Happy Christmas Dad
this and every year

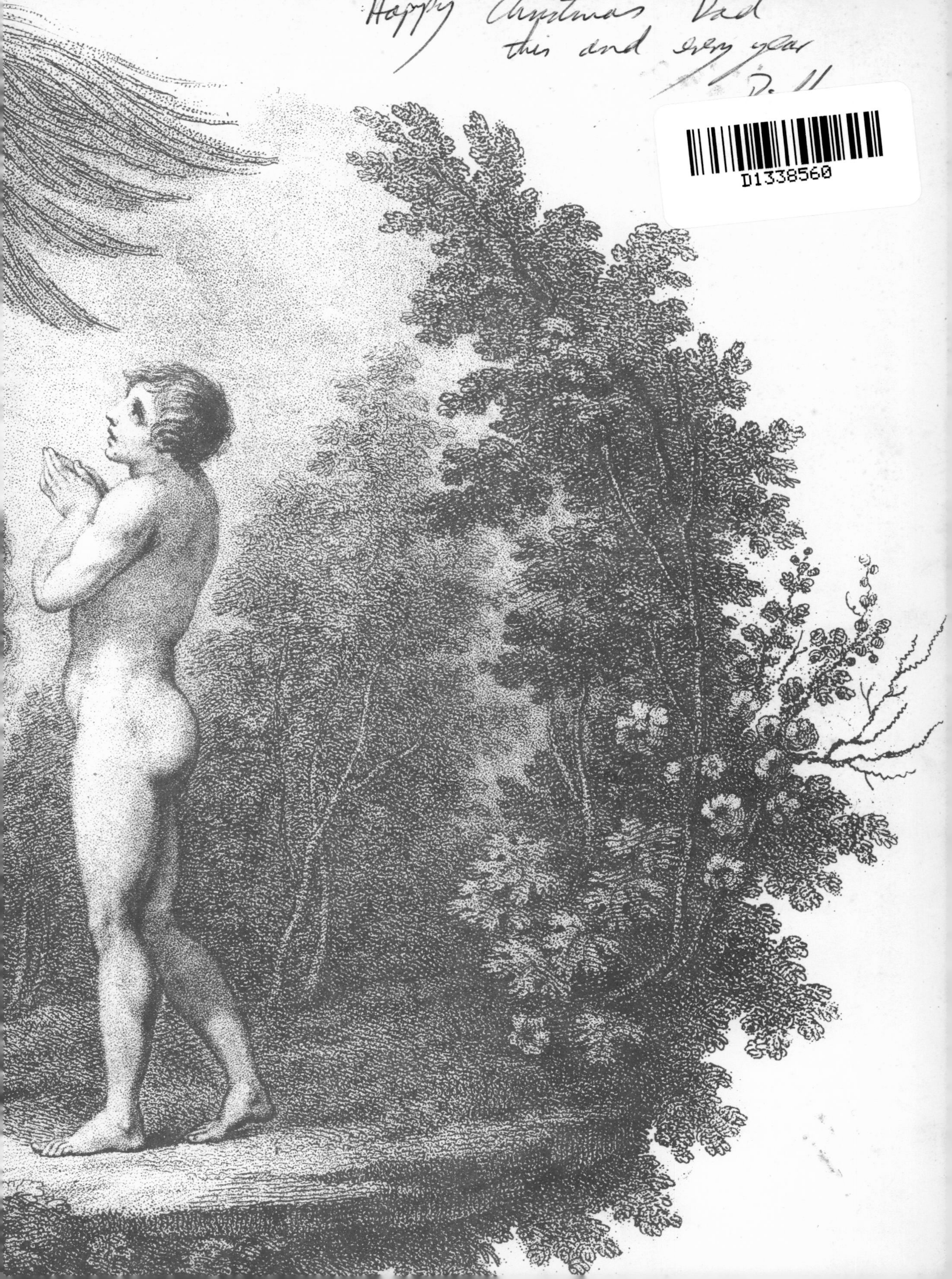

Man, Nature and Ecology

Man, Nature and Ecology

Aldus Books and Jupiter Books

Contents

SBN 490 00166 1
This edition published in 1974 by
Aldus Books and Jupiter Books, London
© 1974 Aldus Books Limited, London
Printed and bound in Spain
by Novograph, S. L. and Roner, S. A.
Crta. de Irún, Km. 12,450. Madrid-34
Dep. Legal: M. 4.226-1974

Contributors

Sir Julian Huxley

Sir Julian Huxley was one of the first scientists to foresee the swingeing population-increase that faces us today and to warn of its consequences. Over 40 years ago he said in a radio talk that ever-increasing population was a danger and that the only answer lay in an increasing use of birth control. Sir John Reith, head of the British Broadcasting Corporation, summoned him to his office and sternly rebuked him for "polluting the ether" with such talk. In 1950 he, with other demographers, was accused of spreading unnecessary alarm when he predicted that the world population would reach 3000 million by the year 2000. In fact, the population passed the 3200 million mark in 1963. We know now that his warnings were justified and that his predictions, far from being alarmist, erred on the side of caution. Today, Sir Julian is still actively concerned with the problems of pollution, of demography, and of ecology generally.

Introduction

We are faced today with a simple question—can man survive on his planet?

The scale of pollution, the rate at which we are using up our finite sources of energy, the growing gap between the amount of food we need and the amount we produce, all add up to a major threat to our very lives. At the root of all these problems is overpopulation.

At the outset of man's existence and all through early prehistory, population growth was necessary. But, once we get to historic times, population increase begins to produce crowded cities, with their slums, their noise, their health hazards, their violence; to invade new and unspoilt land; and to encourage wars of conquest. Today the total world population is about 3700 million. By the end of this century, even if the birth-rate is slightly reduced, it will have almost doubled, to 6500 million. Already this accelerating population growth, coupled with man's technological and scientific advances, harnessed to the profit motive, has led to unchecked industrial growth whose wastes—and sometimes products—pollute the earth, the rivers and seas, and the air we breathe.

Less and less open countryside remains and many plants and animals have disappeared. Cities like London, Tokyo, or Calcutta grow monstrously large. To feed the world's peoples we exhaust the earth's content of natural minerals, then load it with artificial fertilizers. To obtain adequate crops we use pesticides, most of which are seriously damaging to human and animal life. Even so, some countries face starvation and many suffer undernourishment— particularly as regards proteins. The gap between the developing and the developed countries is widening, not closing. Our raw materials—timber, coal, and oil—are being used up at an alarming rate. Our descendants will have to struggle to exist in a denuded planet.

The moral is simple to enunciate, but hard to carry out. We *must* reduce man's increase rate—to zero if possible, because any rate above zero means that population still increases at compound interest. We must also put world ecology above mere profit-making.

It is important, therefore, that all of us should have a sound understanding of our own ecology and that of our planet, because we *have* to find answers to our problems and find them quickly.

This book presents dramatically and unequivocally the grave and ever-worsening situation in which we find ourselves. But it also provides the basic information—starting from simple ecological concepts—that will enable the reader to understand the choices that confront us and the sacrifices we must undoubtedly make if our children and our children's children are to have a life of any quality upon this planet.

Have we the will to make the changes we must make? Have we the power? I do not know. But, if only because of our growing understanding of our plight, I still hope.

Julian Huxley

POPULATION
reaping the whirlwind

In the 3000 to 4000 million years that living things have inhabited the Earth they have undergone—and survived—almost every kind of ecological catastrophe. But the worst crisis of all is now at hand. The web of living relationships that survived millennia of ice and drought, of mountain-building and drowning, of radiation . . . that web is now threatened by the overactivity of just one species: mankind.

Of all living things we are unique in our intelligence and the way we apply it to acquire and exploit the Earth's resources. From the most commonplace—such as water or soil—to the rarest, we consume them all. And the process is self-sustaining. We use the resources to increase our dominance, to gain more resources, to increase yet again. . . . Not only is it self-sustaining, it is accelerating—with effects that are obvious to us all.

For most of the 2 million years we have behaved like this, the Earth's life-support systems have been able to adapt—more or less. True, we have turned many a fertile wilderness into barren desert; but, compared with what we are about to achieve, those were small, local episodes. And at the heart of the problem lies our mushrooming numbers.

Early Man numbered perhaps 125,000 on the African plain. Despite fluctuations due to disease, ice ages, and other hazards, he spread to all parts of the Earth and his numbers rose to about 5 million by 8000 B.C. War, famine, and plague continued to take their toll, but the general upward trend remained—to 250 million by the time of Christ, to 500 million by 1650. Then the takeoff really began. Until then the population had taken a leisurely 1500 years to double. But by 1850, a mere 200 years later, it had doubled again. By 1930 (only 80 years on) it had doubled yet again, reaching 2000 million. By the middle of the present decade (only 45 years on) it will be 4000 million. And by A.D. 2000 . . . ?

2000

1950

1900

1850

1800

1750

1700

Asia (China from 1900)

Central & S. America

N. America

USSR

Africa

Australasia

Europe

Rest of Asia (after 1950)·

Rate of increase of world population in 50-year stages, shown here in diagram form.

YEARS TO DOUBLING DAY

Region	Years
N. AFRICA	23
W. AFRICA	27
E. AFRICA	27
CENTRAL AFRICA	32
SOUTHERN AFRICA	29
JAPAN	63
S. W. ASIA	24
S. E. ASIA	25
E. ASIA	39
CANADA	41
U.S.A.	63
MIDDLE AMERICA	21
CARIBBEAN	32
TROPICAL S. AMERICA	24
TEMPERATE S. AMERICA	39
N. EUROPE	117
W. EUROPE	117
E. EUROPE	88
S. EUROPE	78
U.S.S.R.	70
OCEANIA	35

Handicap race toward the abyss

WHAT DOUBLING MEANS

When we turn from a global view of population growth and look at its effects on individual countries, the prospects become even more menacing. In the developed countries the rate of increase is low. None of them will double its population this century; many of them will not even see a doubling in the next 100 years. But among the countries that are still struggling to build an industrial base the picture is quite different. Most of them will double before 2000; all will double in the next 100 years.

Just think what doubling means. For every gram of daily bread in an already meager supply there will have to be two; for every schoolteacher, doctor, judge, or clerk there must be two. Where each home, school, clinic, or warehouse now stands, another must be built somewhere nearby. And although some amenities need not be doubled (roads, for instance, or deep-water harbors), others, such as police and government personnel, must be more than doubled as increasing scale and complexity calls for new levels of administrative skill. So, in general, each of these countries must double its amenities and operations merely to *maintain* today's miserable standard of life.

Not even the richest of nations with the finest industrial and economic base could seriously plan to double its amenities in just 30 or 40 years. What hope, then, have the wretchedly poor countries of the world—where most of the underfed 2500 million are already to be found?

Left: the number of years before populations of various parts of the world double.

STARVATION
one battle already lost?

The explosive growth of our population is bringing us close to the limits of the Earth's ability to support not just mankind but life itself. Those limits are all around us now—in resources, in unpolluted water and air, in undepleted soils, and above all in food.

If our goal is merely to survive as a species, we face tough times ahead. But if we still cherish the nobler and more edifying goals that have inspired our greatest thinkers and moral leaders, then we can already set down our first major defeat. We have lost the battle to feed mankind as adequately as most other species feed.

As far as food is concerned this is a starving world—a slum where 2500 million people go hungry every day, where another 1500 million barely manage, and where a lucky 500 million, the first-class passengers in Spaceship Earth, need only worry about overfeeding. The tragic fact is that even if all the food mankind now eats were to be distributed with absolute fairness by some cosmic Solomon, everyone would go hungry.

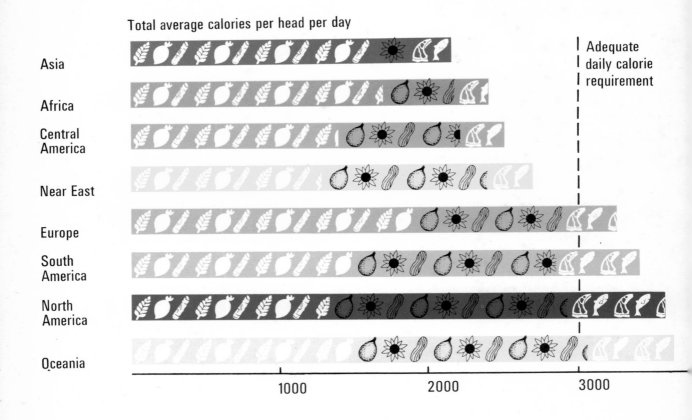

Total average calories per head per day

Asia

Africa

Central America

Near East

Europe

South America

North America

Oceania

Adequate daily calorie requirement

1000   2000   3000

14

The picture on the left shows starvation conditions in the Warsaw Ghetto under Nazi rule in World War II. When it was first published it shook the conscience of the world. Similar conditions, as the picture of Calcutta (above) shows, are common now in many of the world's great cities—too common to create much of a stir.

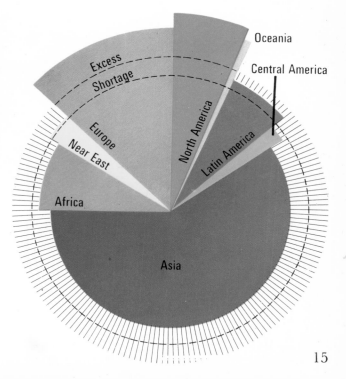

The chart on the left shows the average daily calorie intake in eight major regions of the world; in four of them it is inadequate. The chart on the right shows how the world's available foodstuffs (all the color taken together) are distributed. Anything reaching beyond the disk defined by the radiating bars is excessive. The sad fact is that even if all the excess were redistributed evenly, the colored areas would reach only to the inner broken line. In other words, the whole world would experience a shortage of food.

15

RESOURCES
the storehouse is emptying

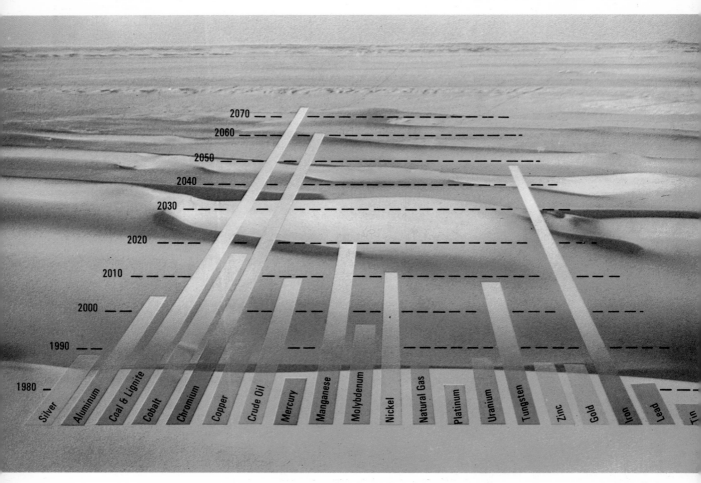

The industrial civilization that supports us depends in turn on an ever-increasing supply of raw materials. Only a few, such as wood and water, are renewable. Most are in the form of fixed reserves locked somewhere in the Earth's crust. They all cost time and money—and energy—to extract and process. The final exhaustion of many is now clearly in sight.

Some, such as iron, copper, and aluminum, are widely distributed and exist in the classic mineral ratio—a few very-high-grade deposits at one end of the scale, massive low-grade deposits at the other end, and every grade and volume proportionately in between. This kind of distribution has led to short-sighted optimism: as cheap-to-win high-grade deposits run out, the price rises and makes lower-grade ores economic to extract . . . and so on until we reach the massive deposits of lowest-grade ore at some happily remote future date.

Above: chart showing estimated lifetimes of known recoverable minerals in the Earth.

Unfortunately this classic ratio does not hold for most minerals. It isn't true of cobalt, gold, lead, manganese, mercury, molybdenum, nickel, tin, tungsten, or zinc. Not only are these patchily distributed but, when the high-grade ores are gone, there are no lower-grade ores within the reach of any acceptable price rise. Yet industry depends absolutely on every one of these metals; and each metal has at least one application for which there is no conceivable substitute. So here, indeed, is an unavoidable limit to the endless growth that has brought us to our present crisis. When these ores have been exhausted, the Earth will support our modern kind of civilization as harshly as a desert supports life itself.

The problem is heightened by the understandable desire of the poorer countries to industrialize. A measure of the hopelessness of their aim can be seen from these figures: to bring today's population to the standards that prevail in the developed countries would call for 30,000 million tons of iron (today's profitable reserves—30,000 million tons), 500 million tons of copper (reserves—220 million tons), 300 million tons of zinc (reserves—85 million tons), 50 million tons of tin (reserves—6 million tons) . . . and so on. To bring the expected population of the year 2000 to that level we should have to double the tonnages required.

The impossibility of that task shows how hollow are any arguments about the *exact* size of economically winnable reserves. The vast differences between what we need and what we know we have make it clear that some of the world will never emerge from the Stone Age and most of it is never going to emerge from the Iron Age.

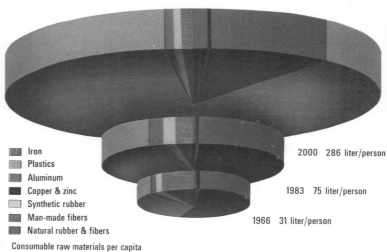

Iron
Plastics
Aluminum
Copper & zinc
Synthetic rubber
Man-made fibers
Natural rubber & fibers

2000   286 liter/person

1983   75 liter/person

1966   31 liter/person

Consumable raw materials per capita

As our civilization grows more complex and demanding, we consume the Earth's abundance at an ever-increasing rate. The chart above shows consumption per person of certain key materials at three equally spaced dates in the second half of this century.

The very act of removing certain minerals can so poison the soil that nothing can grow there for decades after. The Welsh copper mine in the picture on the left was closed over a century ago; the site is still barren of all forms of life.

FROM MOUNTAIN PEAK TO OCEAN TRENCH

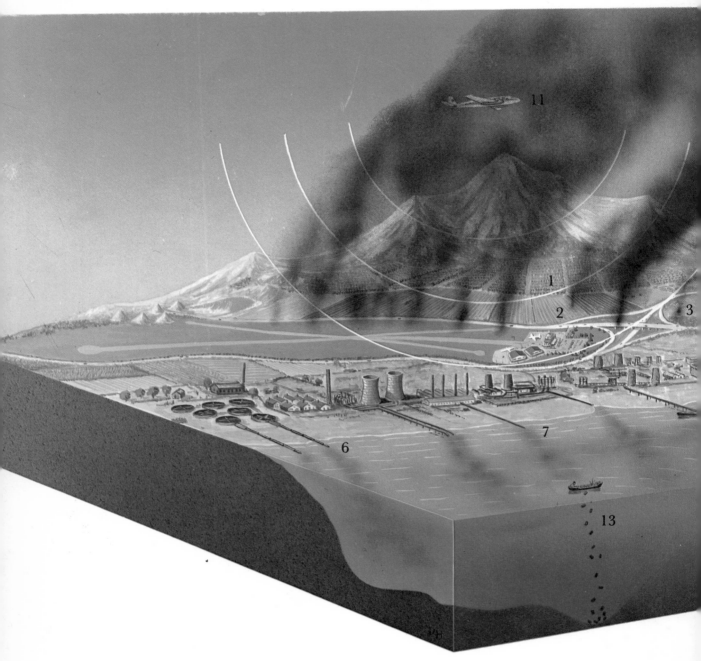

There are at least 13 major areas where we can, and must, control our polluting activities, each identified in the landscape above:

1. Forestry using DDT and other persistent organochlorine pesticides, as well as defoliants and herbicides, damages life far outside the forest.

2. Modern farming reduces genetic variety, destroys wilderness, depletes soil structure, ignores sensible rotations, pollutes with inorganic fertilizers, pesticides, and herbicides—all making for an unstable ecosystem.

3. Vast highway systems encourage wasteful automobiles, consume resources, increase water runoff and so deplete groundwater, and promote the mobility that leads to nomadic, unstable, and unhappy communities.

4. Sprawling suburban communities consume land, deplete city revenues, and add to air pollution with bonfires and small, inefficient furnaces.

the mark of man is everywhere

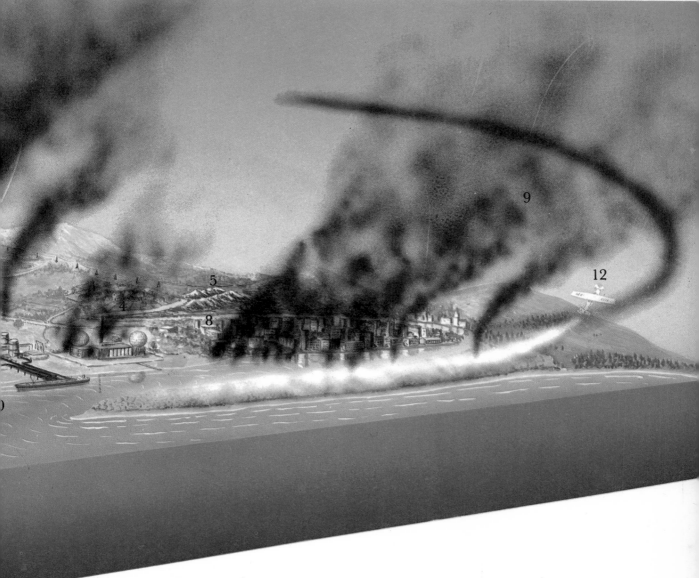

5. Exurban garbage tips sterilize land, pollute neighborhoods, conceal illegally dumped toxic wastes, and harbor nondecaying plastics.

6. Sewage effluent, treated or untreated, overenriches waterways, upsets the plant/animal balance, and squanders a valuable resource.

7. Industrial effluents load the environment with persistent and highly reactive poisons; industrial waste heat is a threat to many river ecosystems.

8. Cities combine the kinds of damage done by suburbs and highways.

9. The inefficient and incomplete burning of fossil fuels raises the concentration of unnatural pollutants to dangerous levels, wastes resources, and endangers the global heat balance by increasing atmospheric carbon dioxide.

10. Spillage from tankers, refueling ships, and offshore drilling operations accounts for some 500 million gallons of oil pollution every year.

11, 12. Airplanes fill the upper atmosphere with particles from inefficiently burned fuels and impoverish the lives of millions with their noise.

13. Dumping of wastes at sea pollutes the water and threatens destruction of its plankton, on which all Earth's life depends.

ANIMALS ENDANGERED

1 Thylacine	**44** Caribbean monk seal	**88** Cyprian mouflon	**133** Kagu	**172** Olivaceous bulbul
2 Rusty numbat	**45** Hawaiian monk seal	**89** Giant pied-billed grebe	**134** Great Indian bustard	**173** Rufous-headed robin
3 Leadbeater's possum	**46** Dugong	**90** Short-tailed albatross	**135** New Zealand shore	**174** Seychelles magpie
4 Sclay-tailed possum	**47** Przewalski's horse	**91** Diablotin	plover	robin
5 Broad-nosed gentlé	**48** Asiatic wild ass	**92** Cahow	**136** Eskimo curlew	**175** Starchy
lemur	**49** African wild ass	**93** Stejneger's petrel	**137** Hudsonian godwit	**176** Omao
6 Mongoose lemur	**50** Mountain zebra	**94** Abbott's booby	**138** Audouin's gull	**177** Teita olive thrush
7 Fat-tailed lemur	**51** Central American tapir	**95** Chinese egret	**139** Grenada dove	**178** Grand Cayman thrush
8 Fork-marked mouse	**52** Great Indian	**96** Korean white stork	**140** Kakapo	**179** Nihoa millerbird
lemur	rhinoceros	**97** Giant ibis	**141** Night parrot	**180** Seychelles warbler
9 western woolly avahi	**53** Javan rhinoceros	**98** Japanese crested ibis	**142** Ground parrot	**181** Eyrean grass wren
10 Verreaux's sifaka	**54** Sumatran rhinoceros	**99** Trumpeter swan	**143** Orange-fronted	**182** Chatham Island robin
11 Indris	**55** Square-lipped	**100** Néné	kakariki	**183** Tahiti flycatcher
12 Aye-aye	rhinoceros	**101** Crested shelduck	**144** Orange-bellied	**184** Tinian monarch
13 Woolly spider monkey	**56** Black rhinoceros	**102** Laysan teal	parakeet	**185** Seychelles paradise
14 Goeldi's tamarin	**57** Pygmy hippopotamus	**103** Brown teal	**145** Beautiful parakeet	flycatcher
15 Tana River mangabey	**58** Wild Bactrian camel	**104** California condor	**146** Paradise parakeet	**186** Piopio
16 Tana River red colobus	**59** Persian fallow deer	**105** Galápagos hawk	**147** Puerto Rican parrot	**187** Kauai oo
17 Orang utan	**60** Brow-antlered deer	**106** Hawaiian hawk	**148** Imperial parrot	**188** Stitchbird
18 Pygmy chimpanzee	**61** Sika	**107** Monkey-eating eagle	**149** Red-necked parrot	**189** Truk great whiteye
19 Mountain gorilla	**62** Père David's deer	**108** Mauritius kestrel	**150** St Lucia parrot	**190** Ponapé great whiteye
20 Ryukyu rabbit	**63** White-tailed deer	**109** Seychelles kestrel	**151** St Vincent parrot	**191** Akepa
21 Volcano rabbit	**64** Western giant eland	**110** La Pérouse's	**152** Prince Ruspoli's	**192** Kauai akialoa
22 Kaibab squirrel	**65** Wild Asiatic buffalo	megapode	turaco	**193** Nukupuu
23 Delmarva Peninsula	**66** Tamarau	**111** Pritchard's megapode	**153** Red-faced malkoha	**194** Akiapolauu
fox squirrel	**67** Anoa	**112** Horned guan	**154** Seychelles owl	**195** Maui parrotbill
24 Utah prairie dog	**68** Kouprey	**113** Prairie chicken	**155** New Zealand laughing	**196** Ou
25 Block island meadow	**69** European bison	**114** Western tragopan	owl	**197** Palila
vole	**70** American bison (wood	**115** Blyth's tragopan	**156** Puerto Rico	**198** Crested honeycreeper
26 Beach meadow vole	bison)	**116** Cabot's tragopan	whippoorwill	**199** Bachman's warbler
27 Cuvier's hutia	**71** Jentink's duiker	**117** Sclater's monal	**157** Narcondam hornbill	**200** Golden-cheeked
28 Dominican hutia	**72** Giant sable antelope	**118** Chinese monal	**158** Okinawa woodpecker	warbler
29 Mexican grizzly bear	**73** Arabian oryx	**119** Imperial pheasant	**159** Ivory-billed	**201** Kirtland's warbler
30 Polar bear	**74** Scimitar-horned oryx	**120** Edwards's pheasant	woodpecker	**202** Semper's warbler
31 Giant panda	**75** Addax	**121** Swinhoe's pheasant	**160** Imperial woodpecker	**203** Seychelles fody
32 Black-footed ferret	**76** Bontebok	**122** White eared pheasant	**161** Bush wren	**204** Tristan grosbeak
33 Giant otter	**77** Hunter's hartebeest	**123** Brown eared pheasant	**162** Small-billed false	**205** Tristan finch
34 Southern sea otter	**78** Swayne's hartebeest	**124** Elliot's pheasant	sunbird	**206** Ipswich sparrow
35 Spanish lynx	**79** Black wildebeest	**125** Hume's bar-tailed	**163** Rufous scrub bird	**207** Dusky seaside sparrow
36 Florida cougar	**80** Beira	pheasant	**164** Noisy scrub bird	**208** Cape Sable seaside
37 Asiatic lion	**81** Slender-horned	**126** Mikado pheasant	**165** Raza Island lark	sparrow.
38 Tiger	gazelle	**127** Palawan peacock	**166** Rothschild's starling	
39 Barbary leopard	**82** Sumatran serow	pheasant	**167** Saddleback	
40 Atlantic walrus	**83** Japanese serow	**128** Japanese crane	**168** Kokako	
41 Ribbon seal	**84** Takin	**129** Whooping crane	**169** Hawaiian crow	
42 Ross seal	**85** Nilgiri tahr	**130** Sandhill crane	**170** White-breasted	
43 Mediterranean monk	**86** Walia ibex	**131** Zapata rail	thrasher	
seal	**87** Markhor	**132** Takahé	**171** Dappled bulbul	

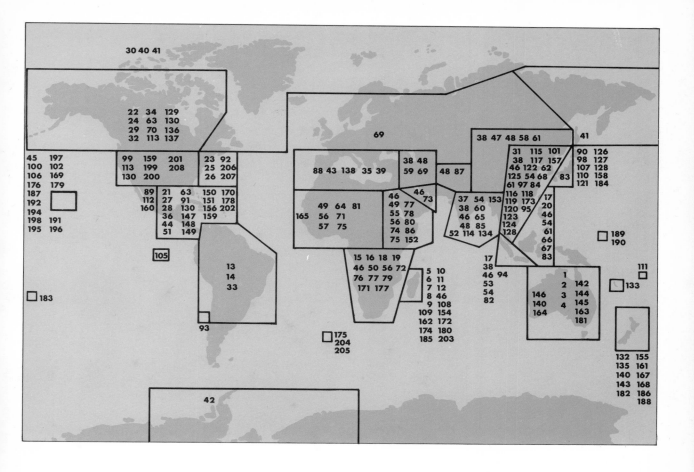

Most of the species that ever lived are now extinct. It is a natural process and extinction is the natural fate of any animal that has specialized too far to change when its environment changes, or of any animal that meets with a better-adapted competitor.

But—as in many other fields—the appearance of man adds a new factor to the natural equation. The threat we pose to our fellow creatures does not result from direct competition with them for territory or food; it is an accidental and sometimes frivolous by-product of our relentless pursuit of our own goals and satisfaction. No one *planned* to wipe out the dodo or the passenger pigeon; no fur-loving movie star deliberately plots the extinction of tigers and leopards. It just happens that sailors looking for easy meat, plainsmen out for fun, and actresses on ego trips achieve these extinctions in passing, as it were.

In the last 2000 years we have helped wipe out about 200 species—many, it is true, on the verge of natural extinction. A third of them have gone in the last 50 years, and the rate is still increasing. At the moment over 350 vertebrate species are on the danger list (most of them named and located on these pages and a few shown in the drawing above). It is not just that their lives are in moral trust to us; we have a much more selfish reason for caring. Their deaths are a barometer to measure our destruction of the world environment. The bell that tolls their passing tolls for us as well.

Each figure on the map represents a species of bird or mammal in danger of extinction, which breeds in part or all of the areas in which the figure appears. The distribution of whales has been omitted to avoid confusing the other references.

FAILURE OF THE CITIES

Everyone knows the disadvantages of cities—the noise, the crowding, the dirt, the slums. . . . Until recently most people assumed that the advantages were more than compensation. They included the rich network of cultural intercourse, encouragement of invention, trade, science, and a host of other activities that traditional rural communities could not support, and the accumulation of wealth to finance every kind of new venture. At more practical levels it was often argued that because cities made more efficient use of such resources as power, transportation, and communications, and because they broadened the tax base, they were cheaper.

Such generalizations are true only as long as the city dwellers rely on mass transit systems. The automobile has changed everything. Its voracious appetite for paved road surface destroys city centers and inner suburbs, separating communities from one another and making the center no longer a desirable

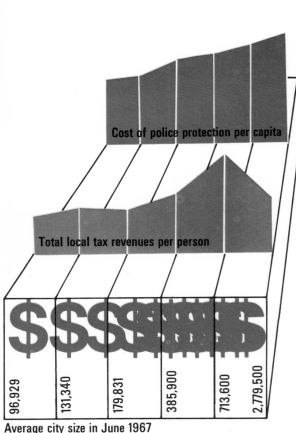

Cost of police protection per capita

Total local tax revenues per person

| 96,929 | 131,340 | 179,831 | 385,900 | 713,600 | 2,779,500 |

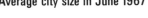

Average city size in June 1967

One argument long maintained in favor of cities is that their efficiency increases with size because they can supposedly deploy resources better. Charts on these pages, based on recent detailed studies of a number of Californian cities, refute that contention. They show that cities get more

expensive to run the larger they grow. They also promote moral deterioration, so that the largest cities are those with the highest rates for crimes against people. This is reflected in the increasing cost of police protection per citizen. Economies of scale are completely outweighed by the cost.

place to live. The rich and middle classes move out. The social structure is upturned, and so, too, are all the formerly valid arguments about cities. Modern communications provide an international network for cultural intercourse; government and business provide all the encouragement for invention, trade, science . . . and so on. The city is superfluous—a slum huddled around a dying shopping and cultural center.

American cities have set the pattern—not deliberately, but simply because that is where private affluence and automobile ownership are greatest. There is now a direct relationship between the size of a city, its population density, the crime rate, the cost of police protection, and (with few exceptions) the tax burden on citizens.

Many of the benefits of cities still remain; but they are dwindling assets in the face of these stark increases in social and financial cost.

Number of crimes per 100,000 people per year

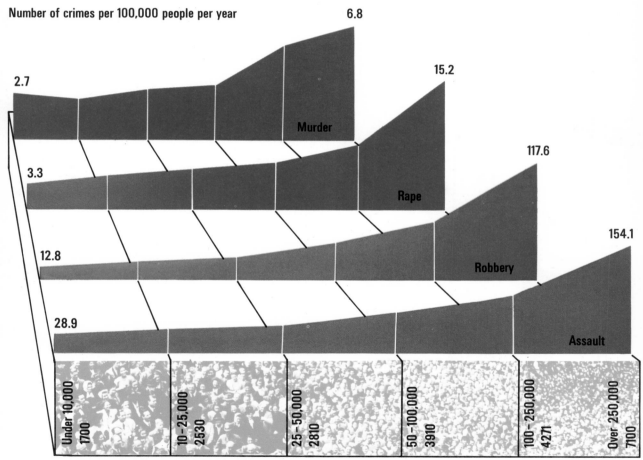

6.8

15.2

117.6

154.1

2.7

3.3

12.8

28.9

Murder

Rape

Robbery

Assault

Under 10,000
1700

10–25,000
2530

25–50,000
2810

50–100,000
3910

100–250,000
4271

Over 250,000
7100

Population size of cities in 1960 & Average number of people per square mile

DAMAGED WORLD

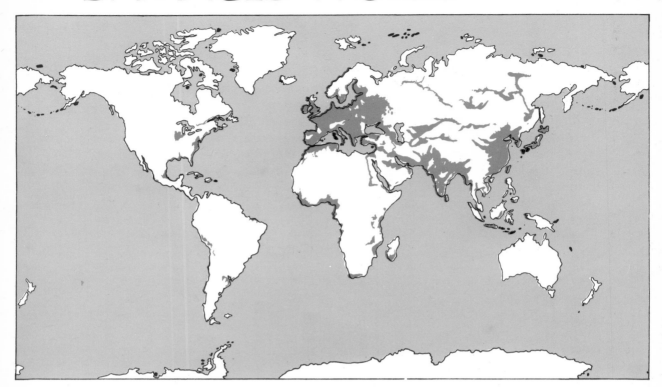

1880

For most of our time on Earth we have modified our environment in ways that can, with a little generosity, be called "natural." We burned or cleared a naturally balanced ecosystem and replaced it with something that was more productive only from our own selfish viewpoint. But this kind of change is very similar to the change *any* dominant animal can make when it invades new territory. And as long as our numbers were small and our technology primitive, the effects on our global ecosystem were marginal. Locally, though, the story could be very different. Despite that primitive technology we long ago managed to make irreversible changes in the Middle East, India, China, Europe, and Central America. Before, these areas had rich and balanced ecosystems, which we depleted, simplified, unbalanced, and finally destroyed. The results can be seen in, for instance, the deserts of Jordan and the parched uplands of Spain.

Even so, these local changes are as nothing compared with the widespread and deep-seated changes we can now make with the help of our powerful technology and the pressure of our mushrooming numbers. In one generation we can achieve (if that is quite the word) what would have taken our forefathers several centuries. True, that same technology can help postpone disaster for a while—we can deplete the soil, then pour on chemicals to cover up the damage, then breed new varieties to raise sagging output, then spray further chemicals to kill off pests that only one-crop farming encourages . . . and so on. Each step leads down to ecodisaster. The twin pressures of population and industry are accelerating us on this course—so that soon only the deserts, mountains, and cold barren regions near the poles will be more or less as Early Man would have recognized them.

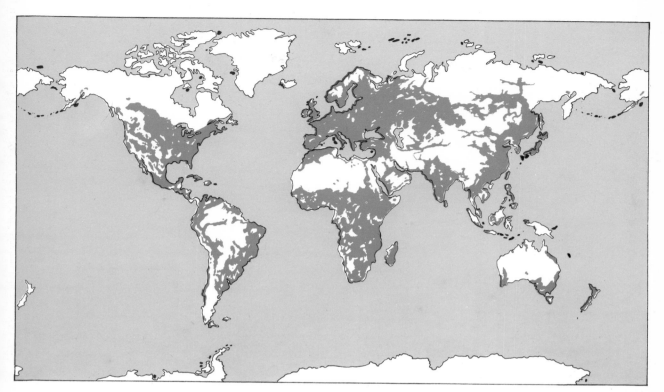

today

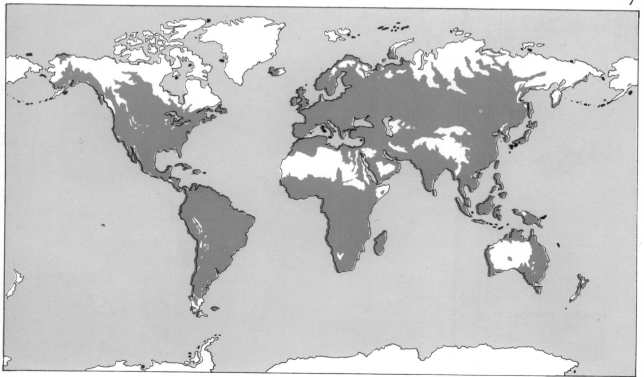

The red in the maps on these pages marks those areas whose natural ecosystems have been, are now, and will soon be, irrevocably altered by man. Current interest in ecological problems is a hopeful sign that these changes will not, in the future, necessarily be ecologically damaging.

2000

Understanding the machine

For over a century and a half now we have been studying our societies with the help of mathematics—census taking, import-export returns, attitude surveys, social security listing . . . almost every aspect of our lives that can possibly be expressed in numerical terms has been.

With the advent of computers, people began putting different sets of figures together and using the tireless computing power of the machine to see how

The chart below demonstrates the main lines of interaction between different elements in our complex social machine. Birth-rate, for instance, affects capital investment, natural resources use, living standards, and, ultimately, death-rate. Each other element has similar, complex repercussions.

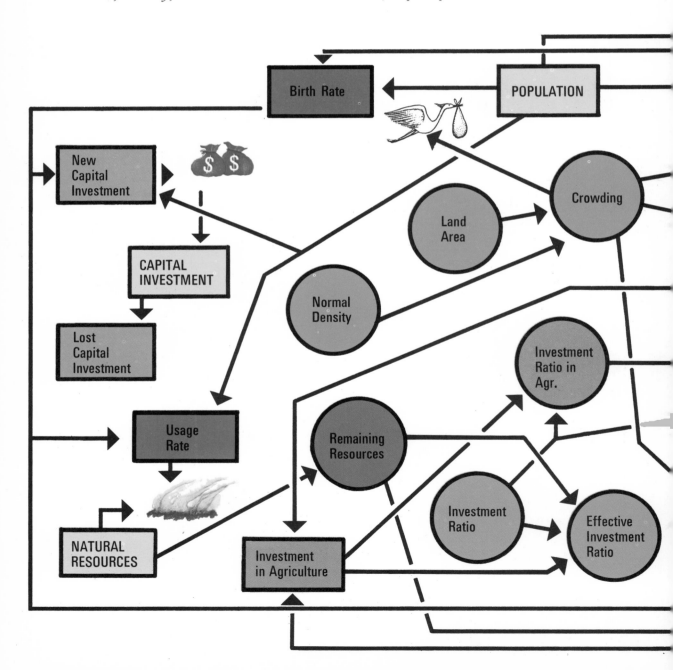

we have made

each quantity varies in terms of every other quantity. Thus we gain a new insight into how the various elements of modern civilization interact and affect one another. With the help of this insight we can go on to build mathematical models of those interactions and then—most important of all—we can embody the model in a computer program to predict the effects of different policies.

Already it is quite clear that the elements of industrial civilization are very closely linked and very sensitive to changes in one another. Therefore the system and its problems must be seen as a whole and tackled as a whole. Piecemeal reform might actually do more harm than good.

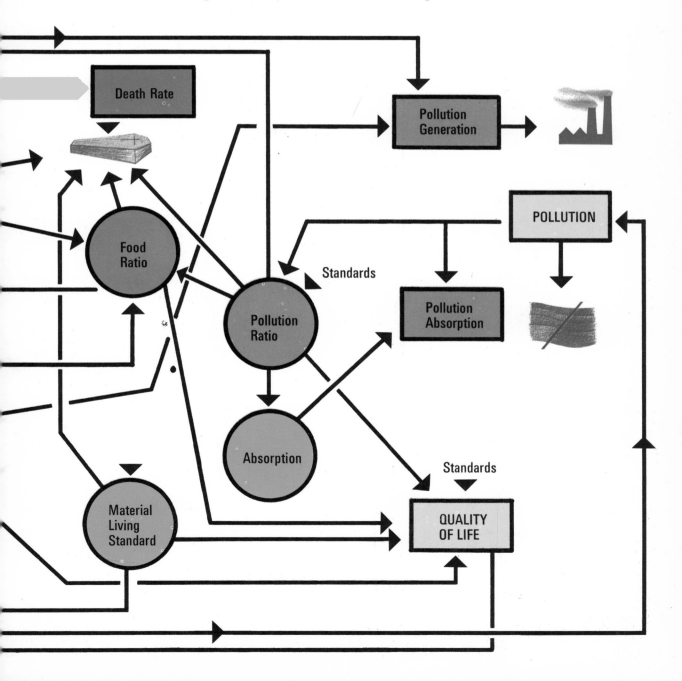

Predator and prey, part of the finely balanced complex of living matter making up a local ecosystem.
By overlapping and interacting with other ecosystems it forms part of a global web of life.

1

Nothing in Isolation

The whole of nature can be compared to a complex household in which a number of people do different jobs, and it is from this idea that the word *ecology* was coined. It comes from the Greek word *oikos*, a house, and means, literally, the science that deals with the home conditions of living organisms. Ecologists are people who study the ways in which organisms or groups of organisms are related to the living and nonliving parts of their environment. Their aim is to provide a complete picture of how these complex relationships work out in any given area.

More than a century ago the English naturalist Charles Darwin instanced this interplay of organisms and their environment when he suggested that there might be a connection between the number of cats in a district and the yield of red clover in the fields. The story went like this: red clover is pollinated only by bumblebees, because no other insect can reach its nectar; so if the bumblebee became extinct, red clover would disappear. But, Darwin continued, bumblebees are the prey of field mice, which eat their larvae and destroy their nests. So, in districts where field mice are on the increase, both bumblebees and clover will suffer. However, the number of mice in a district partly depends on the number of cats. So the more cats there are in the neighborhood, the better the clover crop is likely to be.

Darwin's original story ended here, but there is no reason why we should not extend it to include human beings. For example, we might make cat-owners a fifth link in the argument. If we assume that unmarried women are more likely to keep cats than married ones, we may say that a village where many of the women are unmarried will have an unusually large cat population, and hence a particularly good clover crop; if the marriage rate rises, there will be fewer cat-owners, and the clover yield will decline.

The argument is ingenious, but inconclusive. We cannot say whether it is right or wrong, because it leaves out things that should be included, and assumes others that cannot safely be taken for granted. It takes no account, for instance, of weather and soil conditions, which may play a decisive part in the cultivation of clover. On the other hand, it assumes that the fields are fertile, whereas in fact not even a billion bumblebees could produce a good clover crop if there were too few nutrients in the soil. It also makes the bold assumption that field mice are always the chief destroyers of bee larvae. The only way to determine the extent to which the argument is valid would be to make a careful and quantitative study of the environment, examining the physical and chemical composition of the soil and climate, and the interplay between the chief plants and animals in the area. Such a study would provide us with a picture of the *total* environment, including all the factors affecting the clover crop. Here, as always in ecology, correct conclusions can be drawn only when all the important aspects of the situation have been observed and taken into account. Throughout this book we must steadily remember, even when the reminder is not explicitly included, that any given fact or activity must always be set against this wider—and yet wider—context.

In a subject of such vast scope, however, one person cannot expect to study everything at once. We shall, therefore, find it convenient in this book to look at nature in terms of what ecologists call levels of organization. In increasing order of scale and complexity there are five levels: 1. The *organism*—the individual plant or animal. 2. The *population*—not only groups of people, but also groups of individuals of any one species. 3. The *community*—all the different populations within a given area, whether that area is as small as a pond or as large as the Sahara Desert. 4. The *ecosystem*—the sum of all the communities in an area, together with their nonliving environment. 5. The *biosphere*—the largest unit of all; it includes the sum total of life on earth. In practice, of course, we cannot fully investigate any one of these levels of organization without discovering something about the others. This is because each larger unit includes all the smaller units, just as the organism itself includes organs, tissues, and cells.

Not even an ecosystem can always be studied in isolation, for one often merges into another. But though there may be no sharp boundary between them, we shall find that they are very useful working units, mainly because an ecosystem can be as large or as small as we care to make it. The important thing to remember is that whichever one we choose, it includes *all* the organisms in it, from the bacteria in the soil to the birds and insects in the air; it also includes *all* the factors of the nonliving environment—nutrients, temperature, wind, relative humidity, light intensity, and so on. Each of these factors has some effect on organisms and, as we shall see later, some of them have also an effect on one another.

We may be tempted to think of the world of nature as a state of anarchy and chaos, with every plant and animal fighting for food and survival. Plants compete with each other for light and nutrients, animals eat plants, and other animals feed on the plant-eaters. Even when an animal dies, microorganisms in the air and the soil compete with insects for its remains. But nature's warfare, though intense, is usually waged in an unspectacular way, so that on a country walk, for example, we usually see little evidence of it. It proceeds slowly, steadily, and for most of us almost invisibly. But in spite of the perpetual

For centuries man has dominated many ecosystems; but nature is still to be reckoned with. Right: scouring action of the Colorado River bites deeper and deeper into the Grand Canyon.

Study of a single population, such as the gannet colony (far right), is called population ecology. *But, to make a complete study we have to consider other factors—making a* comprehensive *ecological study.*

conflict within nature, the overall pattern of plant and animal relationships remains precise, balanced, and orderly.

Unfortunately for the would-be ecologist, there is no single branch of modern science that tries to study all the known laws of the natural world. So, if he is to master his subject, he must enlist the help of many scientific disciplines, including chemistry, physics, geography, meteorology, and biology. Further, the aim of ecology—to give a complete picture of nature at work—is an extremely ambitious one. No one has yet even identified all the organisms in an ecosystem of any size, let alone estimated all the ways in which the environment affects them. But there is no need to despair, because it is quite possible to become a good ecologist without first having to master all the other branches of science in detail. There are certain basic principles that not only relate these sciences but also help us to understand the effects of many natural laws in any given ecosystem. When we have understood these basic principles, we can appreciate the orderliness and logic of nature's network without being daunted by its complexities.

Ecology is worth studying for the sheer interest of it, but there is a more compelling reason why we should understand it. Man, as the dominant species on this planet, has more influence over nature than any other living thing, and that influence is increasing all the time. At present, there are already over 3000 million of us; by the year 2000, there may well be more than 7000 million, all competing for food and space. But so far our power of interfering with nature has been greater than our understanding of how nature works: although we are totally dependent on nature for food, we have already turned large areas of the earth's surface into desert, and much more into

unproductive land. It is hardly surprising that man has been making mistakes, because until quite recently he has not known what he has been doing. But now he knows; either he abides by the rules of nature or he risks disaster.

The science of ecology is developing very quickly, but there is a time lag between the discoveries that ecologists make, and the spread of information to people who need to be well informed. Many people are now being called on to make decisions that must affect ecosystems; town planners, for example, may build on farming land without stopping to consider what effects their action may have on the surrounding fields. Such an action, although undertaken with the best intentions, may result in driving insects that were formerly doing no harm onto cereal fields where they become pests. Nobody expects the people who have to take this sort of decision to be expert ecologists, but they should at least know something about the principles of the subject, so that they can weigh their actions and either see what the consequences are likely to be, or recognize that they must call in an expert.

When the balance of an ecosystem is disturbed, whether accidentally or deliberately, the consequences can be serious for man. Top left: Gruinard Island, Scotland, contaminated by germ warfare experiments and now unsafe for humans. Left: dust storm in New Mexico.

The natural cover of plants has been overgrazed by domesticated animals so that strong winds can remove the soil bodily. Above: though not advocating a return to a more primitive age, we must learn to live in equilibrium with nature.

*Energy can enter an ecosystem in only one form:
as light energy from the sun. This energy is
captured and used by green plants, such as trees,
whose leaves contain the pigment chlorophyll,
and is partially converted by photosynthesis into
stored chemical energy.*

2
Energy-the Driving Force

A continuous flow of energy is essential to life. Nothing happens of its own accord, without some change somewhere; and change always means that energy is used up.

Energy is defined as "the capacity for doing work." Matter cannot begin moving, or change its direction once in motion, unless a force is applied to it. When a force is applied to a stationary body and sets it in motion, work is done; and we can express the amount of work by multiplying the force exerted by the distance the body moves. In fact, various units are used to measure work, depending on whether the source of energy is mechanical, electrical, heat, and so on. In ecology, however, it is more convenient to express all forms of energy in the same units, and we can do this if we stick to *calories*, the units employed in measuring heat. The calorie, spelled with a small c, is the amount of heat that will raise the temperature of one gram of water by 1° centigrade. Because one calorie represents a very small amount of heat, it is more convenient to use the kilocalorie, or kcal (often called the Calorie, with a capital C), which is equivalent to 1000 calories.

How is it that different forms of energy—sound, mechanical, light, electrical, and nuclear—can all be expressed in terms of heat energy?

The answer is that all forms of energy can be completely converted into heat, but they cannot be completely converted into any other single form. If we want to measure the heat equivalent of electrical energy, for example, we can do so by passing a known amount of current through a coil of insulated wire immersed in a lagged vessel of water called a calorimeter. By measuring the increase in temperature of the water, we can then work out in calories the amount of heat energy that is equivalent to the electrical energy employed. Of more direct importance to the study of living things, we can also work out the energy content of various foods, and express the result in kcal per gram. To do so, we burn food with oxygen in a vessel called a bomb calorimeter, and measure the resulting increase in temperature. This enables us to calculate the number of calories that are equivalent to the chemical energy contained in the food.

Energy in organisms

We shall see (page 43) that food contains energy, even though that energy may not be doing work; the same is true of fossil fuels, such as coal and petroleum. That is why energy must be defined as the "capacity for doing work." When energy is locked up and idle, it is called *potential* energy; when it is doing work, it becomes *kinetic* energy. We can say, for instance, that all living things store potential energy in the form of food and that this changes into kinetic energy when they do any form of work.

All organisms use food for growth—to increase the size of the body—and to repair worn-out tissues. When this happens, some part of the food actually adds to the bulk of the organism or replaces bulk lost. Some other part supplies the energy needed to take matter from food and build it into the tissues of the organism. Both plants and animals require extra food during reproduction, when they have to form seeds, eggs, or young. Animals also need large amounts of energy for moving about; some also need energy for keeping the heart constantly pumping blood through the arteries and veins. The very intake of food—grazing and chewing vegetable matter, or tearing and chewing flesh—calls for work. Then the food has to be swallowed, processed in the gut, and the indigestible residue expelled. All this demands a supply of energy.

None of these processes can go on unless the plant or the animal has the correct body temperature. The reason is that the speed of all chemical reactions, including those that take place in living things, varies directly with

This cheetah, sprinting at over 60 miles per hour, is converting the potential energy of previous meals into kinetic energy. In the conversion about 75 percent of energy is lost as heat.

temperature. In general, the rate of chemical reactions doubles for every rise of 10°c. So, if the body temperature of an organism falls below a certain level, the chemical reactions that enable it to use the energy contained in food come almost to a halt. This explains why many animals are torpid in winter, while growth in plants comes to a standstill. On the other hand, too high a temperature will rupture the delicate chemical bonds of living matter, and then the organism may die, as we may after a prolonged, very high fever.

Heat is produced automatically in all organisms whenever one form of energy turns into another, and whenever work is done. Sometimes this is not enough to allow chemical reactions to proceed at the pace needed to sustain sufficient activity, so some food is then broken down to produce more heat. The efficiency with which different animals can do this varies greatly. Most animals, for example, are cold-blooded, and their body temperature is much the same as that of their environment. In cold conditions the heat they produce by burning food does little to raise their body temperature above that of their surroundings, and this means that many cold-blooded animals become inactive or hibernate in cold weather. In some warm-blooded animals, including humans, the body temperature is maintained at about 37°c—the most efficient temperature for life activities—because, in the course of their evolution, they have developed ways of preventing loss of body heat.

Many animals also use energy—that of sound—to communicate; and some organisms, such as fireflies, deep-sea fish, and certain toadstools, use it

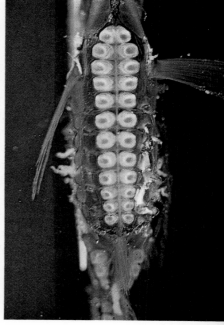

The deep-sea hatchet fish (Argyropelecus) *generates light, converting food energy into light energy in special organs (above right). The process involves the oxidation of compounds.*

37

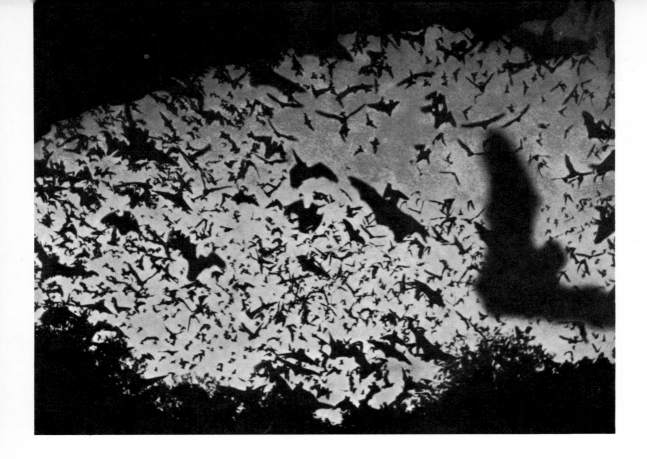

to produce a cold phosphorescent light. Another widespread form of energy in organisms is electrostatic. The chemical bond energy in solid foods is electrostatic, and whenever organisms do work this bond energy is used (p. 43). Also, every movement we make involves electrostatic changes in nerves and muscles. Nerve cells conduct impulses quickly along fibers and, with the help of the brain, they coordinate the activities of most parts of the body. Although nerve impulses can be measured in millivolts, they do not flow as an electric current does in an ordinary circuit. An electric current flowing along a wire is carried by the movement of electrons, whereas an impulse traveling along a nerve is carried by chemical means. Some animals, such as the electric catfish, also use electricity to create an electrostatic field around themselves. When an intruder disturbs this field the catfish can gauge its size and position before deciding whether to flee or to attack. Another fish, the electric eel, actually stuns its prey with electrostatic shocks of about 400 volts.

Sources of energy

A minute amount of the energy available on our planet comes—in the form of heat—from inside the earth itself, but by far the greater part comes from the sun, in the form of solar radiation. It is a mistake to imagine that the sun gives out light and heat in the same way as, say, an immense piece of burning coal. What actually happens inside the sun is a process of thermonuclear fusion, rather like that in an exploding hydrogen bomb, but continuous and

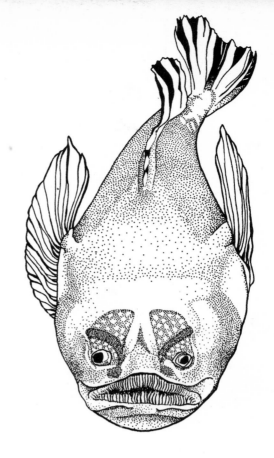

The stargazer, Astroscopus (right), produces large amounts of electrical energy in special nervous tissue near the eyes. When buried in the sand only its eyes protrude. The shock prevents predators from picking at the eyes.

Left: bats, swarming out of a cave at sundown, have specialized in producing intermittent bursts of high-pitched sound energy, as a means of locating each other and their insect prey.

steady instead of momentary and uncontrolled. This process differs from the burning of a piece of coal in that it does not follow either the Law of Conservation of Matter or the Law of Conservation of Energy. In the second quarter of the present century, scientists came to realize that these two laws cannot be separately applied to nuclear reactions. Instead, such reactions obey a single law that combines both of them—the Law of Conservation of Mass and Energy. This new law states that the sum of mass (or matter) and of energy always remains constant, and it implies that mass can, in certain circumstances, be converted into energy.

This is what happens inside the sun, where given masses of hydrogen are constantly being converted into very slightly smaller masses of helium plus vast amounts of energy. The important thing about this reaction is that a minute decrease in mass is equivalent to a prodigious increase in energy. Einstein's famous equation $E = mc^2$ states how we can calculate the amount of energy equivalent to a given mass. In it, E stands for energy, m for mass, and c^2 for the square of the velocity of light, and the whole equation means that the amount of energy gained is equivalent to the amount of mass lost multiplied by the square of the velocity of light. The velocity of light is 300 million meters per second, so, even when the figure for m (mass) is tiny, the figure for E (energy) will have to be colossal to balance the equation.

Even so, the fact that the sun's hydrogen is continuously turning into slightly smaller masses of helium suggests that the sun must be getting smaller and

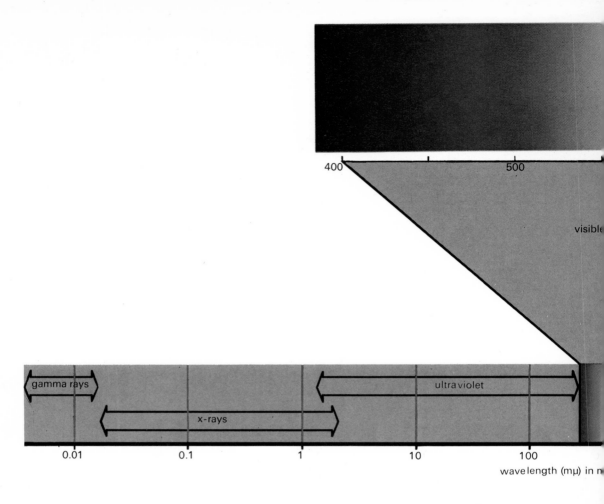

400   500

visible

gamma rays

x-rays

ultra violet

0.01   0.1   1   10   100

wavelength (mμ) in m

will eventually become a mass of inert helium. This is true, but it is estimated that the sun will not exhaust itself for about 5 million million years.

If vast quantities of energy can be produced by the conversion of very small quantities of matter, why can we not get all the energy we need in this way? The answer is that scientists have not yet discovered how to control, much less store, the tremendous energy produced by thermonuclear fusion. Unless and until they do, we must rely entirely on a continuous flow of energy from the sun for heat and light and for the energy in our food and fuels.

Radiation from the sun can conveniently be regarded as traveling in the form of electromagnetic waves, the length of which can be measured in millimicrons (mμ), one millimicron being equal to one millionth of a milli- meter. Some kinds of solar radiation, such as gamma rays, have wavelengths of much less than 0.1 mμ; at the opposite extreme are radio waves, some of which have wavelengths of several km. In between the shortest and longest wavelengths are other radiations, such as visible light and ultraviolet rays. When we talk about a radiation spectrum (diagram above) we include all these wavelengths, but what most concerns us here is the relatively narrow band of radiation called visible light. This narrow band, between about 300 mμ and 760 mμ, is the part to which animals and plants normally respond.

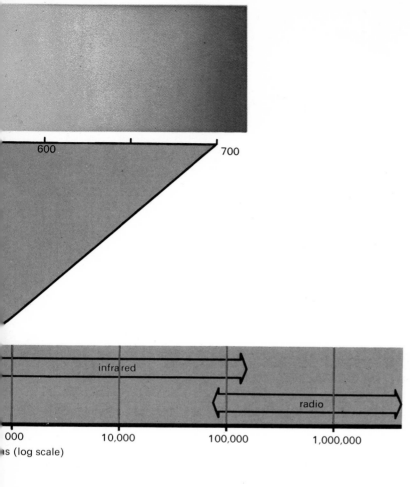

The sun radiates electromagnetic waves over a wide range of wavelengths, shown across the bottom of the diagram. Reading from left to right, the earth's atmosphere intercepts lethal gamma rays and X rays, and most of the ultraviolet rays, which cause sunburn and are harmful to man in large doses. One small section of the total spectrum (enlarged in upper part of diagram), in between ultraviolet and infrared, which we call visible light, provides the energy on which all life processes ultimately depend.

Before life evolved, the earth's surface received a much wider range of radiation at much higher intensities than it does today, including gamma rays, X rays, and ultraviolet rays. All of these are deadly at high intensities, especially those of the shortest wavelengths (gamma rays), for the shorter the wavelength of rays, the greater is their energy. Such radiations are lethal to organisms because they rupture the delicate bonds that hold organic molecules together in three-dimensional patterns. Life as we know it became possible only when the earth's atmosphere had developed in such a way as to filter out the greater part of these lethal radiations. Now the ozone layer of the atmosphere largely keeps out wavelengths of between 290 mµ and 320 mµ; wavelengths of over about 800 mµ are mostly absorbed by the ozone layer and by the atmospheric carbon dioxide and water vapor. Thus the main components of the sun's radiation that now reach the earth's surface are visible light and small quantities of infrared and ultraviolet rays.

In what form does solar energy enter the ecosystem? Forgetting those wavelengths that reach the earth in only small amounts, there are just two contenders: heat and light. In the chemistry of nonliving things, reactions dependent on light are uncommon. Heat is what usually provides the energy for those reactions and creates the conditions in which they take place; and in fact the temperature must be fairly high for many reactions to occur at all. But living organisms cannot use temperatures in this way because their complex molecules can exist and interact only within a very limited temperature range. It therefore follows that the only form of solar energy that remains to be used is the energy in visible light. And even this form of energy can be captured and used only by green plants, which contain the pigment called *chlorophyll*, a very complex compound that gives the familiar green color to vegetation. Chlorophyll, perhaps with the help of certain other pigments, is able to absorb light energy and use it to start the process called *photosynthesis*—the production of organic compounds such as glucose from carbon dioxide and water. Land plants take in carbon dioxide from the air and water from the soil; aquatic plants have water in plenty and dissolved carbon dioxide in the water. The basic equation for photosynthesis is:

$$6CO_2 + 12H_2O \longrightarrow C_6H_{12}O_6 + 6O_2 + 6H_2O$$

carbon   water   glucose   oxygen   water
dioxide

Green plants are thus able to build up organic foods from simple inorganic substances. The ingredients of photosynthesis outside the plant itself—water and carbon dioxide (and light)—are all free in that they exist in large quantities on earth, and existed even before life came into being. Green plants are thus not dependent on other living things for their "food" and it is for this reason that solar energy enters the ecosystem by way of photosynthesis. We shall follow this story further in the next two chapters.

Because photosynthesis provides the only supply of energy for the ecosystem, it is worth examining how the process works. It occurs in two stages—the *light reaction*, which requires light, and the *dark reaction*, which uses the products of the light reaction—and proceeds in both light and darkness. These stages of photosynthesis are shown in more detail in the diagrams on pages 44-45.

In the light reaction, light activates pigment molecules, including chlorophyll, so that electrons are expelled from them. The energy from these electrons is then transferred to reactions in which two important compounds— ATP (adenosine triphosphate) and $TPN.H_2$ (triphosphopyridinenucleotide) —are formed; without energy, these reactions do not occur.

$$ADP + P + Energy \longrightarrow ATP$$

adenosine   phosphate   adenosine
diphosphate   group   triphosphate

It follows that if ATP were to revert to ADP and P, this energy would be liberated. Both ATP and $TPN.H_2$ may be compared to tiny batteries capable of providing energy for work at any time, and their energy is required in the dark reaction. $TPN.H_2$ has another function: it contains hydrogen, which comes from water, and this too is required in the dark reaction. The water's oxygen is the source of the oxygen set free in photosynthesis.

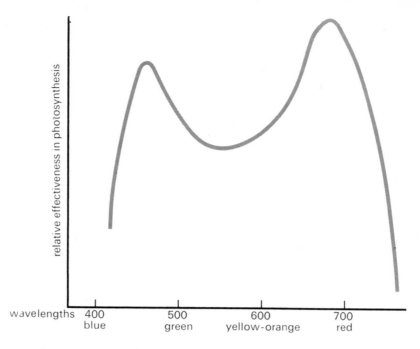

Right : diagram shows the relative effectiveness of different wavelengths of light in promoting photosynthesis in the green leaf of a higher plant. The curve shows that photosynthesis is most active in blue-green and red light. This kind of graph is called an action spectrum.

relative effectiveness in photosynthesis

wavelengths 400 500 600 700
blue green yellow-orange red

In the dark reaction, carbon dioxide finally combines with hydrogen from $TPN.H_2$ and is transformed into the carbon compounds of solid foods. As we have seen, the energy to drive these reactions comes from the breakdown of ATP and $TPN.H_2$. This energy is the chemical bond energy of all solid foods—and, indirectly, of fossil fuels, which are themselves the product of photosynthesis that took place millions of years ago.

By combining the organic compounds of photosynthesis with nutrients from the soil or water, plants form all the carbohydrates, proteins, and fats needed to build protoplasm. They obtain energy for protoplasm-building by the process of respiration, in which a part of the food formed in photosynthesis is broken down step by step with the help of oxygen. The process is gradual, and does not produce too much heat at any one stage; it liberates the potential bond energy in the food and locks it up again in ATP molecules. This process is basically the reverse of photosynthesis, as we can see from the equation that summarizes it:

$$C_6H_{12}O_6 + 6O_2 \longrightarrow 6CO_2 + 6H_2O + ATP$$

ATP is the main substance in both plants and animals that can store energy in a form in which it can be used directly for work. Substances such as glucose are storehouses of potential energy, but the living cell cannot use them to do work until the process of respiration has produced ATP. Without ATP no organism could emit sound or light, contract muscles, or transmit electrical impulses. But whatever the importance of ATP, the fact remains that the source of all the energy used by living things is light that enters green plants in photosynthesis. Without this energy and without green plants, the living world as we know it could not exist.

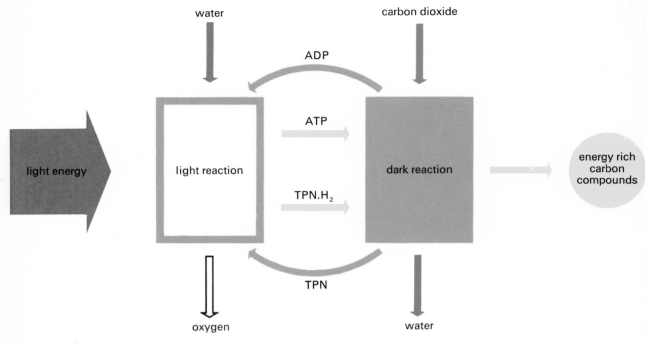

water   carbon dioxide

ADP

ATP

light energy

light reaction

TPN.H$_2$

dark reaction

energy rich carbon compounds

TPN

oxygen   water

$$6\,CO_2 + 12\,ATP + 12\,TPN.H_2 \longrightarrow C_6H_{12}O_6 + 6\,H_2O + 12\,ADP + 12\,P + 12\,TPN$$

Above : diagrammatic summary of the light and ark reactions of photosynthesis. In the light reaction TPN and ADP acquire energy by conversion into TPN.H$_2$ and ATP. This energy is used in the dark reaction to synthesize carbon-rich compounds such as glucose. ATP loses energy by reverting to ADP; TPN.H$_2$ gives up energy and becomes TPN. Both compounds are re-energized in the light reaction.

Below : the cyclic part of the light reaction. Light expels an electron from a chlorophyll molecule ; the electron's energy passes to an energy carrier at A; this carrier then transfers the energy to a reaction in which ADP and P form the molecule ATP. The electron that has spent its energy returns to the chlorophyll molecule. Expulsion of electrons is continuous during daylight.

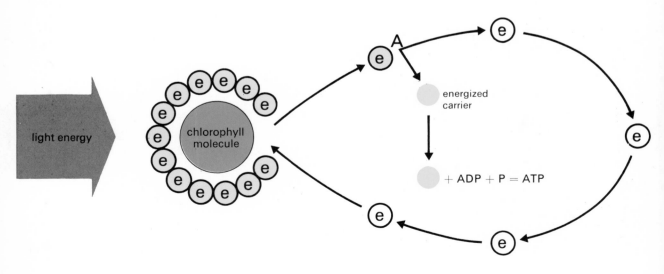

light energy

chlorophyll molecule

A

e

energized carrier

+ ADP + P = ATP

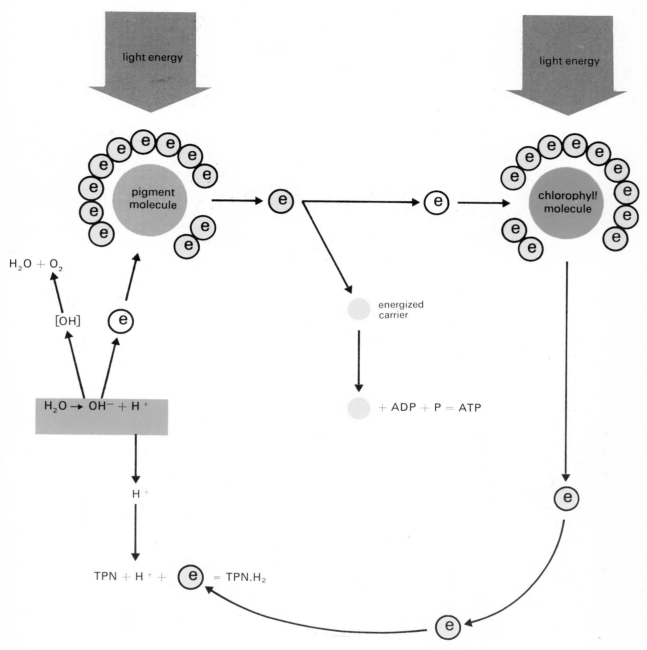

Above: the non-cyclic part of the light reaction. Light displaces an electron beam from a pigment (nature unknown), and carriers transfer its energy to a reaction in which ADP and P form ATP. The electron lost from pigment is replaced by one from a hydroxyl ion formed from the natural ionization of water. Loss of this electron turns the hydroxyl ion into a hydroxyl radical (OH), *several of which combine to form water and oxygen. Meanwhile light displaces an electron from chlorophyll, and a carrier transfers its energy to a reaction in which TPN and H ions from the ionized water form TPN.H₂. The electron lost from chlorophyll is replaced by the one previously lost from the pigment.*

The first forms of life on sterile ground are plants; animals arrive later, when plants have created suitable soil conditions. Here boulders are covered with various types of lichens—often one of the first plants to appear on barren ground.

3

Ways of Life

The world of living things is divided into two main groups. The first group consists of organisms that make their own food and build themselves up from simple inorganic materials such as water, carbon dioxide, and salts. These are the *autotrophs* (literally "self-feeders"). Another name for them is *primary producers*. The term is slightly misleading, because there are no *secondary* producers, but it does serve to emphasize the importance of the autotrophs in nature: all other living things depend directly or indirectly upon them. The second group comprises all living things that are not autotrophs. These are the *heterotrophs* ("other-feeders"); another name for them is *consumers*. Because many of the primary producers are familiar green plants and many of the consumers are large animals, the situation is conveniently summed up in the biblical phrase, "All flesh is grass."

Autotrophs

All the self-feeders contain chlorophyll, or chlorophyll-like substances, and all of them make organic matter from inorganic ingredients present in soil, rock, air, and water. To do so they need energy, and they do not have to obtain this by breaking down organic material. Not all of them rely on the same energy source. Some, called *phototrophs*, use light. Others, called *chemotrophs* (mainly certain bacteria and blue-green algae), obtain energy by oxidizing inorganic substances; for example, the soil bacterium *Nitrosomonas* obtains energy by oxidizing ammonia to nitrite. Chemotrophs are thus of great importance in the soil, particularly during the establishment of the first life on barren ground. In established ecosystems on land, however, the phototrophs play by far the larger part. They include mosses, ferns, conifers, and flowering plants such as grasses, shrubs, and trees. They also include certain plants called *algae*, and photosynthetic bacteria, although these are of only minor importance in established ecosystems on land.

In water, however, algae play a dominant ecological role. *Phytoplankton*—the plant component of the multitude of minute living things (plankton) that float near the surfaces of seas and lakes—includes a very high proportion of single-celled algae called *diatoms*; it also includes many chlorophyll-containing flagellates, which can also be regarded as single-celled algae. Although these organisms are not usually visible to the naked eye as individuals, their density in some parts of the sea may be as high as that of plant life on land. The only large plants in the sea are the multicelled algae called seaweeds. The largest British species, *Saccorhiza bulbosa* (the sea furbelow), grows up to 4.5 meters, but the American *Macrocystis* may measure as much as 60 meters.

Large plants of other kinds exist in fresh water, around the margins of lakes, ponds, and rivers. Some of them are rooted to the bottom, but others float freely at the surface. In tropical South America, which is their native habitat, the water hyacinth and the water fern occur as floating mats on lakes

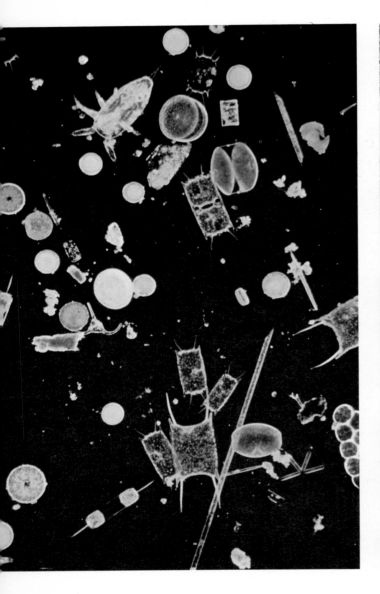

Microscopic marine phytoplankton, the pasture of the sea; mixed diatoms and dinoflagellates (left); living dinoflagellates (above). Large plants seldom occur free-floating at sea, in contrast with fresh water. Right: river choked with water-hyacinth.

and rivers; but where these plants have been accidentally introduced into other regions, they have run wild. In parts of Africa, for example, thousands of miles of rivers have been choked with water hyacinth, making navigation difficult, while in Sri Lanka (Ceylon) water fern has smothered many lakes and rice fields. This provides a good illustration of the way organisms may become uncontrollable and harmful when they are transferred without sufficient forethought to a new environment.

In order to make food, aquatic plants, like land plants, must have both nutrient salts and light. Fresh water usually contains less salts than sea water, and so it usually supports less plant life, but there are exceptions. Shallow estuaries and marshes that receive nutrient salts drained down from a wide area of surrounding land can be very fertile indeed. So, too, can rivers and lakes that receive sewage effluents, or agricultural fertilizers that have been washed off the land.

Because plants need light, most aquatic plant life is found near the surface of the water. By the time light has passed through 5 meters of water, roughly half of it has been absorbed, and at a depth of about 25 meters only three percent of it remains. Some sea plants have developed special pigments that enable them to utilize those wavelengths that remain at greater depths, and red algae, for instance, can live at depths of 100 meters. But in rivers and lakes the water is usually too opaque to support plant life at a depth of more than about 6 meters.

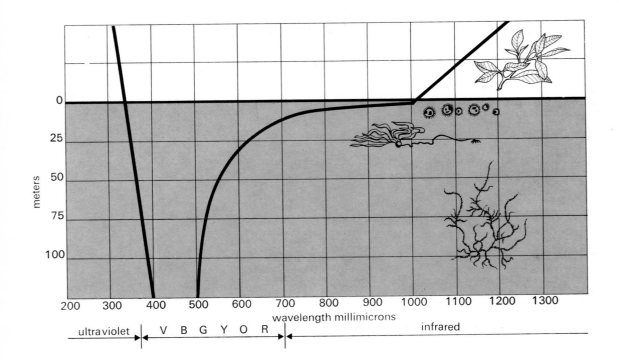

When sunlight, already filtered by the atmosphere, penetrates water, its spectrum is still further narrowed until the predominant color is blue. The solid blue lines show wavelength boundaries inside which 90 percent of solar energy is concentrated in the lower atmosphere and in the ocean. On right of diagram are shown the kinds of vegetation that flourish at different levels: in air, green leaves; near water surface, green algae; at 25 meters, brown algae; at 100 meters, red algae.

At certain times of the year a shortage of light limits the growth of many aquatic plants; and a shortage of nutrient salts can produce the same effect at any season. These two factors combined can produce some strange results. In some seas, for instance, there is an explosive burst of phytoplankton in the spring and fall, when strong winds mix the lower waters with the upper waters, bringing nutrient salts to the surface. But in the summer, when there is plenty of light, there is actually a decline from the spring abundance of phytoplankton because the essential salts in the surface water have already been used up.

Primary productivity, or the rate at which autotrophs develop in a particular region, is governed by light intensity, temperature, nutrients, water, and the ability of the entire ecological community to use and circulate materials. If all these factors are present to the right degree, productivity will be high, whether the region is grassland, forest, or water. But if just one factor is deficient, the region will be unproductive. Many deserts, for example, have sufficient nutrients and sunlight, as well as the right temperature, to support abundant vegetation; absence of water alone makes them infertile.

What, then, are the regions of highest primary productivity in the world? The American ecologist Eugene P. Odum recognized three main levels of fertility in the world. He considered that the most productive regions are certain shallow water areas, moist forests, alluvial plains that support natural communities, and areas of intensive agriculture devoted to productive crops such as sugar cane. The second most productive regions are grasslands, coastal seas, shallow lakes, and ordinary agricultural land. The least productive are certain oceans, very deep lakes, and deserts.

There are, however, some important exceptions. Even fairly deep seas can be highly productive where currents bring nutrients to the surface. For instance, part of the Pacific along the coast of Peru is densely populated with phytoplankton, and hence with fish and fish-eating birds, because the Humboldt current continually brings nutrient-rich water to the surface. Another exception is the coral reef, which, though surrounded by relatively unproductive tropical seas, is as productive as a tropical rain forest.

Heterotrophs

The autotrophs, or self-feeders, stand at the beginning of the story of food, in the role of primary producers. The heterotrophs, or other-feeders, include *all* other organisms, every one of which is a consumer and depends directly

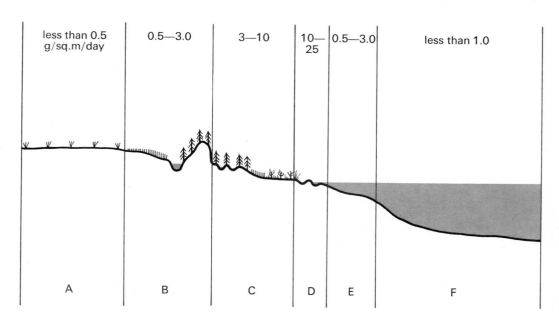

less than 0.5 g/sq.m/day	0.5—3.0	3—10	10—25	0.5—3.0	less than 1.0
A	B	C	D	E	F

The diagram shows the distribution, in six main groups, of primary production measured in grams per sq. meter, through a typical cross section of the biosphere: A=deserts; B=grassland, deep lakes, mountain forests, some farmland; C=moist forests, shallow lakes, moist grasslands, most farmland; D=some estuaries, coral reefs, springs, alluvial plains, intensive agriculture; E=shallow seas, continental shelves; F=deep oceans. Note by far the highest production is in section D.

or indirectly on autotrophs for its food supply. Heterotrophs must, in fact, feed on organic matter from plants or animals; and this organic matter may be in the form of living plants or animals, semidecomposed plants or animals, or organic compounds in solution. Biologists divide heterotrophs into three types—saprophytes, parasites, and holozoic organisms—according to their method of obtaining food.

Saprophytes (none of which are animal) feed on soluble organic compounds from dead animals and plants, most of them simply absorbing compounds that are already dissolved. However some fungi—molds and toadstools—and many bacteria also break down undissolved foods by secreting digestive enzymes on to them. *Parasites* are more difficult to define and we shall have more to say about them later on in this chapter. But for the time being we may describe them as organisms that at some time during their lives make connection with the living tissues of a different species, on which they rely for their food; this may take the form of soluble compounds, insoluble compounds, or both. There are many animal parasites, many parasitic fungi, and a few parasitic flowering plants, such as mistletoe. The *holozoic organisms* are all animals, and have the typical animal method of eating—by mouth. They absorb large food particles as well as soluble compounds.

Quite apart from their method of obtaining food, the heterotrophs can also be divided into different categories according to the number of removes at which they stand from the autotrophs. Those that live directly on autotrophs are collectively called *primary consumers*, or herbivores. Primary consumers

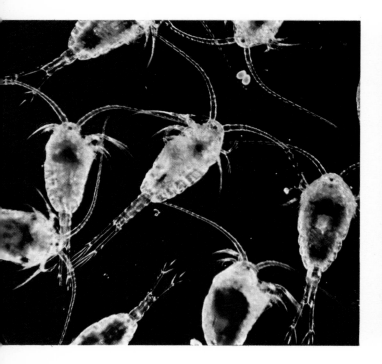
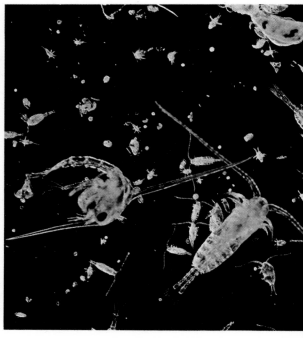

Primary consumers of phytoplankton in the sea are numerous species of zooplankton. Left: copepods of the genus Calanus. *Right: mixed zooplankton. Marine grazers are small, as befits their diet of microscopic phytoplankton.*

that feed on living plant material are called *grazers*. On land they include sheep, which feed on growing herbage; giraffes and goats, which feed mainly on leaves and young shoots; and also the many insect larvae that feed on stems, roots, and leaves. In the open sea the chief grazers are crustacea called *copepods*, which are only a few millimeters long. In coastal waters, common grazers include worms, mollusks, and some echinoderms, such as starfish and sea urchins. The small animals and animal larvae that graze on phytoplankton are called *zooplankton*. Most of them obtain their food by filtering the phytoplankton out of the water by means of bristly appendages.

Many organisms live on debris from decaying plants. They, too, are primary consumers. Such organisms, called *detritus feeders*, are important in the ecosystem because they help to return to the soil what growing plants remove. Plants take up nutrients from the soil, and if these were not replaced it would soon become sterile. As it is, a host of detritus feeders decomposes the dead plants, breaking down the complex molecules of which they are made into simple molecules that can be absorbed by other, living plants.

The soil contains an astonishing number of organisms. In one gram there may be millions of bacteria, and thousands of fungi, algae, protozoa, and flagellates. Most of these are consumers, and many are primary consumers. Bacteria are probably the most important decomposers in grassland, and fungi in forests. Soil organisms are more dependent on one another than communities of larger animals are, in that each of them can break down only part of the plant body. Some decomposers require cellulose or lignin, others

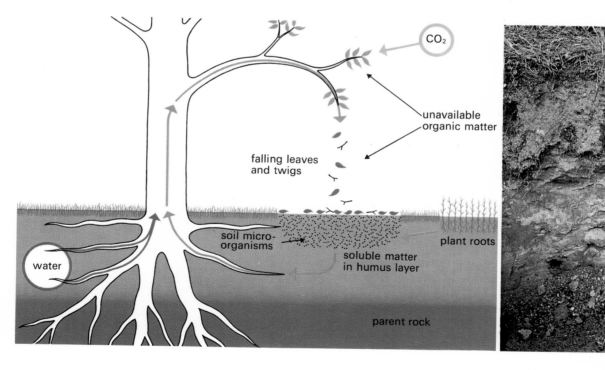

The nutrients required by this tree are drawn from the soil and replaced by the activities of detritus feeders working on the surface and below, in the humus layers (gray in the soil profile at right). Litter is converted to humus and nutrients.

require sugars as well. The sugars cannot start work on the plant until the cellulose or lignin has done its part.

Other soil detritus feeders include many small invertebrates such as earthworms, woodlice, millipedes, mites, springtails, and roundworms. All of these, although they are not usually considered as decomposers, consume dead plant material and convert much of it into food remains or feces that consist of smaller particles with a proportionately larger surface area. These small-volume, large-surface-area particles can then be more easily attacked by microorganisms. In northwest Europe, earthworms are especially important in this respect. Some invertebrates probably also make chemical changes to the food that passes through the gut and back into the soil.

Land detritus feeders, of course, have their aquatic counterparts such as worms, mollusks, and bacteria. Also, a few fish are detritus feeders—for example, those species of catfish that shovel sediment into their mouths.

Primary consumers are eaten, sooner or later, by *secondary consumers*, or carnivores (flesh-eaters). On land, the lion eats the wildebeest, the wolf eats the deer, and the hawk eats the fieldmouse; soil organisms that decompose herbivore remains are also secondary consumers. In water, many fish feed on zooplankton; for instance, the herring feeds on copepods. Many fish, however, feed on other fish, as the cod does on the herring. The cod and all other animals that feed on secondary consumers are called *tertiary consumers*.

An omnivorous (all-eating) animal may belong to all three consumer categories. A man, for instance, is a primary consumer when he eats lettuce, a secondary consumer when he eats mutton, and a tertiary consumer when he eats herring. Omnivores are very common in nature; even a full-blooded carnivore such as the tiger occasionally eats a little plant food.

In short, a plant is eaten by one organism, which is eaten by another organism, which in turn is eaten by yet another, and so on. This linear sequence of events, from the plant to the last carnivore, is called a *food chain*, and in general terms is written thus:

Plant ⟶ herbivore ⟶ carnivore 1 ⟶ carnivore 2

Alternatively, we can write:

Primary producer ⟶ Primary consumer ⟶ Secondary consumer ⟶ Tertiary consumer

Here are some examples of food chains:

1. Seawater diatom ⟶ copepod ⟶ herring ⟶ cod ⟶ man
2. Freshwater diatom ⟶ mayfly larva ⟶ caddis fly larva
3. Grass ⟶ vole ⟶ weasel

Food chains usually consist of only three or four links; we shall see in Chapter 4 why this is so. Strictly speaking, however, we should extend the chain to include the microorganisms that feed on the remains of the last carnivore, thus adding a few more links.

While it is convenient, and not incorrect, to talk of food chains, it is important to remember that isolated food chains do not exist in nature. This is because each link in one chain interconnects with many other chains. Take, for example, the food chain: diatom—copepod—herring—cod. The copepods feed on diatoms and flagellates, but the same diatoms and flagellates are also eaten by other zooplankton. The copepods in their turn are eaten, not only by herrings, but also by larval sand eels.

There is a special term to describe the whole complex of feeding relationships within an ecosystem: a *food web*. The diagram on this page, which includes the diatom—copepod—herring food chain, shows that a community's food web is extremely complicated. Even so it is worth trying to unravel its most important links. We may then estimate what would happen in the ecosystem if we interfered with one of the main links.

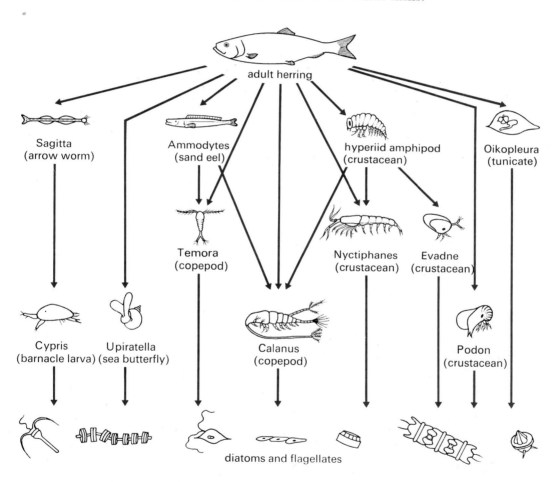

Diagram of interconnected food chains ending in a herring, and called a food web. It is as simple as possible; but if we introduced other species of plankton and fish, the food web would become extremely complicated, as it is in nature.

Associations between organisms

We can sum up what has been said so far about the feeding habits of organisms in the following way. All living things belong to one of four *trophic levels*: the primary producers (autotrophs), the primary consumers (herbivores), the secondary consumers (carnivores), and the tertiary consumers (also carnivores). All organisms that belong to the last three trophic levels are heterotrophs; some of them feed saprophytically, some feed parasitically, some feed holozoically.

However, many organisms obtain all or some part of their food supply, or derive other important benefits, by living in more or less close association with organisms of other species. Such associations vary in the degree of intimacy they entail, in the extent to which they affect either or both parties, in their degree of permanence, and so on. It is clear that if we are to discuss associations of different kinds, we need different names for them, and here a difficulty arises. For many years biologists have used such words as symbiosis, commensalism, and parasitism to denote various kinds of association, but as time has passed these terms have somehow come to be differently defined by different workers in different countries, and at present there is no universal agreement as to their precise meaning. But many biologists, especially in America, have now reached agreement on logical definitions, and these are the ones we shall use in this book.

The term that embraces all close associations is *symbiosis*, which comes from the Greek word "living together." It is used to denote any kind of association that involves two or more different species living together for all or part of their lives. Under this broad heading come three other terms, each of which denotes a different kind of living together. The first is *commensalism* (literally, "eating at the same table"); it refers to any association from which

Above: the remora fish, or shark sucker, hitchhikes to the scene of a kill, and picks up scraps. The remora and shark do not interfere with each other. This relationship is commensal. Right: Spanish moss is epiphytic, using trees and even *telegraph wires as an above-ground support. Far right: the egret and the rhinoceros form a symbiotic partnership for mutual benefit. The egret feeds on ticks that infest the rhino, and the rhino has the benefit of regular "de-lousing."*

one participant derives feeding benefits, or other important benefits, without affecting the other in any significant way. The second term is *mutualism*, which simply means living together to mutual advantage. The third is *parasitism*, which is used to describe the kind of association in which one organism derives its food from another, being more or less harmful to it and certainly never beneficial.

A good example of commensalism is the relationship between the two-foot-long remora fish and certain sharks. The remora fish has a sucking disk, developed from the anterior dorsal fin. With this disk it attaches itself to a shark (or, less commonly, to some other animal—even to a skin-diver). The shark, with the remora as passenger, swims to its own feeding ground, where the remora detaches itself and feeds on the scraps that the shark leaves. Afterwards it fastens on to the shark once more, and is transported to another feeding area. Clearly only one partner benefits, while the other remains virtually unaffected.

In some commensal associations one of the partners acquires ready-made housing, and is thus protected from the predatory world outside. The many organisms living in the cavities and canals of sponges are a good example—in a single loggerhead sponge as many as 17,000 organisms have been found. Some commensals require other organisms for support. Plants such as certain orchids and Spanish moss can be considered as commensals of the trees on which they grow because such plants rely on the tree for support, and for the special environmental conditions that exist at that level. The tree neither suffers nor benefits from the association.

Mutualism always confers benefits on all who share in it. One such association is that between several species of African birds and the rhinoceroses, elephants, and other large mammals on whose backs they commonly perch.

The bird enjoys a peaceful meal, feeding on the lice and ticks in the mammal's skin. In return, the mammal gets a delousing; in addition, it may receive a useful warning if the bird suddenly takes off in fright. British farmers have reported the same kind of association between starlings and sheep. Such associations are of a fairly casual kind, but there are other kinds of mutualism in which the two organisms cannot live apart.

What we call lichens are in fact dual organisms, consisting of algae and fungi living together in very close contact and utterly dependent on each other. The alga relies on the fungus to supply it with water, while the fungus utilizes food that the alga produces by photosynthesis—and that it could not produce without water.

Algae are also mutualistic with corals—small marine animals that secrete around themselves a hard limestone skeleton. The benefit that algae receive has long been known; they find a safe home by embedding themselves in the tissues and skeletons of the corals. But when the association was first investigated in aquaria, it was not clear what benefit the corals received, because they seemed capable of living quite happily without algae. When studies were made on corals in the sea, however, it became clear that there was not enough food (zooplankton) present to account for their yearly growth. Eventually the use of radioactive tracers revealed that in the relatively sterile conditions of tropical seas, the corals were indeed taking up organic substances produced by the algae. This shows, incidentally, how easy it is to come to wrong conclusions when ecosystems are simulated in the laboratory.

On land a particularly important example of mutualism is to be found between nitrogen-fixing bacteria and leguminous plants such as alfalfa and clover. The bacteria live in nodules on the roots; they convert atmospheric nitrogen gas into nitrogenous compounds that can be used by the legume. In return, the bacteria receive from the legume carbohydrates and safe housing. Most nonleguminous plants, which harbor no such bacteria, have to rely on nitrogen salts in the soil, and are unable to make direct use of the abundant nitrogen in the atmosphere. We shall see in Chapter 5 how this mutualism makes legumes so important in agriculture.

Mycorrhiza, or "fungus root," is the name given to an association between certain fungi and the roots of higher plants. It occurs among many food crops, but there we do not yet know its full significance. However, given certain soil conditions, most species of trees have also been found to contain mycorrhiza, and it is probably very common in natural forests. Here we do know that the association between tree and fungus is a truly mutualistic one, benefiting both partners. The soil must be well aerated and have a fair supply of organic matter, but it must also be deficient in nitrogenous compounds and other minerals. Where such conditions exist, trees tend to have a high carbohydrate content and a low protein content because of the lack of nitrogen. But fungi use sugars (carbohydrates) as their source of energy, and those that do not form a close association with other organisms must make sugars by

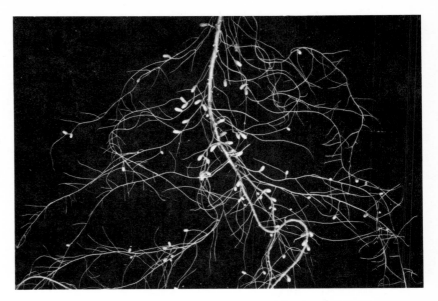

Right: part of the root system of a legume showing the nodules in which mutualistic bacteria are fed by the plant and in return fix atmospheric nitrogen into compounds useful to the plant. Photomicrograph (middle) shows a cross section of nodules packed with bacteria. These bacteria in legumes are different from free-living nitrogen-fixing bacteria in the soil.

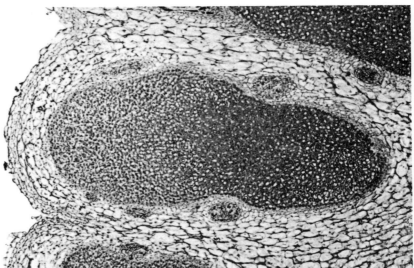

Below: mycorrhiza—a mutualistic association between fungus and root. The fungus receives carbohydrate from the root and gives nitrogen compounds in return.

An unusual mutualistic relation exists between the "cleaner" goby fish and the tiger rock fish. The cleaner rids the tiger of diseased sores and encrustations, thereby securing a square meal. The cleaner does not, apparently, depend on the tiger fish for nourishment, but the tiger fish becomes very unhealthy if deprived of regular cleanings.

breaking down the lignin and cellulose skeletons of plants. But those fungi that live in or on the roots of trees rich in carbohydrates can obtain their sugar supply far more easily by drawing on the dissolved sugar in the tree sap. In return, the fungus breaks down the organic matter in the soil into nitrogenous salts and passes them on to the tree, thus increasing its ability to synthesize proteins. In this association, part of the fungus is in the soil and part branches out inside the root of the tree. Some mycorrhiza exist only *inside* roots, but we do not know what their function is.

A particularly interesting mutualistic relationship exists between various species of cleaner fish and certain other fish, which we may conveniently call customers. The cleaner does the customer a good service by freeing it from external parasites and diseased or damaged tissue; in so doing, the cleaner obtains a square meal, and a change from its usual diet of plankton. As it swims toward a customer, the cleaner draws attention to itself—sometimes by performing a dance; meanwhile the customer lies still, often in a trance. The cleaner then scours the customer so meticulously that no part of it is left untouched. It even cleans the gills, teeth, and inside of the mouth, and one cleaner was actually seen to free the cornea of the eye from encrustations. If the customer moves during this operation it may receive a sharp jab from the cleaner. Though some customers are large predatory fish, they rarely eat or even harm the cleaner.

Most customers come across cleaners by accident, while traveling from one feeding area to another, but certain fish go for regular checks at special cleaning areas, and some even go regularly to one particular cleaner fish. The analogy of a visit to the barber is irresistible. At cleaning stations, fish customers wait their turn, often jostling for attention, and though many of them are incompatible or predators, they behave rather like animals at a waterhole, rarely attacking each other while waiting. In some coral reefs there are thousands of cleaners and customers, but research has shown that such reefs are only one type of cleaning station in the sea, and that a number of profitable fishing grounds are in fact cleaning stations. Before this was known such areas were a mystery, because only grazing and spawning grounds were believed to attract such large numbers of fish.

Parasitism, unlike mutualism, benefits only one party. The parasite obtains food, and sometimes shelter, while the other party—the "host"—gains nothing and usually suffers to a greater or lesser extent. Nevertheless, parasitism is a true form of symbiosis, or living together, and is very different from the relationship between predator and prey, which culminates in the death of the prey. Indeed a predator lives by killing its prey, whereas it is in the interests of a parasite that its host should remain alive, for if it happened to kill the host it would also destroy the feast. Thus, in parasitism, if any deliberate killing is done, it is normally done by the host trying either to rid itself of a nuisance or to repair injuries it has received. In ordinary speech most of us use the word

Hookworm is today reckoned to be the biggest single cause of ill-health, over 600 million people being affected. Right: life history starts with larvae penetrating bare feet or wrists, traveling to lungs, coughed up, swallowed, taking up residence in the gut (below, see photomicrograph of head, and cutting teeth colored brown). Hookworms live on blood, causing severe anemia; they lay their eggs, which pass out with feces. The only control is to wear shoes and ensure a proper disposal of feces.

parasite as a term of abuse, but we should remember that it is an extremely widespread way of life, and one that considerably extends the range of habitats available to living things.

Sometimes the harm a parasite does is so slight that the health of the host is virtually unaffected. Few of us, for example, are aware of the roundworms (nematodes) that live in the gut where our food is digested. Other parasites, such as the viruses that parasitize our respiratory passages and cause the common cold, have a mildly adverse effect on us, the hosts. Still others, of course, do immense harm; malaria, bilharziasis, pneumonia, polio, cholera, and dysentery are all caused by parasitic microorganisms. Yet, though we tend to think that harmful parasites predominate, they are actually in the minority. Most of the harmful ones have entered into association with the host relatively late in the course of their evolution, so that host and parasite have not yet had enough time to adapt to each other.

Man himself has been afflicted with a great variety of dangerous parasites because his travels have brought him into contact with parasites to which he has not already become adapted. Domesticated European cattle suffer in a similar way when they are introduced into Africa. Once infected by the protozoan that causes sleeping sickness, they die quickly, but the native game, which has become adapted to the protozoan, often survives. If parasitism is to succeed, adaptation is essential, because a parasite that wipes out a whole host community will itself become extinct.

There is a great variety of habit and structure among parasites, and the degree of association is also very variable. Endoparasites, which live inside a host, are closely associated with it, and leave it only on dispersal to another

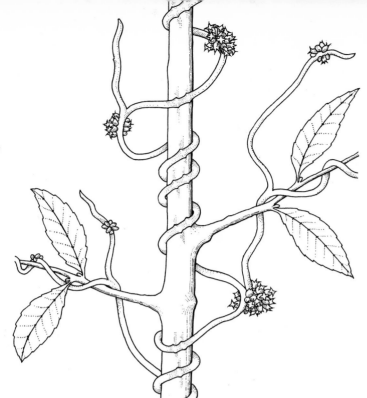

Left: a flea sucking blood from a man's thumb. Because it lives outside its host it is called an ectoparasite, *in contrast to the hookworm (previous page), which lives inside the host and is called an* endoparasite.

Right: the dodder is an unusual flowering plant; it lacks roots and leaves, and also chlorophyll, so it cannot synthesize its own food. Instead, it twines around other plants and penetrates the host stem to draw on the host's nutrients.

host. Ectoparasites, such as lice and fleas, live on the surface of a host, and establish a connection with it only when they suck blood, even though they may live on it for most of their lives.

Mosquitoes, on the other hand, can be called *temporary* parasites, because their direct association with the host is limited to the time spent on it during periodic bloodsucking visits. Finally, there are the organisms that become parasites only when opportunity arises: these are *occasional* parasites. Maggots of blowflies and bluebottles, for example, normally feed on decaying flesh and vegetable matter, but the adults may lay eggs in septic wounds. The maggots then become parasitic and feed on the dead flesh of a living host.

Most animal parasites are small invertebrates, but a few, such as the hagfish and the bloodsucking bats, are vertebrates. Animal parasites either parasitize other animals, as do the tapeworms that live in pigs, or parasitize plants, as do gallflies, which lay their eggs inside leaves. Most plant parasites are fungi, and most of their hosts are other plants; among the common plant diseases they cause are mildew, potato blight, and rusts. But some fungi also infect animals; ringworm and athlete's foot are two fungal diseases of man. There are also some parasitic higher plants—such as dodder (a member of the Convolvulaceae), which lacks chlorophyll. It twines around other plants, its suckers penetrating the host stem and drawing on the food circulating within it. Mistletoe by contrast is a *partial* parasite, supplementing its own photosynthesis with food from the tree upon which it grows.

All parasites have evolved from free-living ancestors, and in the process many have developed special adaptations that make them quite different from their nonparasitic relatives. There is, for instance, a parasite called

Sacculina that afflicts crabs; its adult form consists of no more than a bag of reproductive organs, which is nourished by tubes that branch out all through the crab's body. Although *Sacculina* is a crustacean, it is recognizable as such only in its free-swimming larval stages.

We have seen that plants have a very wide range of feeding methods. Most of them are autotrophs and make their own food; but some live as saprophytes, relying on decaying plant and animal substances, and some maintain a symbiotic association with other organisms. Other possibilities still remain: certain fungi, for instance, may live either as saprophytes or as parasites, while a few kinds of algae may be either autotrophs or saprophytes. One group, however, fits into none of these categories and is, in a sense, carnivorous—the insect-eating plants. These plants often grow on soils poor in nitrogen, and they have

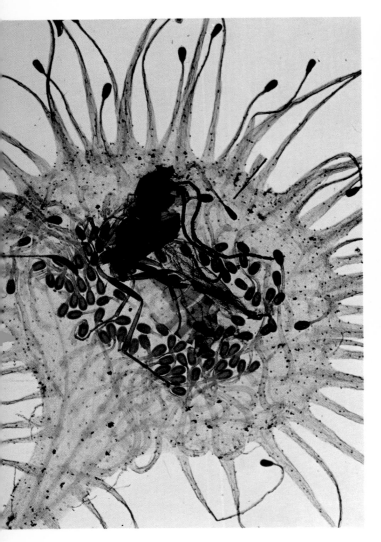

Three carnivorous plants that supplement their normal food by devouring insects. Glistening droplets on the leaves of the sundew (above) attract insects, which get stuck and enfolded by tentacles (left). The Venus's-flytrap (right top) has bilobed leaves with sensitive sticky hairs. Insects are trapped by bristles when the halves of the leaf close on the victim (far right).

The aquatic bladderwort (right) has small bladders fitted with trapdoors that spring open when an organism touches them. The bladder expands creating an inrush of water and victim.

solved this problem by feeding on protein-rich insects. Pitcher plants, for example, have some leaves constructed like cups, which contain a watery fluid. When an insect is attracted to the rim of the cup by sugar, it tumbles over and slides down the slippery surface and into the fluid. In some pitchers, downward-pointing spines prevent the insect from climbing out. Protein-digesting enzymes in the fluid, and possibly also bacteria, then digest the insect, and the products are absorbed by the plant. The Venus's-flytrap is another well-known insect-eater. Its leaves have two lobes and are fringed with stiff bristles. When an insect touches the sensitive hairs on the leaf, the lobes snap shut so that the bristles interlock, leaving the trapped insect to be digested. The Venus's-flytrap makes an interesting but macabre house plant, and if flies are scarce it accepts small pieces of meat.

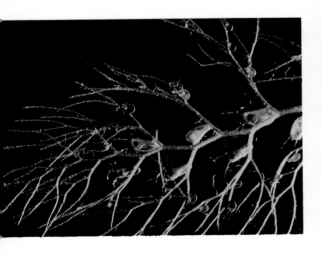

*Though animals eat either plants or other animals,
the end product of the food chain is the return of
soluble material to the soil. Here two carnivorous
stages of a food chain are shown: vultures eating
the remains of a dead tiger.*

4

Energy Flow

In Chapter 2 we explained how light energy from the sun enters the world of living things by way of green plants, and how they use it in the process of making food from inorganic ingredients. That food then becomes the source of energy for all life processes, including growth, not only for the primary producers or autotrophs themselves, but also, indirectly, for all other trophic levels—primary consumers (herbivores), secondary consumers (carnivores), and tertiary consumers (other carnivores).

This passing-on of energy in the form of food from one trophic level to the next involves many energy transformations, and we have seen that at every transformation some energy is dissipated as heat. There are thus losses of usable energy at each stage. So in any *self-contained* ecosystem we should expect to find most food in the primary producers, less in the primary consumers, less still in the secondary consumers, and least of all in the tertiary consumers.

This is, in fact, what we do find, but that is not the whole story. Energy losses at each stage vary considerably from one ecosystem to another. By studying these variations we may discover if it is possible to increase our own food supply sufficiently to provide for the huge increase in human population that is likely to come about in the near future. But before we turn to this all-important topic, we need to look at some of the devices that ecologists employ to represent the flow of energy in ecosystems.

Ecological pyramids

In all ecosystems many food chains are interlinked to form highly complex food webs. These involve many different autotrophs, many different herbivores, and many different carnivores. To show the flow of food (and energy) through such a web is extremely difficult. Attempts have therefore been made to devise diagrams that simplify matters by dealing with each trophic level as a whole, placing all the primary producers in one group, all the primary consumers in another group, and so on. Such devices are called *ecological pyramids*, and several examples are shown in the figures on page 68.

The figures on page 68 are constructed by counting all the individual primary producers in a given ecosystem and drawing a rectangle of fixed depth

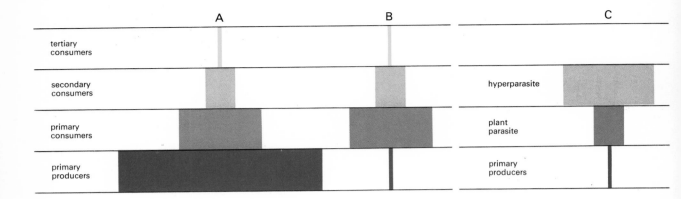

whose length represents that number; the same thing is done for the other trophic levels, in sequence. Finally, all the rectangles are placed one on top of another, in correct trophic order. In figure (A) the result is an upright pyramid, which is what we might well expect, because we know that the amount of food available decreases at each step. Indeed, an upright pyramid of numbers of this kind is obtained whenever there are more autotrophs than herbivores, and more herbivores than carnivores, in an ecosystem. It is the kind of result we should get for a blue-grass field where a few birds and moles feed on larger numbers of beetles, spiders, and ants, which in turn feed on larger numbers of herbivorous insects, which in their turn feed on even larger numbers of individual grass plants.

In ecosystems where the autotrophs are small, large numbers of them are needed to support comparatively few primary consumers. However, where the autotrophs are large, they may easily be outnumbered by the herbivores, and the pyramid will then look like figure (B), where the two lowest rectangles might represent insects feeding on a few large trees. With parasitic food webs, we actually find ourselves drawing inverted pyramids, as in figure (C). Such a pyramid might represent a tree that supports many mistletoe plants, which are themselves parasitized by even more fungi.

It is obvious that if an ecological pyramid is constructed merely according to the numbers of organisms at each trophic level, it forces us to compare organisms of very different sizes, such as a diatom and a tree, or a mouse and an elephant. It is usually more instructive to compare the *weight* of organisms in each level. The weight per unit area, or *biomass,* can be expressed in one of two ways: either in terms of the gross weight of organisms, including the weight of their water content, or in terms of dry weight—that is, the weight without their water content. When our main purpose is to make comparisons, dry weight is the better measure, because the proportion of water in different organisms varies considerably.

Because it is impossible for a given weight of producers to support a greater weight of consumers, it looks as if pyramids of biomass must always be upright, like figure (A) on page 69. Yet inverted pyramids of biomass do sometimes

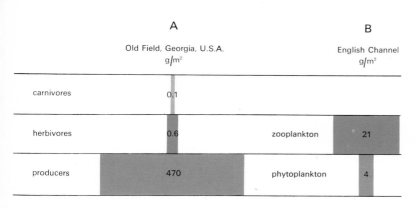

A

Old Field, Georgia, U.S.A.
g/m²

B

English Channel
g/m²

carnivores	0.1		
herbivores	0.6	zooplankton	21
producers	470	phytoplankton	4

Far left: pyramids of numbers; the areas of the rectangles denote numbers of organisms at each trophic level. In A, primary producers are small and more numerous than the large primary consumers. In B, primary producers are large and one plant supports many herbivores. In C, a parasitic system produces an inverted pyramid. Left: pyramids of biomass; numerals are grams dry weight per square meter. A is a field in America; B is the English Channel, where fast-breeding phytoplankton are grazed so fast by zooplankton that biomass is small.

occur, as figure (B) shows. This anomaly arises when the time factor is left out. Pyramids of biomass denote the *standing* crop—that is, the biomass at any particular time, or during some arbitrary period; that period may be as little as a month for the trees in a forest, or a day for algae. But a tree may take five years to grow to a certain size and to reproduce, whereas an alga, during the same period, may produce millions of progeny that together weigh more than the tree. So the standing crop of the tree represents material accumulated over a very long period of time, whereas that of the alga represents an accumulation over a very short period. A similar difficulty arises with herbivores and carnivores. Their weight in an ecosystem may also need to be measured for longer or shorter periods, because of differences in growth rate. Pyramids of biomass,

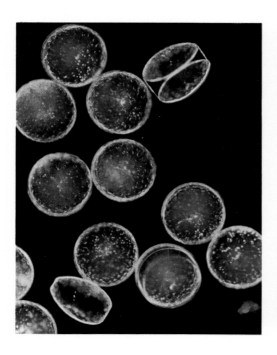

The marine algae (left) are smaller, more numerous, and weigh less, than the beech trees on the right. Algae take only a few days to reproduce, whereas beech trees do not become fruitful until approximately 35-40 years old.

then, do not indicate the total amount of food produced or made available to the next trophic level in a fixed time, such as a year; that is, they do not indicate the rate at which it is produced.

We now have an explanation for the anomalous shape of figure (B), which represents the phyto- and zoo-plankton of the English Channel. The phyto-plankton reproduce much faster than the zooplankton and if they were sampled over a whole year, the standing crop would appear much larger. But most of them are eaten as soon as they appear, and, as a result, it is possible to find a smaller biomass of phytoplankton than of zooplankton at certain times of the year.

As these examples show, to consider trophic levels solely in terms of numbers or biomass is rather like budgeting for a year on the basis of one month's accounts. In ecology it is useless to estimate one month's income and expenditure of energy and then multiply by 12, because plants and animals receive and expend different amounts of energy at different times of the year. In nature, as in economics, it is the long-term view—that is, the rate of accumulation, or *productivity*—that really matters.

As far as animals are concerned, food counts only as a supply of materials for growth and energy. It is not really important whether it consists of many small organisms or a few larger ones. Neither does it much matter whether it is heavy or light, since one food may contain 3.5 kcal per gram dry weight, while another food may contain 7 kcal. In other words, the weight of food consumed can give a very misleading impression of calorific intake. The usual formula, 1 gram $= 4$ kcal, is an average, and may be quite wrong in specific cases.

So if we want to trace the flow of energy through an ecosystem, neither pyramids of numbers nor pyramids of biomass prove wholly satisfactory. Recently, however, ecologists have learned to construct far more revealing diagrams, called *pyramids of energy*. In these, each trophic level still has its own rectangle, but each rectangle denotes the energy available per unit time. (This is usually, though not always, expressed in kilocalories per square meter per year, or $kcal/m^2/yr$.)

Pyramids of energy are always upright, as can be seen in the figure on page 71, but they do not always taper in the same way. If we compare such pyramids for two different ecosystems, the differences in tapering show at a glance the differences in energy losses at various trophic levels. As we shall see, it is these differences that may point to new methods of increasing man's food supply.

However, it is easier to begin by examining the flow of energy in a general way, by looking at what happens in an unspecified ecosystem rather than trying to follow in detail what happens in any particular one. Let us assume that light energy reaching our ecosystem from the sun is 300,000 $kcal/m^2/yr$. Of that total, the plants will use for growth as little as one percent. In other words, about 3000 $kcal/m^2/yr$ will end up in the bond energy of plant proto-plasm. The amount of protoplasm formed per unit area per unit time is called the *gross production*.

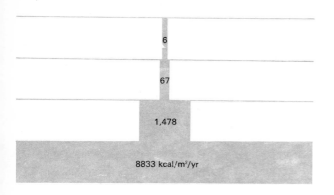

6

67

1,478

8833 kcal./m²/yr

Much of the great loss between light energy received and energy turned into plant protoplasm can be attributed to the inefficiency of photosynthesis. Indeed, the process is about as efficient as James Watt's early steam engines and bears no comparison with the efficiency of a diesel engine—about 45 percent. But neither machines nor organisms can be considered in terms of efficiency alone; just as important is what they are capable of doing, how accurately they can work, how long they can last without repair, and so on. Photosynthesis may seem to us to be relatively inefficient, but because no man-made machine can yet use light energy to synthesize foods in enormous quantities, no practical comparison can reasonably be made.

All the autotrophs in a self-contained ecosystem are eventually eaten by herbivores—either by grazers or by detritus feeders that live on plant remains. However, herbivores do not eat all the material formed by photosynthesis, because the plants themselves use at least one sixth of it in respiration, when it is broken down into carbon dioxide and water in order to supply energy. This energy, as we saw in Chapter 2, is used for growth, active transport within the plant, and so on. Each time work is done inside the plant, some of the converted energy escapes as heat, and this heat raises the temperature of the atmosphere, though the rise is negligible. Because a sixth of the 3000 kcal absorbed in the original plant protoplasm is lost in respiration, only about 2500 kcal are left for the primary consumers, or herbivores. This figure, the gross production minus the respiratory losses, represents the *net production*.

Plants lose energy in two processes—most in photosynthesis, some in respiration—whereas animals lose it mostly through respiration. The muscular work that animals do demands a considerable amount of energy, and so the herbivores' loss of energy through respiration is very large—often as much as 90 percent. Thus, of the 2500 kcal consumed by herbivores, not very much more than 10 percent goes to form flesh—about 250 kcal. This is the amount of energy available to the secondary consumers, or carnivores that feed on herbivores. Of the 250 kcal taken in by these secondary consumers, only about 25 kcal go to form flesh—again an efficiency of 10 percent. And of the 25 kcal consumed by tertiary consumers, the very small figure of 2.5 kcal remains as flesh—once more an efficiency of 10 percent.

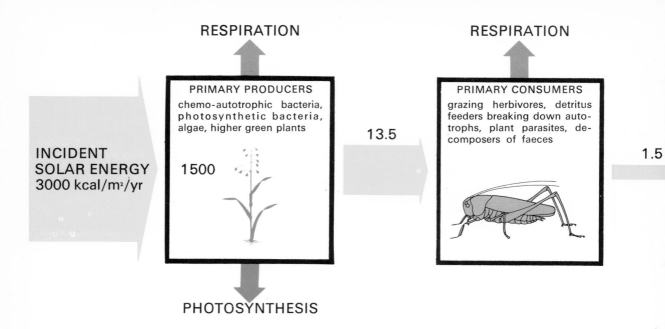

RESPIRATION

RESPIRATION

PRIMARY PRODUCERS
chemo-autotrophic bacteria, photosynthetic bacteria, algae, higher green plants

1500

13.5

PRIMARY CONSUMERS
grazing herbivores, detritus feeders breaking down auto-trophs, plant parasites, decomposers of faeces

1.5

INCIDENT
SOLAR ENERGY
3000 kcal/m²/yr

PHOTOSYNTHESIS

The numerals in the food chain shown above represent the amount of energy present in protoplasm at each stage of the average food chain; in other words, the net production. This

figure is also the amount of energy that is consumed by the next trophic level. Much of this energy, however, is lost in respiration—mainly as heat—and so much less energy is available for the

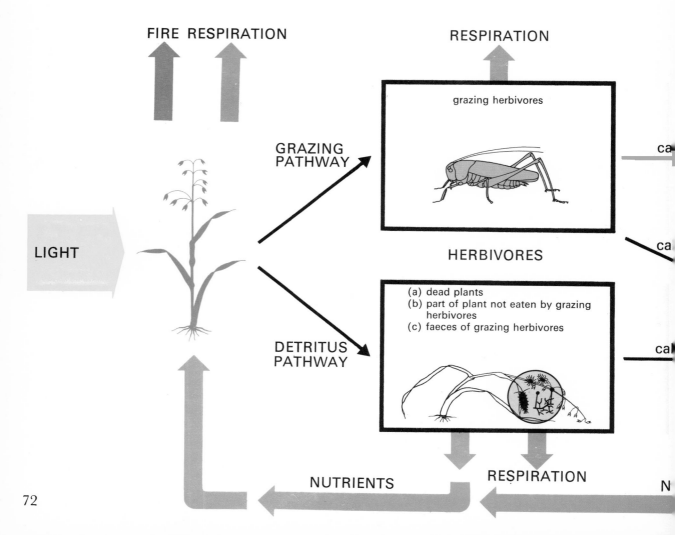

FIRE RESPIRATION

RESPIRATION

grazing herbivores

LIGHT

GRAZING
PATHWAY

ca

ca

HERBIVORES

(a) dead plants
(b) part of plant not eaten by grazing herbivores
(c) faeces of grazing herbivores

DETRITUS
PATHWAY

ca

NUTRIENTS

RESPIRATION

N

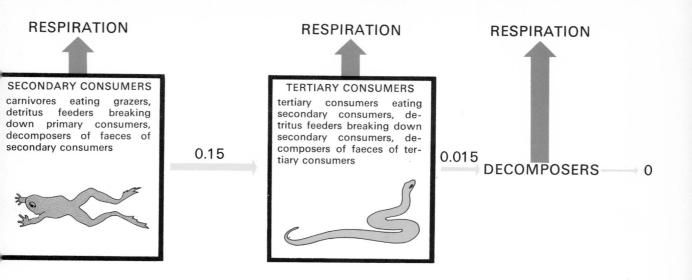

RESPIRATION

RESPIRATION

RESPIRATION

SECONDARY CONSUMERS
carnivores eating grazers, detritus feeders breaking down primary consumers, decomposers of faeces of secondary consumers

0.15

TERTIARY CONSUMERS
tertiary consumers eating secondary consumers, detritus feeders breaking down secondary consumers, decomposers of faeces of tertiary consumers

0.015

DECOMPOSERS

0

next stages; in fact, at the last stage only a minute fraction of the original solar energy remains. Below: when studying the main food chains in an ecosystem, it is better to separate the grazers, which feed on living organisms, and the detritus feeders (decomposers), which feed on dead organisms, into distinct pathways, both of which are equally important in ecosystems.

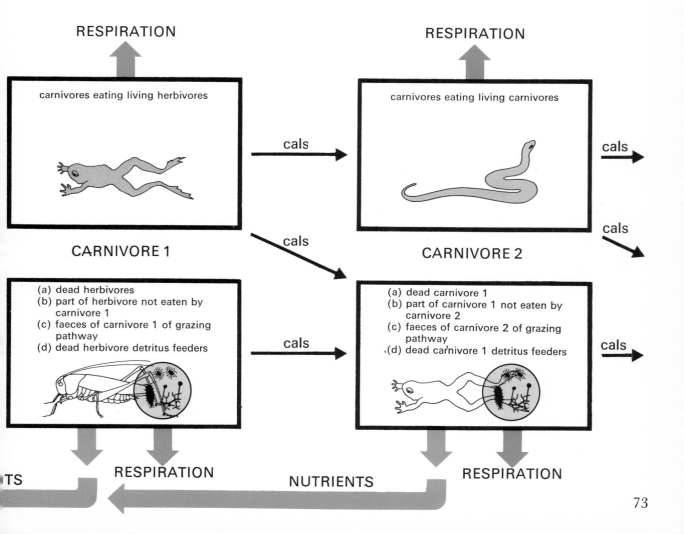

RESPIRATION

RESPIRATION

carnivores eating living herbivores

cals

carnivores eating living carnivores

cals

CARNIVORE 1

cals

CARNIVORE 2

cals

(a) dead herbivores
(b) part of herbivore not eaten by carnivore 1
(c) faeces of carnivore 1 of grazing pathway
(d) dead herbivore detritus feeders

cals

(a) dead carnivore 1
(b) part of carnivore 1 not eaten by carnivore 2
(c) faeces of carnivore 2 of grazing pathway
(d) dead carnivore 1 detritus feeders

cals

TS

RESPIRATION

NUTRIENTS

RESPIRATION

73

Our story is summarized in the top figure on pages 72–73, which shows how the energy in a self-contained ecosystem is lost at each stage of the flow from plants to the last carnivores. Within such an ecosystem, the total quantity of materials in circulation remains more or less constant, and is used over and over again; but the amount of usable energy decreases sharply from one trophic level to the next, so that by the time we reach the tertiary consumers very little of it remains. This explains why food chains seldom have more than four links; beyond the tertiary consumers too little energy normally remains to support further links. The great run-down of energy between primary producers and the last carnivore is made good only by the continual supply of solar energy that enters the ecosystem by way of green plants.

In this figure we have not distinguished at each trophic level between detritus feeders and the organisms that eat living organic matter; the distinction is of little importance when we are concerned only with the circulation of energy and the passage of nutrients in a single self-contained ecosystem. It is usual, however, to chart detritus feeders and other organisms separately, as in the bottom figure. When we compare two or more ecosystems, as we shall now do, we shall find that making this distinction helps to show up the major differences between the ecosystems.

Energy flow in autotrophs

The flow of energy through the plant trophic level in general is shown below (left). The central rectangle represents the stock, or standing crop; the channels from it represent the flow of energy and organic matter. Using diagrams of a similar kind, the other figures compare the energy flow for five different plant systems. (One need not pay much attention to the standing

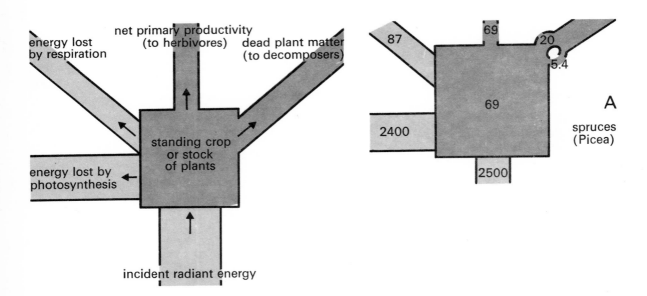

The energy flow through five plant trophic levels. The energy paths use the Standard Nutritional Unit, where one unit equals 10^6 kcal per hectare (2.47 acres) per year. The central rectangle represents the calories in the standing crop and its *relationship with the other channels is arbitrary. The width of channel denotes their energy contents and is logarithmic. Where values could not be found experimentally, it was assumed that 1 gm. dry weight organic matter contains 4 kcal.*

crop figure in the central rectangle, because we have seen that the standing crop bears no relation to the total productivity over a whole year.)

The plant systems in figures (A) to (E) have two features in common: most of the light energy reaching the plant is lost—mainly through the inefficiency of photosynthesis; and the net production is between 80 and 90 percent of the gross production. In every other respect, all the examples differ, often markedly.

The most productive of the primary systems shown is the managed forest of *Picea* (spruce) in Britain, where man cuts down and removes the timber every 21 years and then replants and refertilizes the soil. In a natural forest, of course, most of the calories locked up in the timber would eventually be eaten by detritus feeders, when branches and trunks died and fell to the ground. The main difference between natural and managed forests thus lies in the proportion of the standing crop that passes through the detritus pathway. We can also see that in this managed forest less energy is lost through photosynthesis than in other examples, but that there is a greater loss from respiration. In the grassland example, we can see that a greater proportion of its standing crop is consumed by detritus feeders.

The most striking point about the algal examples—figures (C), (D), and (E)—is the minute size of the standing crop. Aquatic algae absorb and produce about the same amount of energy as do terrestrial plants. They multiply rapidly, but they are grazed so quickly by herbivorous plankton that the standing crop remains small. We have not even charted the detritus pathway for the marine algae shown—(D) and (E)—because so few of them live long enough to die naturally. By contrast, the algae in the salt marsh at Sapelo Island—figure (C)—live in the mud, out of reach of herbivorous plankton, and are eaten mostly by detritus feeders.

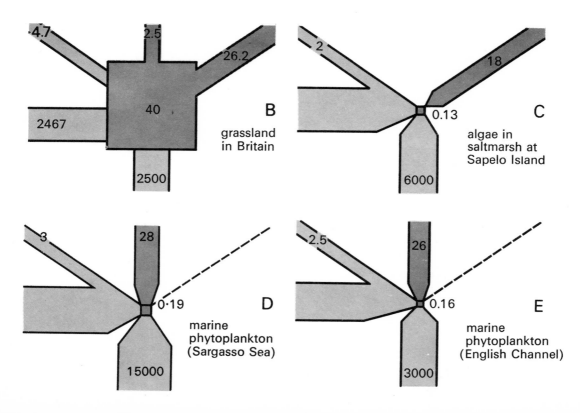

B
grassland
in Britain

C
algae in
saltmarsh at
Sapelo Island

D
marine
phytoplankton
(Sargasso Sea)

E
marine
phytoplankton
(English Channel)

It is a general rule that in land ecosystems the detritus pathway is greater than the grazing pathway, and the standing crop of plants is larger than that of animals; in aquatic ecosystems, on the other hand, the grazing pathway is greater, and there is a larger standing crop of animals (zooplankton, fish, bottom fauna) than of plants (phytoplankton). There are, however, many exceptions to the general rule so far as land ecosystems are concerned. In some heavily grazed grasslands in temperate climates, for example, as much as 50 percent of the annual net production passes through the grazing herbivores, such as sheep, cows, and goats.

Having compared the flow of energy through various autotrophs in various ecosystems, we may ask whether such comparisons can be put to practical use. We shall see that they can, but we must bear in mind that ecological considerations are not the only ones that matter: social, economic, and even aesthetic factors must also be taken into account.

ECOSYSTEM	grams/sq. meter year	grams/sq. meter day	grams/sq. meter day during growing season
NET PRIMARY PRODUCTIVITY IN VARIOUS ECOSYSTEMS AS DRY WEIGHT ORGANIC MATTER			
Wheat, world average	344	0.94	2.3
Wheat, average in areas of highest yield (Netherlands)	1,250	3.43	8.3
Rice, world average	497	1.36	2.7
Rice, average in areas of highest yield (Italy & Japan)	1,440	3.95	8.0
Sugar-beet, world average	765	2.10	4.3
Sugar-beet, average in areas of highest yield (Netherlands)	1,470	4.03	8.2
Sugar-cane, world average	1,725	4.73	4.7
Sugar-cane, average Hawaii	3,430	9.4	9.4
Sugar-cane, maximum Hawaii under intensive cultivation	6,700	18.35	18.4
Mass algae culture, yield under intensive cultivation and added nutrients, Tokyo	4,530	12.14	12.4
Natural club-rush in temperate climate	4,674		
Natural pine forest, average during years of most rapid growth (20-35 years old) England	3,180	6.0	6.0
Natural deciduous forest, England	1,560	3.0	6.0
Natural tall grass prairies (Oklahoma & Nebraska)	446	1.22	3 0
Natural desert, 5 inches rainfall, Nevada	40	0.11	0.2
Natural seaweed beds, Nova Scotia	358	1.98	1.0
SECONDARY AND TERTIARY PRODUCTIVITY OF FISH AS WET ORGANIC MATTER (TO COMPARE FIGURES WITH PRIMARY PRODUCTIVITY DIVIDE BY ONE HALF)			
Herbivores and carnivores in North Sea	1.68		
Herbivores and carnivores in African Lakes	0.16- 25.2		
Stocked carnivores, US fish ponds (sports)	4.5 - 16.8		
Stocked herbivores, German fish ponds (carp)	11.2 - 39.0		
Stocked carnivores, US fish (sports), fertilized waters	22.4 - 56.0		
Stocked herbivores, Philippine marine ponds, fertilized waters	50.4 -101.0		

Right: since the Industrial Revolution, farmers have learned to produce heavier crops per unit area. Top left: the Romain steam cultivator (1857) was versatile compared with specialized modern machines, such as the pea and bean cultivator (other photographs). With this cultivator one man can deal with a field more quickly than 20 of his predecessors. But increased productivity means a greater drain on the nutrients in the soil and hence the need to add synthetic fertilizers.

One practical way of increasing the amount of plant food available to us is to find out which plants show the highest productivity and cultivate accordingly. It is known that many plants that grow in natural ecosystems show higher productivities than many of our agricultural crops (table on p. 76). There is nothing to prevent us from cultivating such plants, but there is no guarantee that people would want to use them as food if we did. Man's eating habits are extremely conservative—of all the 300 different species of plants used for food during the history of agriculture, only 15 are now eaten in any quantity! The most immediate task, therefore, may be to improve existing crops. There is still room for improvement in cultivation and in harvesting techniques, and in most parts of the world the yield of existing crops could, by such means, be doubled. Indeed, with fertilizers, a tenfold increase in productivity is sometimes possible. Also in some parts of the world, including large areas of India, at least half the produce is lost by disease, spoilage, and

predators—especially by rats. If we could increase the usable yields of existing crops there would be less need to turn to others.

Another possibility is to look to highly productive natural ecosystems that have rarely been utilized at all, such as the rain forests in the tropics, and the fertile reed swamps in temperate regions. The British scientist N. W. Pirie, working at the Rothamsted agricultural research station, has shown that protein extracted from the leaves of trees is as good as that from fish meal; it could be successfully mixed into cattle fodder, and even into human food. We might also tap the enormous detritus energy-paths in such land ecosystems as grasslands and forests. In other words, we might use the tough cellulose and lignin skeletons of plant cells as human food. Because, unlike ruminants, we cannot turn cellulose and lignin into digestible form, we should need machines to do it for us. Several such machines have already been designed in the laboratory, including one that makes milk from grass. But most foods produced by such means, though digestible, do not yet have the same flavor as their natural counterparts and much education will be needed to persuade people to accept them as a normal part of their diet. Nevertheless, as world population increases, it will doubtless become more common to turn to unusual plant sources, especially to the detritus pathways, as well as to concentrate on growing more productive crops.

Energy flow in herbivores

The figure below (left) represents the flow of energy through the herbivore trophic level, and shows that any particular herbivore eats only a small proportion of the plant food available to it. There are five main reasons for this. First, most herbivores, whether detritus feeders or grazers, select their food carefully.

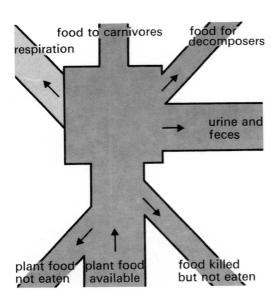

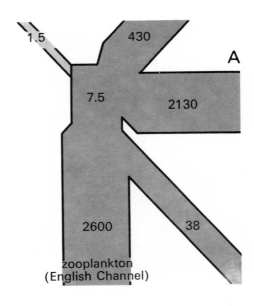

The flow of energy at the herbivore trophic level (left) and three specific examples. Almost all the animals eaten by man are herbivores, and a detailed study of the differences between them can be of considerable value to food producers.

Some herbivores, for example, eat only the roots, stalks, leaves, or fruits of the food plant. Second, the food plant is not always present in the right form or at the right time. Third, wild grazers often instinctively avoid overgrazing, which damages or kills grass and foliage, and overgrazing by domestic cattle is often prevented by man. Fourth, some of the available grass is usually trampled underfoot by grazers and destroyed. Fifth, some of the plant food apparently available to a particular herbivore is in fact consumed by other herbivores.

The food that is left by herbivores, and their feces, accumulate in the surface layer of the soil and are broken down slowly. On grassland the detritus feeders do most of this work, and they limit productivity because they are slow in releasing the nutrients that they produce from the dead grass. The productivity of temperate forests is also limited by the detritus feeders, but tropical forests are very different, as we shall see later.

From 40 to 90 percent of the food a herbivore eats is never actually assimilated. This proportion of ingested plant food passes straight through the gut without being digested, and is voided as feces. Of the remainder—the food that is actually assimilated—up to 90 percent is used during respiration, and there is a further loss in the form of excretory products voided in the urine. A considerable amount of the assimilated food is also used to produce eggs and offspring; but this is not strictly an energy loss, as it increases the herbivore stock, and it is thus not reflected in the figure.

Having looked at the energy losses of herbivores in general, we can now examine three specific herbivore systems (below). Figure (A) shows zooplankton: we see that unlike grazers on land they eat a very high proportion of the food available to them (phytoplankton); we see also that a high proportion of what they eat is voided as feces. Most of the small amount actually assimilated

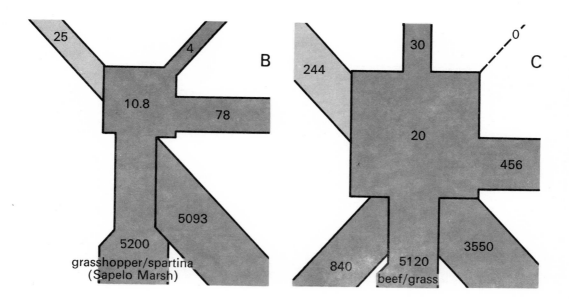

becomes eventually available as food for carnivores. Figure (B) shows the herbivorous grasshopper, which is a very selective eater and assimilates a high proportion of its food—about 25 percent. However, a great deal of the assimilated energy is lost through respiration. In grass-fed bullocks—figure (C)—the food made available to carnivores represents only 30 kcal/m²/yr from an initial grass production of over 5000 kcal—that is, an efficiency of only 0.6 percent. The bullocks eat only about one seventh of the available herbage; they discard roughly two thirds of that as feces, and lose around 90 percent of the assimilated food in respiration. Furthermore, if farmers do not keep their stock in good health, much of the 30 kcal yield of meat will disappear.

In the 1890s mechanization was just beginning to be used in the rearing of herbivores. Left: cowshed with tramline and feed trolley. Since then many technical devices have been used to increase productivity. Below: cows grouped in pens according to their grade of milk. Right: a conveyor belt feeding system operated by one man. Far right: a mechanized milk and feeding parlor. Bottom left: cowman meters out food to cows being milked according to the colored tag hung around the cow's neck (center). Bottom right: analyzing and recording quantity and quality of milk from each cow.

Most of man's domesticated food animals are herbivores, so it is important to do what we can to minimize the large energy losses at this level. The first step is to improve the productivity of our present domestic animals, and much work has already been done in this direction, based on quantitative measurements of energy flow. First of all, it is possible to increase the amount of plant food available to these animals by isolating them from other plant-eaters, such as insects and rabbits. When cows, chickens, or pigs are kept under cover, more food—and food with added nutrients—can easily be brought to them without, at the same time, supplying it to unwanted grazers. Further, it can be brought at the right time and in the right amounts. What is more, by keeping animals

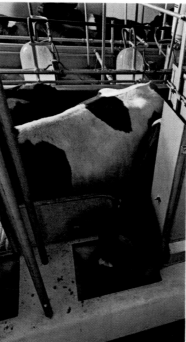

Table shows how one steer and 300 rabbits gain the same weight by eating one ton of hay, but the rabbits gain the extra weight four times faster.

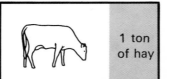

ANIMALS	1 STEER	300 RABBITS
Total body weight	1300 lb	1300 lb
Food consumption/day	16⅔ lb	66⅔ lb
Duration of 1 ton of hay	120 days	30 days
Heat loss/day	20,000 kcal	80,000 kcal
Gain in weight/day	2 lb	8 lb
Gain from 1 ton of food	240 lb	240 lb

off the pasture, a great deal of which would otherwise have been destroyed by trampling, food is saved. Thus the cut-and-carry method reduces wastage through the detritus pathway.

Another way of increasing herbivore productivity depends on the fact that growth efficiencies vary throughout an animal's lifetime. There are two ways of expressing growth efficiencies in animals: gross efficiency and net efficiency. *Gross efficiency* is calculated by dividing the calories of growth (those actually ending up in animal protoplasm) by the number of calories consumed; the result is usually between 4 and 37 percent. A high gross efficiency shows that relatively little of the food is lost as feces. *Net efficiency* is calculated by dividing the calories of growth by the number of calories actually assimilated; this lies between 5 and 60 percent. A high net efficiency indicates that relatively little of the assimilated food is lost in respiration. Now the gross growth efficiency of a young calf is about 36 percent, compared with a mean value of 5 percent over the whole life of a cow. Among other herbivores, too, gross growth efficiency is much higher in early life than later on. We obtain maximum efficiency, therefore, by slaughtering herbivores before the growth efficiency starts to decline, which is usually just before reproductive age. Pullets in broiler houses, for example, are killed at three or four months—about two months before they are due to start laying.

Some herbivores that are not very popular as food are highly efficient in terms of meat production, and this suggests a second step we might take to improve productivity. The speed of production could be increased, for example, by breeding rabbits rather than bullocks. Rabbits are much more efficient than beef cattle in terms of meat production per unit time. It has been shown that if 1 ton of hay is given to 300 rabbits, and the same amount is given to 1 steer, the rabbits gain in weight four times as quickly as the steer. However, since 300 rabbits also consume a ton of hay four times as quickly as one steer, the problem would be to ensure a constant supply of food.

In his search for food, civilized man has put palatability before efficiency, so it is hardly surprising that he has neglected the most efficient herbivores of all, such as certain small aquatic organisms. The net growth efficiency of the water flea, *Daphnia,* for example, is more than five times as great as that of beef cattle, and its gross growth efficiency well over twice as great. If man's sole concern were to obtain the maximum conversion of solar energy into animal protein, he would do well to eat animals such as *Daphnia.* But again we come up against his unwillingness to try out strange foods, even when starving. Yet it is worth emphasizing that many food prejudices are merely the result of habit. Some African tribes are quite happy to eat a porridge made from midges or termites, simply because they have always eaten it.

In spite of his long-ingrained feeding habits, man is making a slow but sure start to harness the food energy of neglected herbivores, as well as neglected plants. In Britain, some food manufacturers have already started to incorporate unconventional foods into popular ones, such as sausage-meat. People do not complain, because they do not realize where the addition comes from, and this strongly indicates that the objection to unconventional foods is often aesthetic. For aesthetic reasons, the Food and Drug Administration in the United States long held up a scheme to produce high-protein flour from certain fish, but finally, in 1967, agreed to subsidize it. Fish flour contains about 80 percent protein, plus vitamins and salts, whereas the protein content of wheat flour is only about 12 percent. It has been estimated that if the U.S.A. were to make use of relatively uncommercial fish in this way, it could double its catch, and make it possible to balance the diet of 1000 million people for 300 days at a cost of only 0.5 cents a day per head.

In certain countries that breed European livestock, there is a particularly good reason for turning to other herbivores for food. In East Africa, for example, European livestock has been raised for a long time, and has been continually afflicted by such parasitic diseases as trypanosomiasis and rinderpest, while the native herbivores, such as zebra and antelope, have remained relatively immune. What is more, the native stock have a higher productivity, and do not damage their habitat to the same extent when grazing. For these reasons, thought is now being given to the cropping of zebra in Africa, and Australia is even exporting canned kangaroo meat.

Because there is an average 90 percent loss of energy when plant food is turned into herbivore meat, we may well wonder if it might not be better for man to stop eating herbivores altogether. If he were to eat only plants, he would at least obtain more food per unit area of land. Unfortunately, however, the plants that contain most calories generally contain least protein, which is an absolute essential in man's diet. To avoid protein deficiency, which is the most common form of malnutrition, and keep healthy, a person needs about 29 gm of protein for every 1000 kcal consumed—that is, about 50 gm a day if he has an office job. Now plant protein from any one species of plant—unlike animal protein—does not contain all the amino acids essential for human growth. Nevertheless, by eating a certain combination of plant food, including a legume, at each meal, it is possible to obtain a balanced diet of protein. In those parts of the world where people cannot obtain enough meat, dairy produce, or fish, this seems to be the best answer to the problem of malnutrition; and even in rich countries steps are being taken to use autotrophs to better advantage. For example, an American firm in Minneapolis has managed to produce synthetic meat that includes protein extracted from vegetable matter.

Energy flow in carnivores

Energy losses in carnivores are broadly similar to those in herbivores; the former, like the latter, consume only a proportion of the available food, and some of the food apparently available to a particular carnivore is in fact consumed by others. In any case, part of the food (the herbivore's bones, for example) is generally left. However, carnivores assimilate more of the food they eat than do herbivores, the proportion usually being 30 to 50 percent, and sometimes as high as 75 percent.

Man eats far fewer carnivores than herbivores, and indeed the only ones he consumes in large quantities are carnivorous fish. This is inevitable, because carnivores represent the last of the trophic levels; their energy content, in any given ecosystem, is far less than that of the herbivores. Nevertheless, carnivores have a much greater effect in ecosystems than is apparent from a study of energy flows alone. A reduction in their numbers, for example, may increase the number of herbivores available as human food, but this increase may upset the balance of nature. A rise in the number of herbivores often leads to overgrazing, which damages plants and occasionally destroys all plant life over a wide area. This, in turn, can lead to soil erosion, and when that happens, only the richest countries can afford the high cost of recolonizing the land.

It is not too difficult to sketch a rough-and-ready outline of the flow of energy through ecosystems in general, as we saw on pages 72–73; but it has rarely been possible to measure in detail the actual energy flow in a particular ecosystem at all the trophic levels, for this is a very lengthy, costly, and difficult job. Even so, it is important to try, because this is the only way in which we can test general assumptions, and equip ourselves to assess the consequences of interfering with the flow of energy at one particular stage. It is commonly assumed, for example, that one ecosystem is isolated from all others, whereas the findings of the American ecologist J. M. Teal show that this is not necessarily true. Teal's energy-flow chart for Root Spring is remarkable in that only 655 of the 2300 kcal consumed by herbivores were produced within the ecosystem; the remainder entered in the form of debris from outside. Another common assumption is that all organisms can be assigned to one particular trophic level. Some algae, however, are partially autotrophic and partially heterotrophic, and omnivores, such as men and bears, are both primary and secondary consumers. Again, it is often assumed that the ecosystem is in a steady state, and does not change greatly from year to year; but, as we shall see in Chapter 8, steady-state ecosystems occur only in certain situations.

As the necessity of finding enough food becomes increasingly urgent, investigations into the energy relations of whole ecosystems will probably become more common, and we may well find that a surprising number of them differ considerably from the kind of general ecosystem depicted on pages 72–73. We have already seen how ecosystems vary in efficiency and productivity, and how this realization is enabling man to increase food production. In the future, information about a greater variety of ecosystems will almost certainly reveal further differences, and enable us to discover more and better ways of turning nature to our benefit.

The New Zealand scene above shows the severe overgrazing and consequent erosion caused by introducing red deer and opossum (right). The severity of defoliation was mainly due to the absence of native carnivores.

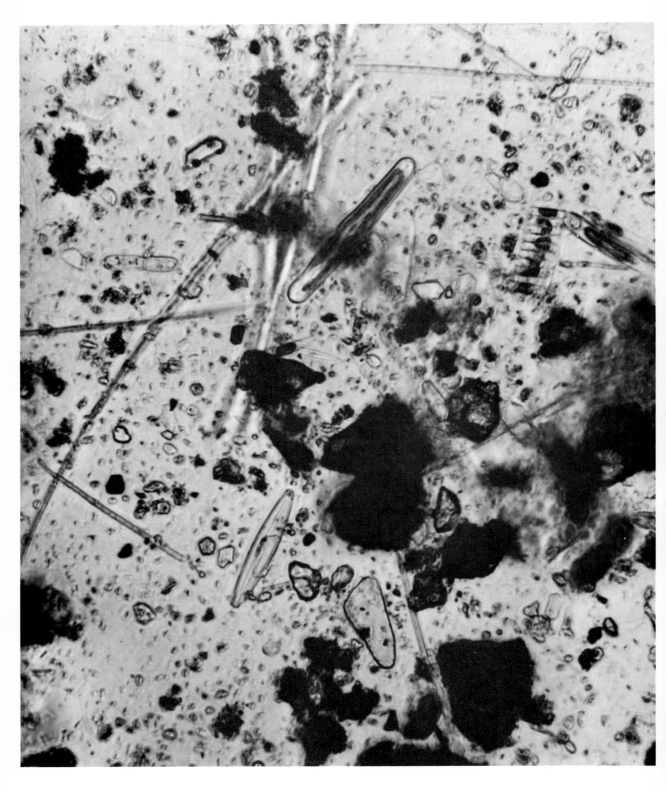

This photomicrograph of a grain of soil shows a variety of algae and heterotrophic microorganisms. The latter play a vital role, for they turn complex organic matter into simpler compounds that plant roots can absorb.

5

The Chemical Environment

We have seen that although the energy that enters green plants and passes from them through herbivores to carnivores is gradually dissipated as heat, the loss is made good by the continual influx of new solar energy. Nature provides the earth with no such influx of new materials: the sole supply is that which is already here, and it is used over and over again. So atoms of common elements such as carbon, oxygen, and nitrogen may at one time form part of our bodies, and at another time be part of the air around us.

Since the main intake of abiotic (nonliving) materials is at the plant level, we can best explore the chemical environment by discovering first how plants obtain these materials and build them up into protoplasm.

Macronutrients

We can get some idea of the chemical make-up of plants from the table on pages 88–89. This records the result of heating five corn plants to $100°c$ to drive off their water, and then identifying and weighing the various chemical elements that remain. The most striking point this table reveals is that oxygen, carbon, and hydrogen together account for about 94 percent of the dry weight of the plants. In fact these three elements form all the carbohydrates—such as sugars, starch, and cellulose—that were originally synthesized in plants during photosynthesis. Because these elements make up the greater part of their dry weight, plants must obviously have access to large supplies of oxygen, carbon, and hydrogen. These elements, together with the six others that head the list, are therefore called *macronutrients*, or "large-scale" nutrients.

In discussing photosynthesis in Chapter 2, we noted that the plants get oxygen, carbon, and hydrogen from carbon dioxide and water. But water plays such an important part in all life processes that more needs to be said about it before we go on to consider other nutrients.

Water dissolves an extraordinary variety of compounds. It is impossible to over-emphasize this, because most chemical reactions in both plants and animals occur only when substances are in solution; furthermore, solids cannot normally enter cells unless they are first dissolved. Sometimes water

ELEMENT	WEIGHT (grams)	PERCENTAGE OF TOTAL	RELATIVE NUMBER OF ATOMS
Oxygen	371.4	44.43	4.640
Carbon	364.2	43.57	6.060
Hydrogen	52.2	6.24	10.440
Nitrogen	12.2	1.46	174
Phosphorus	1.7	0.20	11
Potassium	7.7	0.92	39
Calcium	1.9	0.23	9
Magnesium	1.5	0.18	12
Sulfur	1.4	0.17	9
Iron	0.7	0.08	2
Silicon	9.8	1.17	70
Aluminum	0.9	0.11	7
Chlorine	1.2	0.14	7
Manganese	0.3	0.03	1
Undetermined elements	7.8	0.93	

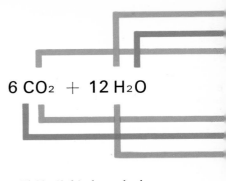

$$6\ CO_2\ +\ 12\ H_2O$$

Table (left) shows the dry-weight composition of five corn plants. The figures reflect the plants' demand for nutrients. Elements below gap are micronutrients.

itself enters into chemical reactions, and sometimes it serves as a catalyst to speed them up. Gases, too, have to go into solution before they can move around inside living organisms. In fact, one of the main functions of water is to act as a conveyor belt. Water also plays a large part in determining the physical environment in which all organisms live, including such important aspects as temperature, humidity, and light. The figure above shows that, of the three macronutrients used in photosynthesis to build up glucose, two—carbon and oxygen—enter green plants directly as atmospheric carbon dioxide; the third, hydrogen, comes from water.

Next on the list of macronutrients is nitrogen, which accounts for 1.46 percent of the dry weight of the corn plants in our sample. Only by combining the products of photosynthesis with nitrogen (and often with small amounts of other elements too) can proteins be built up within cells. These proteins include enzymes—the complex organic catalysts that speed up chemical reactions in living things—and the chromosomes that regulate the synthesis of proteins and govern heredity.

Because four fifths of the air we breathe is nitrogen, we might expect plants to draw an abundant supply of it from the atmosphere. But nitrogen is an extremely inert gas; it does not easily react with other elements. Green plants cannot absorb it directly from the air and incorporate it into organic compounds; instead, they must take in dissolved nitrogenous compounds, such as nitrates, through their roots.

So far we have dealt with four of the macronutrients. The remaining five are phosphorus, potassium, calcium, magnesium, and sulfur. All of these commonly occur in combination with other elements in the form of mineral salts, many of which are soluble in water. The part that each plays in plant life is summarized in the table.

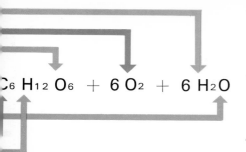

$$C_6H_{12}O_6 + 6O_2 + 6H_2O$$

Equation for photosynthesis (above left) shows that the carbon, hydrogen, and oxygen in glucose, and in the fats and proteins derived from it, come from carbon dioxide and water.

Table right shows the function of five mineral macronutrients in promoting plant growth.

Phosphorus	Part of the nucleic acids (DNA and RNA), which govern heredity and control protein synthesis. Occurs in cell walls and forms energy bonds in ATP.
Calcium	In plants it rigidifies middle lamella of the cell wall. In animals it occurs in bones and teeth, and is essential for muscle contraction.
Magnesium	In plants it is a vital component of the chlorophyll molecule and forms part of some enzymes. In animals it is essential in bones and teeth.
Sulfur	In both plants and animals it is an essential part of some amino acids, which make up the proteins for enzymes and growth.
Potassium	In plants it activates certain enzymes. In animals it helps to maintain osmotic pressure of body fluids and the potential differences in nerves and muscles.

How plants absorb nutrients

We have seen that plants take oxygen, carbon, and hydrogen from carbon dioxide and water. They must absorb all other nutrients in solution through their roots. The simplest way to understand what this involves is to examine what happens when we grow plants hydroponically—that is, without soil.

Soil normally provides the roots of plants with four essentials: anchorage, oxygen (contained in air pockets between the soil particles) for respiration, water, and minerals. In the increasingly common system of cultivation known as hydroponics, anchorage and air for respiration are supplied by beds of sand, or some similar inert material. These beds are flooded at regular intervals, usually once every day or two, with water in which nutrients are dissolved. The nutrient solution must not be allowed to flood the beds for too long at a time, or it would deprive the plant roots of air and literally drown them. So after each flooding it is drained off into storage tanks, where it can be topped up with added nutrients to maintain the right concentration, and used again. Between immersions, the plant roots and their anchorage remain moist and are always adequately aerated.

Plants grown hydroponically get all the oxygen, carbon, and hydrogen they need from air and water, just like any others: but the remaining macronutrients—nitrogen, phosphorus, calcium, magnesium, sulfur and potassium—must be supplied to them in the nutrient solution. In fact all six could be provided by dissolving only three selected salts, for instance calcium nitrate, $Ca(NO_3)_2$; potassium phosphate, KH_2PO_4; and magnesium sulfate, $MgSO_4$. In the dry state these salts, and indeed all other salts, consist of molecules that are quite stable. This might suggest that if a plant needed, say, a supply of nitrate from calcium nitrate, it would have to absorb the calcium as well, whether or not it needed it; but this is not so. When a salt dissolves in water, it undergoes a drastic change, breaking down into separate electrically charged particles called *ions*. A molecule of calcium nitrate, for instance, breaks down into one ion of calcium with a double positive charge (Ca^{++}) and two ions of nitrate (NO_3-) each with a single charge. The other salts break down in a similar way: potassium phosphate becomes K^+ and H_2PO_4-, and magnesium sulfate is converted into Mg^{++} and SO_4-- ions.

We now have six groups of ions all moving about freely and independently in the water, each group containing a different macronutrient. It is only in ionized form that these nutrients can pass through the cell membranes of a plant root; and, to become ionized, salts *must* be dissolved.

Here we have the answer to the problem of how a plant can take in, for example, nitrate from calcium nitrate without also taking in an equivalent amount of calcium.

When the salt is dissolved in water, the nitrate and calcium ions are independent of each other. The membranes surrounding the cells of the plant roots act as a selective barrier, and enable the plant to take in the precise amount of each ion that it needs at any given time. Ionization has another important consequence, too. The nature of the salts from which the ions are formed can thus vary without affecting the final result—the supply of nutrient ions to the plant. For instance, a nutrient solution containing calcium nitrate and potassium phosphate could just as easily be replaced by another solution containing potassium nitrate and calcium phosphate: either solution will yield a supply of the same ions, namely Ca^{++}, NO_3^-, K^+, and $H_2PO_4^-$. This explains why many plants are able to grow equally well on a wide variety of different soils, each containing different kinds and combinations of salts dissolved in the soil water.

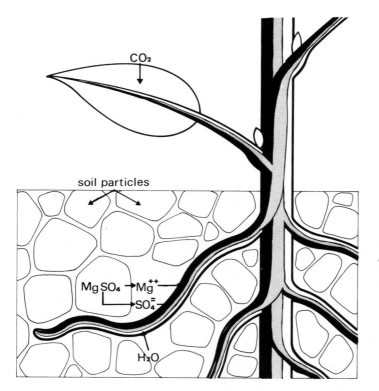

Soil particles contain salts, which ionize in the thin watery film that surrounds them. The ions then enter roots (right), after which more soil salts ionize to replenish the soil solution. Water has therefore two main functions: it is a nutrient, and it is responsible for converting salts, which cannot be used by plants directly, into ions, which can be absorbed and used for growth.

Micronutrients

The table on page 88 shows that corn plants contain several elements besides the nine macronutrients. These are the *micronutrients*—elements that plants need in only small quantities, and often in very minute quantities indeed.

In the first experiments on plants and nutrients, in the middle of the 19th century, scientists placed plants with their roots in flasks of water containing three commercial salts that together provided them with all the macronutrients (other than carbon, oxygen, and hydrogen, which they could get from air and water). These early experiments with commercial grades of salts worked well; but when salts of higher purity became available and were used to make more precise experiments, the plants failed to grow normally. It was soon realized that the small quantities of impurities in commercial salts, unlike those used later, were just as important to plant life as the macronutrients. Further research led to the identification of several other elements essential to plants—the micronutrients—all of which can penetrate into the roots of plants only in ionized form, just like the macronutrients.

The principal micronutrients are iron (as Fe^{++} or Fe^{+++}), manganese (as Mn^{++}), zinc (as Zn^{++}), copper (as Cu^+ or Cu^{++}), molybdenum (as MoO_4^-), boron (as BO_3^{---}), and chlorine (as Cl^-). All of them form part of various enzyme systems. Certain other elements that occur in plants, such as cobalt and vanadium, may also some day be shown to form part of enzymes. Both are already known to play such a role in animal tissues.

There are yet other micronutrients that some species of plants require for specific purposes. The one-celled aquatic plants called *diatoms*, for example, use silicon to build up their elaborate and beautiful external skeletons. Grasses, too, use silicon as a stiffener, and this fact shows up in our table of the

The first experiments in growing plants in nutrient solutions were made by Knop and Sachs around 1865. The culture jars (left) were filled with distilled water plus selected nutrients. The effect of each nutrient was determined by observing the plant's progress when deprived of one or other of the nutrients in succession. From this work has grown the practice of hydroponics. The technique consists of growing plants, usually under cover (right), in trays of inert material such as gravel, to support the roots. These trays are regularly flooded with a balanced nutrient solution, and are regularly emptied so that roots are nourished and aerated.

composition of corn, which is itself a modified grass. Other elements, though not essential to survival, promote increased health and vigor in certain plants; beet plants, for example, produce larger roots in the presence of sodium ions.

However, some plants take up elements for which they seem to have no use at all. One of the most extraordinary instances of this kind is provided by the locoweed of the western United States, which is so named because animals that eat it are liable to become "loco" or crazy. It grows best on soil from which it can take in the uncommon and highly poisonous element selenium. This element seems neither to benefit it nor to harm it, but it may account for the effect of locoweed on grazing animals. Today the occurrence of locoweed is put to practical use. Selenium is known to occur in association with uranium ore, and the presence of those species known to contain selenium is such a good indication of uranium that prospectors use the plant as a useful lead to potentially rewarding uranium lodes.

Nutrients in animals

Nutrients have much the same function in animals as in plants; but animals have several needs that are different from those of plants. For instance, they need potassium and sodium to maintain electrical conductivity in nerves and muscles; many need calcium for bones and teeth, as well as for muscular contraction; and many also need iron to form part of the hemoglobin molecules on which their respiration depends. In general, as we saw in Chapter 3, animals get their nutrients from plants, but there are a few exceptions.

In hot climates, where animals regulate their temperature by sweating, there is an extra demand for both sodium and chlorine in the form of common salt (NaCl). So grazing animals often supplement the small amount of salt that they take in with their vegetable diet by gathering at dried-up pools of water where a certain amount of salt has crystallized out. These "salt-licks" are most important to many herbivores. The carnivores are more fortunate—they get enough salt from the flesh and blood of the herbivores they eat.

The animal with the most highly developed capacity for temperature regulation is man. His need for additional salt has played a large part in history and in the development of trade routes. So long as men lived almost entirely on milk and meat (as the Bedouin of the Hadhramaut in Southern Arabia do to this day), there was no need for extra salt. But as agriculture developed, and the proportion of vegetable to animal food increased, additional supplies of salt became essential. Today the human race, with few exceptions, depends for its existence on salt, which is either mined from ancient deposits (rock salt) or produced by evaporation of sea water. Cakes of salt have even been used for money. Roman legionaries were given a daily ration of salt called a *salarium*; later this was converted into an allowance of money for buying salt, and it is from this that our word *salary* comes.

Iodine is another important nutrient; it is an essential part of thyroxin, a hormone secreted by the thyroid gland. Animals and man usually get enough of it from their food and drinking-water; but there are some places—for instance, in Brazil and in the Peak District of Derbyshire, England—where the drinking-water lacks iodine. For centuries people in these places suffered from an enlargement of the thyroid gland, which grew to more than twice its normal size in a completely futile effort to make thyroxin. This condition is known as a goiter, and in England it was often called "Derbyshire Neck."

We began this chapter by noting that while energy runs down as it flows through an ecosystem, and is replaced by a constant influx of solar energy, materials are used over and over again. It is true that in hydroponics, materials—water and nutrient salts—also run down and have to be replaced, but they are replaced only from the earth's own stock of materials, not from any external source. In all ecosystems dependent on natural soils or natural expanses of water, the "cycling" of materials for repeated use occurs naturally and demands equally important roles to be played by living organisms, the physical environment, and the chemical environment. The processes by which various materials circulate are known as *biogeochemical cycles*, which means "life-earth-chemical cycles." Of these materials, six—water, oxygen, carbon, nitrogen, phosphorus, and sulfur—are particularly important. Let us look at each in turn.

Both animals and man need fairly large amounts of common salt, and it is often necessary for man to supplement his diet by taking it "neat." Although most salt is mined it is often obtained by evaporating sea water in salt pans.

93

Goiter is caused by the lack of a single nutrient—iodine. A nutrient deficiency rarely shows such dramatic effects as this (left); more often a lack produces tiredness or general ill-health, which is difficult to diagnose but is undoubtedly responsible for much debility among humans.

Below: up to 70 percent of all fresh water is locked up in the icecaps and glaciers of the world.

The water cycle

The many essential functions of water in the maintenance of life have already been outlined. Here it is necessary to add only that in many parts of the world, water is also the main *limiting factor* for growth and survival: that is to say, however favorable all other factors may be, the chances of survival and the rate and extent of growth depend mainly on whether or not the right amount of water—not too little, not too much—is available at the right time.

All the water that falls on land and sea comes from and eventually returns to the oceans, which cover two thirds of the earth's surface and contain 97

percent of all its water. The water cycle is in essence a gigantic distillation process, in which the sun's heat evaporates sea water, and the vapor, free of salt, rises into the atmosphere, where winds carry it over long distances. At the lower temperature and pressure of high altitudes the water vapor condenses to produce rain, hail, or snow—known collectively as *precipitation*. Most of this precipitation falls directly back into the ocean; only one eighth falls on land, and an appreciable amount of this goes out of circulation for an unpredictable period by falling as snow on the great icecaps of Antarctica and Greenland, which lock up some 70 percent of the world's fresh water.

Not all the fresh water falling on land that can support life is available to plants and animals. About 30 percent of it evaporates from soil, rock, and plant surfaces before it has had time to become involved with living matter. (That figure, of course, is a world average; in hot, dry regions, evaporation removes virtually all the water before it can benefit life.) The water that does not evaporate penetrates the soil, and is then available to soil microorganisms and plant roots. Some of the absorbed water is built into plant protoplasm—which is mostly water—but a very large proportion travels up the stems of plants and evaporates from the small holes, or *stomata*, of the leaves—a process called *transpiration*. A tree may transpire well over 200 liters a day, and, on average, transpiration accounts for the loss of about 40 percent of the world's rainfall on land. Thus evaporation and transpiration together remove on average 70 percent of the total land precipitation, returning it to the atmosphere, where it eventually condenses and falls on sea or land once more in further precipitation.

The water that is not built into protoplasm, or removed by evaporation and transpiration, percolates down through the soil until it reaches an impervious layer, along which it moves more or less horizontally until it finds some outlet. This leads it back to the sea, either directly or indirectly via lakes and rivers. Where precipitation occurs on waterlogged or impermeable land, the water runs over the surface in sheets or channels, and is called *runoff*. As with ground water, this also finds its way directly or indirectly to the sea. The important point is that all the fresh water that falls on land ultimately returns to the sea, even though the process may take tens of thousands of years for some ground water, or for the vast quantities of water locked up in the icecaps.

Only a minute proportion of the world's water is contained in living matter. We have already seen some of the reasons for this, but there are others. In many parts of the world light showers interspersed with dry spells are of no use whatever to plant life. Even though rain gauges may record a fairly high monthly precipitation, every drop of water can be lost by evaporation. At the opposite extreme, torrential rain is also wasteful—most of it just flows into lakes and rivers before plants can use it. Except in oddities such as cacti, nature has found no way of storing excess water, as it can solid food, for use during long dry periods. Organisms must therefore have a fairly continual supply of water; otherwise they become inactive, stop growing, or die.

The oxygen cycle

Oxygen is not only a constituent of all carbohydrates, fats, and proteins; it is also taken in by organisms to "burn" foods to produce energy, in respiration. It is true that there are some microorganisms that produce energy without oxygen, but no higher organism can do so all the time. What is more, all higher plants and animals require *free* oxygen—that is, oxygen not combined with any other element. Because oxygen, unlike nitrogen, readily combines with most other elements, that requirement is an exacting one.

The fact that oxygen is extremely reactive explains why many elements are seldom found in their pure state but occur quite commonly as oxides. But living organisms protect themselves from violent oxidation by controlling the use of oxygen in the step-by-step process of respiration, mentioned in Chapter 2. By contrast, dead, dry vegetation burns fiercely once it has been ignited—that is to say, raised to a temperature at which oxidation is both spontaneous and self-sustaining.

HYDROLOGICAL CYCLE

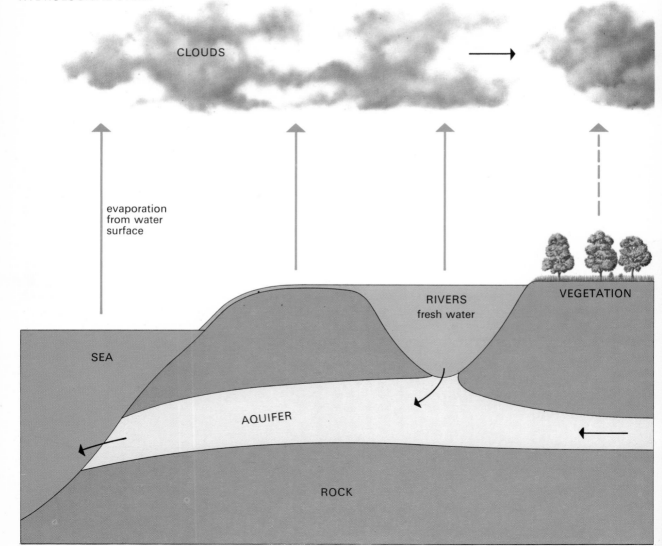

CLOUDS

evaporation
from water
surface

VEGETATION

RIVERS
fresh water

SEA

AQUIFER

ROCK

Every act of plant or animal respiration removes free oxygen from the atmosphere, combining it with other elements and thus temporarily locking it up in compounds; every time a fire burns, free oxygen is removed and locked up in similar fashion. In the process of photosynthesis (page 42), however, green plants release free oxygen into the atmosphere, and they do so on a prodigious scale.

We cannot tell if the amount of oxygen released at any given moment into the atmosphere precisely balances the amount being removed. But we do know that the present proportion of oxygen in the atmosphere—nearly 21 percent by volume—is vastly greater than when the first forms of life emerged. Some 2000 million years ago the only free oxygen in the earth's atmosphere was that formed by the action of ultraviolet radiation on atmospheric water vapor. But that process (which still continues) cannot account for more than one thousandth of the present amount of oxygen. In time, however, photosynthetic organisms evolved and multiplied, using carbon

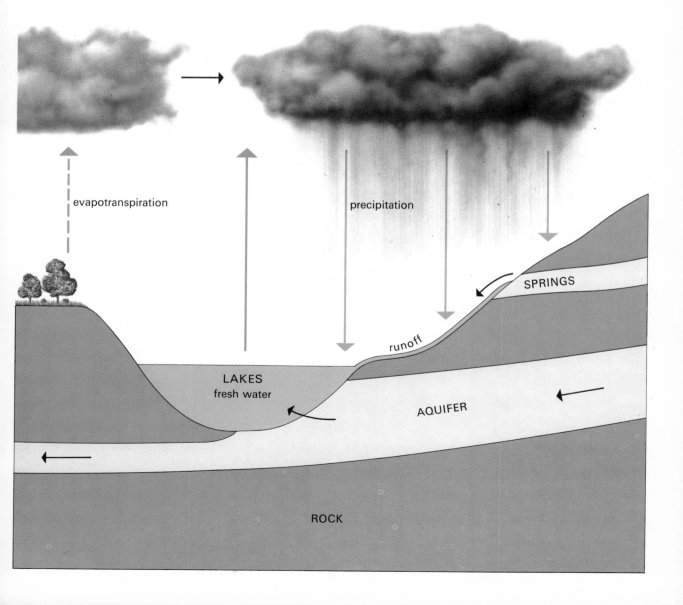

evapotranspiration

precipitation

SPRINGS

runoff

LAKES
fresh water

AQUIFER

ROCK

dioxide and water to produce carbohydrates and free oxygen. The present high concentration of atmospheric oxygen is thus due to green plants—a dramatic example of how living things affect their own nonliving environment.

The carbon cycle

Carbon is the characteristic element of all organic compounds. Among the most abundant of these are cellulose, which makes up a very large part of the skeletal cell walls of plants, and sugars such as sucrose and glucose, which organisms commonly store for future use in the form of starch, including the animal starch called *glycogen*.

Carbon, linked to oxygen in the form of carbon dioxide, enters living systems by way of green plants. There, in the process of photosynthesis, it combines with water in chlorophyll-containing cells to make carbohydrates, some of which are used in subsequent processes to build up proteins, fats, and other organic substances.

When carbon dioxide from the air enters the leaves of terrestrial plants, it dissolves in the water film around the cells and forms carbonic acid (H_2CO_3), which breaks down into hydrogen ions (H^+) and bicarbonate ions (HCO_3^-). In seas, lakes, and rivers, carbon dioxide also dissolves and gives rise to bicarbonate ions, which can diffuse into the cells of aquatic plants. Such plants therefore obtain the carbon dioxide necessary for photosynthesis directly from the water in which they grow.

Plants have been estimated to manufacture 200,000 million tons of organic compounds a year—all, of course, containing carbon. Yet, compared with nitrogen and oxygen, the supply of carbon dioxide in the atmosphere is very

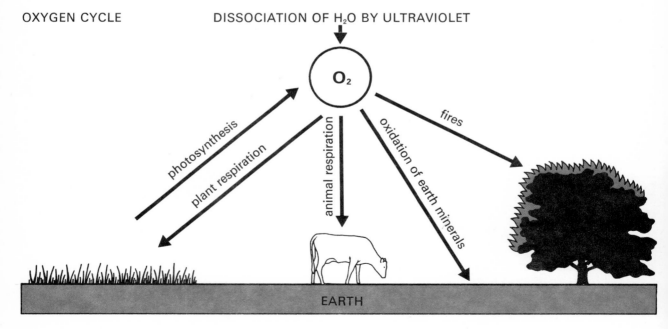

OXYGEN CYCLE   DISSOCIATION OF H₂O BY ULTRAVIOLET

O₂

photosynthesis

plant respiration

animal respiration

oxidation of earth minerals

fires

EARTH

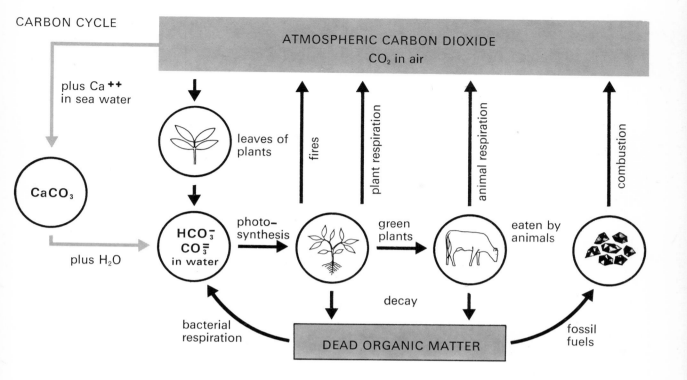

CARBON CYCLE

ATMOSPHERIC CARBON DIOXIDE
CO_2 in air

plus Ca^{++}
in sea water

$CaCO_3$

plus H_2O

leaves of plants

HCO_3^-
$CO_3^=$
in water

photo-synthesis

fires

plant respiration

green plants

animal respiration

eaten by animals

combustion

decay

bacterial respiration

DEAD ORGANIC MATTER

fossil fuels

small—only 0.03 percent by volume, or about 2,400,000 million tons. The sea contains about 30 times as much carbon dioxide as the air, in the form of bicarbonates and carbonates.

The quantity of carbon dioxide in the air is so small that photosynthesis by land plants would exhaust it in a few years if it were not replenished. Constant replenishment is brought about mainly by living organisms, through respiration. This process, being in essence the reverse of photosynthesis, produces carbon dioxide thus:

$$C_6H_{12}O_6 + 6O_2 \rightarrow 6CO_2 + 6H_2O$$

In addition, countless millions of microorganisms in the soil release carbon dioxide by breaking down the carbohydrates of dead plants and animals. Natural fires, domestic fires, and the age-old practice of burning vegetation to improve grazing also produce huge amounts of carbon dioxide (and exhaust an equal amount of oxygen).

Since the Industrial Revolution, man has added considerably to the carbon dioxide content of the air by burning ever-increasing quantities of fossil fuels, and so putting back into circulation carbon that had formerly been locked up for millions of years in coal or oil deposits. It has been estimated that in the last century alone there has been an increase of 14 percent in atmospheric carbon dioxide, and if this trend continues it is probable that photosynthesis will be more intensive in years to come. This is not a rash statement, because

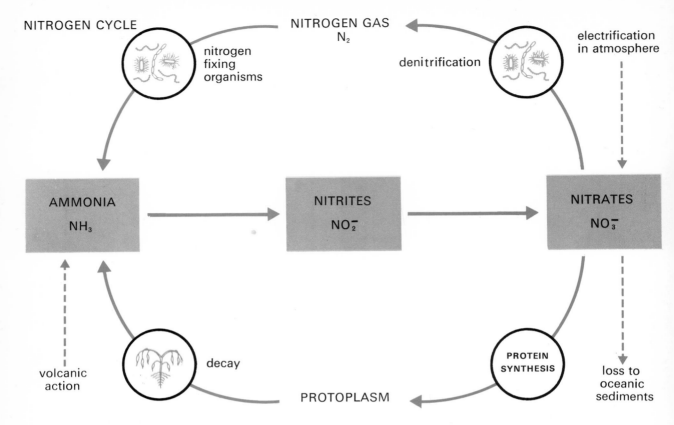

NITROGEN CYCLE

NITROGEN GAS
N_2

nitrogen
fixing
organisms

denitrification

electrification
in atmosphere

AMMONIA
NH_3

NITRITES
NO_2^-

NITRATES
NO_3^-

volcanic
action

decay

PROTEIN
SYNTHESIS

loss to
oceanic
sediments

PROTOPLASM

*One of the many natural fires
that yearly help to return to the
atmosphere the carbon dioxide
that plants remove by photosynthesis. During the last 30
years man has burned immense
quantities of fossil fuel, and this
has raised the concentration of
carbon dioxide in the
atmosphere by 30 percent. It is
quite possible that in the years
to come this will result in
increased photosynthesis.*

we already know that many plants under glass grow faster if extra carbon dioxide is added to the atmosphere of the greenhouse.

The nitrogen cycle

About half the dry weight of protoplasm is protein, and about one fifth of this is nitrogen. Protein is especially important in those parts of the cell concerned with heredity—the nucleic acids—and in enzymes, which catalyze most chemical reactions in living things. In many plants, protein is concentrated mainly in the fruits and the seeds—the parts that man crops even when he leaves the rest. In animals, protein occurs mainly in the muscles, and in general the importance of protein is underlined by the fact that the most common form of human malnutrition is protein starvation. As we saw (p. 84), adults require about 50 gm. of protein a day to keep healthy.

We have seen that the higher plants cannot make direct use of the plentiful supply of nitrogen in the air, but must instead secure it from the soil, mainly in the form of nitrogenous salts. This raises the question: how does atmospheric nitrogen become available to plants in these forms? Part of the answer we have already seen in Chapter 3, where the existence of nitrogen-fixing bacteria in the roots of legumes was given as an example of mutualism. These bacteria "fix" atmospheric nitrogen by combining it with hydrogen to form ammonia; later the ammonia is converted to nitrites and nitrates, which help the legume plant to synthesize proteins.

For the past 2000 years, farmers have improved their fields, and secured heavier crops by first growing legumes such as clover and alfalfa and plowing them back into the soil. They had no idea why this practice improved the land, and indeed it is only in the last hundred years that the mutualistic nitrogen-fixing bacteria were found. At first, these bacteria were thought to be the only source of available nitrogen; but now we know many free-living micro-organisms in soil can fix nitrogen and make it available for higher plants. These include certain saprophytic bacteria, many photosynthetic bacteria, and the blue-green algae. The blue-green algae that flourish in shallow water and high temperatures, are particularly useful in fertilizing paddy fields—so much so, that fields have been successfully seeded with extra algae to raise their general fertility and improve their rice-bearing capacity.

On average, the microorganisms we have mentioned fix between 0.5 and 3 kg. of atmospheric nitrogen per acre per year; in the most fertile regions the figure sometimes reaches 100 kg. In addition, some nitrogen combines with oxygen, forming nitrogen oxides; these then combine with water vapor to form nitrous and nitric acids, which are brought down with the rain. In the soil these acids are converted into nitrites and nitrates, thus making the nitrogen available in a form that plants can use.

Yet the main source of available nitrogen in the soil is organic matter itself. When plants and animals die, most of the great store of nitrogen locked up in them returns to the soil; much is also returned in the form of animal

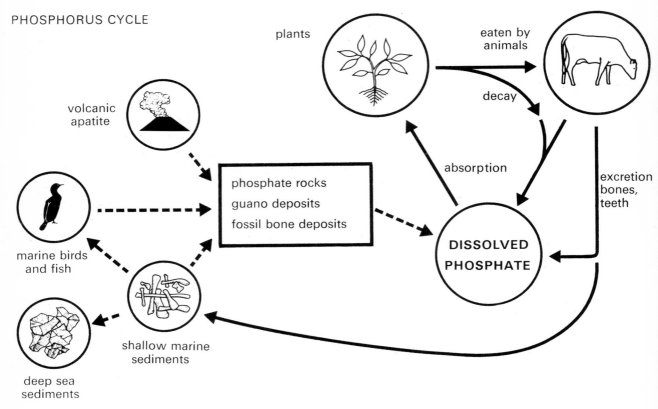

plants

eaten by animals

volcanic apatite

decay

phosphate rocks
guano deposits
fossil bone deposits

absorption

excretion
bones,
teeth

marine birds
and fish

DISSOLVED
PHOSPHATE

shallow marine
sediments

deep sea
sediments

feces and urine. Myriads of soil microorganisms then turn the proteins and other complex compounds back into simpler nitrogenous compounds that living plants can absorb.

There are, however, other bacteria that spend their lives breaking down nitrogenous compounds and releasing free nitrogen to the atmosphere; this process is called *denitrification*, the very reverse of nitrogen fixation. There are other sources of loss, too. When soil is very wet, nitrogenous compounds can easily be washed, or *leached*, down into the subsoil, beyond the reach of plant roots; eventually, dissolved in ground water, they find their way via rivers and lakes to the sea. Fires, too, can remove soil nitrogen by driving it off as ammonia, which rapidly disperses in the air.

In natural, fertile ecosystems, the losses of nitrogen by plant absorption, denitrification, fires, and leaching are more or less balanced by decay and nitrogen fixation. In many ecosystems, however, the losses are greater than the gains, and despite the vast store of nitrogen in the atmosphere, the low rate of nitrogen fixation often keeps primary productivity at a low level. In other ecosystems, man removes large quantities of protein by intensive harvesting, and returns insufficient nitrogen, in the form of fertilizers, to balance the loss. Further, in most civilized countries, a high proportion of nitrogen voided by humans in feces and urine is no longer returned to the soil as fertilizer, but is eventually washed or dumped into the sea.

Perhaps the biggest hope of increasing soil nitrogen lies in the fact that man now fixes atmospheric nitrogen on an industrial scale, and can thus synthesize simple nitrogenous compounds to nourish his crops and make up for the losses due to intensive cultivation.

The phosphorus cycle

Phosphorus, as we have seen, is a component of the nucleic acids (DNA and RNA), and of ATP; it also forms part of cell membranes. To build up such complex organic compounds, plants absorb phosphorus, mainly as ortho-phosphate ions ($H_2PO_4^-$); when organisms die, microorganisms in the soil break down these organic compounds and release phosphates once more.

The phosphorus cycle is not quite as simple as that, however. Land eco-systems tend to lose phosphates by erosion, which washes them down into the sea, where they eventually react with other minerals and form phosphatic rocks. These are only rarely uplifted on to land again, and then not in all parts of the world. By deforestation, and other activities that increase erosion, man accelerates the flow of phosphates to the sea. Every year 3,500,000 tons are lost in this way.

Other factors compensate, at least in part, for this loss. The weathering of phosphatic rock by water adds large quantities of phosphates to the soil; some phosphorus, though not as much, is also added by wind-borne salt spray and the fall-out from volcanic gases. Furthermore a good deal finds its way back from the sea to the land. Upwelling ocean currents bring phosphates from the depths to the photosynthetic zone, where they are incorporated into organic compounds by phytoplankton; these compounds eventually nourish fish and, in turn, fish-eating birds. Fishing by man restores 60,000 tons of phosphorus to the land each year; and fish-eating guano birds annually excrete on to their nesting grounds many tons of phosphorus compounds, much of which man uses to fertilize his fields.

Even so, the balance sheet shows a net loss of phosphorus from land to sea. Man can make good part of it by quarrying phosphatic rock and incorporating it into fertilizers. In fact, between one and two million tons of fertilizers are so produced each year, but much of this is also washed off the land and ends up in the sea. The solution to the problem may well come from oceanic research, which is already looking into the possibility of dredging the ocean depths for minerals, including phosphates.

The sulfur cycle

Sulfur, which organisms need to build up certain proteins, is fairly plentiful. There are very large reservoirs of it in the form of sulfurous sediments lying at the bottom of seas and deep lakes, and deep down in the soil. From these reserves the sulfur slowly circulates to and from the photosynthetic upper waters and the upper layer of the soil, where plant life is abundant, and where sulfur is rapidly cycled.

Plants take in sulfur mainly in the form of sulfate ions (SO_4^{--}). When terrestrial plants and animals die and decay, the sulfur-containing compounds in them are converted by soil bacteria into sulfides and then into sulfates, so replenishing the supply of sulfur available to a new generation of plants. But dead, aquatic organisms decay mainly in airless conditions in the depths of seas and lakes; their anaerobic ("without-air-living") bacteria convert their sulfur-containing compounds into hydrogen sulfide, a poisonous, foul-smelling gas. This usually means that man can make no use of the deepest waters of the lakes. However, it often happens that the hydrogen sulfide reacts with iron compounds to form iron sulfide; this in turn acts on insoluble phosphorus compounds on the bottom and converts them into soluble forms that organisms can use.

Here, then, the sulfur cycle influences the phosphorus cycle. It is becoming increasingly apparent that several other mineral cycles affect one another in similar ways. The use of artificial fertilizers therefore calls for knowledge of how different minerals interact in different circumstances if it is to prove a success.

Temporary and local shortages

When we talk about "cycles" of various nutrients we need to be on our guard. The word suggests that if there is enough of any particular nutrient to start with, there will always be enough, because the supply will simply be used over and over again. Over short periods of time—and sometimes over very long ones—this is not necessarily true. Fire may temporarily deplete the

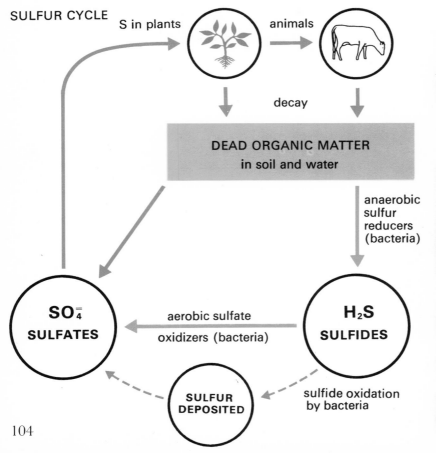

SULFUR CYCLE

S in plants   animals

decay

DEAD ORGANIC MATTER
in soil and water

anaerobic
sulfur
reducers
(bacteria)

$SO_4^=$
SULFATES

aerobic sulfate
oxidizers (bacteria)

H_2S
SULFIDES

SULFUR
DEPOSITED

sulfide oxidation
by bacteria

Left: most of our phosphate fertilizer is mined from phosphatic rock; but phosphate washed into rivers and oceans is continuously depleting the world's supplies. Above: epiphytes, such as these tropical club-mosses and ferns, have abandoned the forest floor to reach the sunlight they need for photosynthesis. They derive their nourishment and water from the air—as rain, and as dust that collects to form humus in the crevices of the trees where they anchor their roots.

soil of nitrogen, and other nutrients can be leached into the subsoil and eventually washed into the sea. Carbon can be removed from the carbon cycle for millions of years, as much of it was in the Carboniferous Age, when forests of giant tree ferns became buried and slowly converted into coal. Over the millennia, however, things even out. Nitrogen-fixing bacteria, for example, replace nitrogen lost from the soil; geological uplift restores to the land minerals that were long since washed into the sea; burning coal and oil puts carbon back into the carbon cycle.

Even so, an overall world sufficiency of a given nutrient does not guarantee that there will be enough of it in any one place. Essential elements are rarely distributed evenly over the face of the earth; most ecosystems are deficient in one or more, and the particular deficiency from which the soil suffers plays a large part in determining what kinds of flora, and hence fauna, it can support.

If a region is fortunate enough to contain all the essential elements, even this does not ensure that productivity will be high, for productivity depends not merely on the presence of essential materials in sufficient quantity but also on the rate of turnover. And the rate of turnover of a nutrient commonly depends on the form in which it is present. For instance, if plants are to flourish, the soil they grow in must contain a certain amount of phosphorus in a form they can absorb. Such phosphorus is likely to have a quick rate of turnover. But phosphorus may also be present in other forms, which most plants cannot absorb. Though this will certainly have a slower rate of turnover, it

Left: these plants show the effect of a lack of phosphorus (left), potassium (middle), and molybdenum (right). A shortage of a particular nutrient often causes a characteristic appearance in plants. An agriculturalist skilled in diagnosis can thus rectify the particular deficiency in his soil by providing the missing nutrients, hence greatly increasing his crop production.

The virtue of mixed cropping in many parts of the world lies in the fact that plants have different requirements. Whereas one crop may be able to tap only certain soil salts, another may be able to utilize salts that the first crop could not use. Together they can often make maximum use of soil nutrients.

cannot be written off as useless. By the process of weathering and chemical reactions in the soil, some of the seemingly unavailable phosphorus is gradually turned into usable forms. Furthermore, there are a few species of plants that can make use of it just as it stands; and when they die they return phosphorus to the soil in a form that other plants can absorb.

This process is just one example of how different plants in the same ecosystem can affect one another, often to mutual advantage. It also points to the value of mixed cropping in agriculture. This enables some plants to utilize compounds that others cannot, and so makes the best use of the soil. But, on the other hand, the presence of two different compounds in the soil does not always mean that plants can make use of both. Sometimes, indeed, it means that they cannot use either, for the presence of one may inhibit the absorption of the other. Furthermore, such *antagonism*, as we call it, may occur for different reasons, in different soils, and in different climates.

Microorganisms in the soil, then, play an extremely important part in the cycling of nutrients. But the microbiology of the soil itself remains one of the least understood branches of biology, just because it involves such a host of interacting organisms and such a host of variables in the chemical and physical environment—all of which can differ greatly even from one part of a field to another. It is therefore no surprise to find that computers are being increasingly employed to help investigators evaluate the complex interacting factors that affect the fertility of the soil.

Above: close-up view of mixture of rye-grass and white clover. Above right: cocksfoot (a grass) grown in alternate strips with alfalfa.

*The physical environment of an area determines
the nature of its ecosystem. Most environments
are fairly constant, but occasionally disasters such
as tornadoes (top) and earthquakes (bottom)
damage the environment and destroy living things.*

6

The Physical Environment

Twentieth-century man, with all his command of science and technology, is still very much at the mercy of the climate. This is especially true in the vast underdeveloped parts of the world, where agriculture and water supplies depend so precariously on climatic change that hundreds of thousands of people may die in a bad year. But climate has many more effects on the lives of most other organisms. Climate can control their shape, color, reproductive cycles, and seasonal behavior. Among the most dominant features of climate that affect animals are light, temperature, wind, rainfall, and humidity. All these factors are physical ones, in contrast to those of the tangible chemical environment, and they all vary from region to region and from one time of year to another. In this chapter we lump all these factors under the word *climate*, which comes from the Greek *klima*, meaning zone or region. *Klima*, however, has the additional meaning of *slope*, and when we talk of the physical environment we shall also have in mind topography, which concerns the irregularities of the earth's surface.

Although we can give separate definitions of the living environment, the chemical environment, and the physical environment, it is worth repeating that they are not isolated from each other in nature. For instance, in some regions temperature affects soil nutrients and so helps to determine what plants can grow and how well they grow. The canopy of a tropical rain forest greatly affects not only the concentration of carbon dioxide down to the forest floor, but also the temperature. In this chapter we shall see how physical factors interact with the chemical and living environment.

Light

Strictly speaking, light is only that part of the sun's radiation to which the human eye responds by way of vision, but for convenience, when we talk of light we shall here include the ultraviolet and infrared radiation that reaches the earth's surface.

Besides being the source of energy on which life depends, light influences living things in several other ways. Year after year plants germinate, grow,

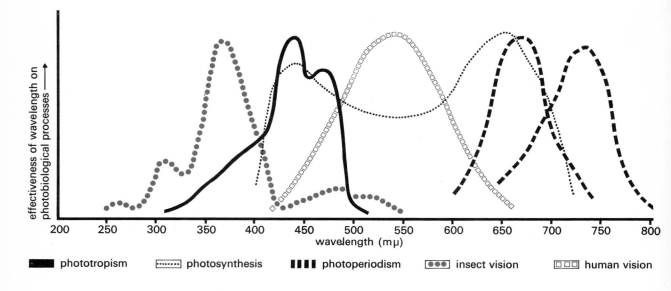

flower, and fruit at predictable times, and one of the main factors controlling this biological clock is light. Light is the most predictable factor of the physical environment. Given the latitude of a place, we can say what the average intensity and duration of light will be at any time of the year.

The name given to the effect of light on flowering and germination is *photoperiodism*. In many plants, flowering depends on the relative length of day and night. Soybeans and chrysanthemums, for example, flower only when the hours of daylight fall short of a certain value, and are called *short-day* plants. Spinach, barley, and petunias, on the other hand, flower only when the hours

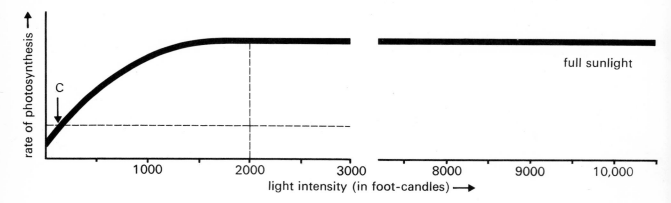

110

The rate of photosynthesis increases with light intensity up to a maximum, after which it levels off (in this particular plant at a light intensity of 2000 foot candles). The point marked C is called the compensation point, *at which the loss of carbohydrate by respiration is balanced by the rate of photosynthesis. At any rate above C the plant can gain weight and grow taller.*

of daylight exceed a certain value; these are *long-day* plants. *Day-neutral* plants, such as the tomato, are unaffected by day length. Actually, short-day and long-day plants do not respond to the length of daylight but rather to the length of uninterrupted darkness. This is a fact that can easily be demonstrated. If we give a short-day plant a brief flash of light during the night, it does not flower; it has been robbed of the long period of uninterrupted darkness it needed. On the other hand, breaking up long nights with brief flashes of light induces long-day plants to flower. The pigment that controls flowering in this way is *phytochrome*, and it responds only to wavelengths of between about 660 and 730 mμ (red and infrared).

Light also affects the seeds of many crops and weeds. Those of the Grand Rapids variety of lettuce, for example, will not germinate at all in the dark, but minute quantities of red light induce them to do so. Other seeds, such as those of the Californian poppy, germinate only in darkness.

Another widespread response of plants to the stimulus of light is called *phototropism*, which means light-turning. The main stems of most plants grow toward the light, which in normal conditions means more or less vertically. This vertical stance—due partly to phototropism and partly to a tendency to grow away from the pull of gravity (called *negative geotropism*)—not only places plants in the best position to receive light for photosynthesis but also enables more of them to exist in a given area than if they grew more or less horizontally. In many plants, too, especially trees, the elevated position helps the wind to disperse pollen and fruit. Many plants are oriented so as to receive a maximum of light, and their leaves are often arranged horizontally so that

Left: time-lapse (exposure once every four minutes) of a fungus "stem" growing toward two light sources separated by 30°. The "stem" hunted between the two sources. Above: wind disperses the fruits of the dandelion.

111

the upper ones cast the minimum of shade on the lower, so creating what is called a leaf mosaic. This can be well seen in the familiar house plant called the fig-leaf palm (*Fatsia japonica*). But many other plants, such as the cacao tree, grow best in shade, and they may even become diseased under full sunlight.

The red wavelengths, which are so important in both photosynthesis and photoperiodism, play no part in phototropism. This depends on pigments called *carotenoids*, which absorb light from the violet, blue, and green regions of the spectrum—that is, wavelengths of below 550 mμ.

The effects of light extend to animals as well as plants. Some very simple marine invertebrates that are attached to the ground or some other solid object by a stalk, bend toward the light in much the same way as plants. Other simple free-moving animals, such as flatworms, are equipped with photosensitive spots that enable them to move toward or away from the light. But only three major groups of animals have evolved complex image-forming eyes. They are the mollusks (such as squids and octopuses), insects, and vertebrates. In each of these groups, eyes have evolved independently, yet in all, the chemistry of the visual process is much the same. The light-absorbing pigment concerned is vitamin A, which is closely related to the light-sensitive carotenoids in plants. In fact, animals cannot make vitamin A except by converting the carotenoids

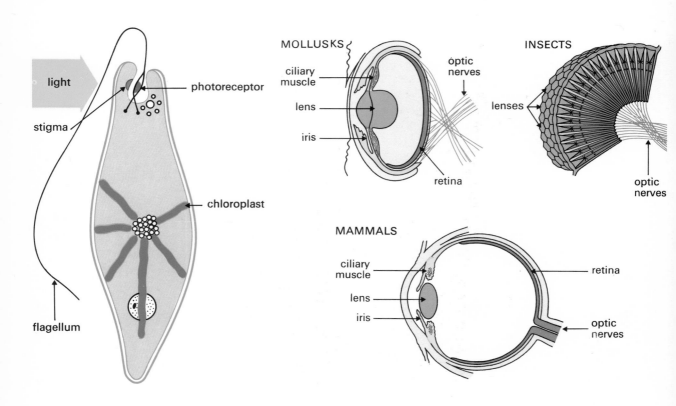

The one-celled Euglena *has the simplest of "eyes"—a photosensitive area shielded on one side by a pigment spot. As it swims with a spiral motion its photoreceptor is alternately illuminated and shaded, except when it points toward the light source. The eyes of octopus, insect, and mammal (above) all evolved separately, but are based on a common plan with a lens or multiple lenses to project an image onto a receptor (retina).*

that they get from plants. So it is not surprising that the chemistry of vision is similar to that of phototropism.

In general, eyes respond to radiation of wavelengths between about 380 mμ (violet) and 760 mμ (red). Vertebrate eyes fail to respond to ultraviolet radiation, not because the retina is insensitive to it, but because the yellow lens acts as a filter and prevents it from reaching the retina. However, the eyes of some insects are exceptional in that they do respond to ultraviolet.

Many animals, like many plants, are affected by the relative lengths of day and night. Reproductive cycles, migration, and seasonal changes of color in the feathers and fur of some birds and mammals are all geared to light intensity and day length. Finally, one effect of moderate ultraviolet radiation on some animals, including man, is that it enables the skin to synthesize vitamin D.

In addition to showing various responses to light, many bacteria, invertebrates, and fish respond to darkness by actually producing light, a process known as *bioluminescence*. Glowworms generate light flashes to attract the opposite sex; fish living in the blacked-out world of the deep seas emit light probably for recognition, and certainly in some cases to attract their prey.

Temperature

One of the most important factors affecting both terrestrial and aquatic life is temperature. On land it varies with latitude, altitude, the changing seasons, the nearness of oceans, the direction in which the land slopes, and so on. In the

One of the responses of animals to changing seasons is a change of color that makes the animal less conspicuous against its background. The brown summer and white winter fur of the stoat (above) is a good example of camouflage.

113

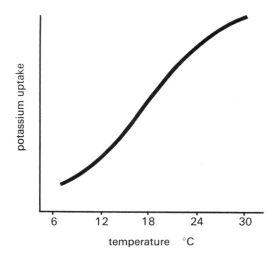

Temperature—a physical factor—also affects the chemical environment. Left: graph shows the effect of temperature on the uptake of potassium by barley roots. Most organisms cannot survive temperatures above 40°C for long, but certain algae live in the near-boiling geyser (below).

sea it depends both on latitude and on ocean currents that transfer warm or cool water over enormous distances. But because water has a far higher specific heat than soil and rock, and so takes longer to heat up and cool down, temperature fluctuations are far smaller in the sea than on the land.

Most organisms are active between 0°c and 40°c, but for many species the range is considerably less. There are two main reasons for this. First, all chemical reactions, including those in plants and animals, proceed faster at higher temperatures and more or less stop when the temperature falls below a certain minimum. This is why plant productivity falls in temperate regions during winter, and also why some animals hibernate. Secondly, most chemical reactions in living organisms are catalyzed by enzymes, and these are destroyed at temperatures over about 45°c. So organisms that are unable to keep their temperature below this figure cannot live in very hot regions. However, some forms of life exist at temperatures well outside the 0°-40°c range. The dormant, resting stages of organisms, such as the seeds of plants and spores of some microorganisms (protozoa and bacteria, for instance) can withstand extremely low temperatures down to −75°c, and some bacteria and algae can survive in hot springs at temperatures around 92°c.

The body temperature of cold-blooded animals approximates to the temperature of their immediate surroundings. In cold weather they are therefore inactive, because the body's chemical reactions proceed too slowly. In very hot weather they may also become torpid as the enzymes begin to overheat, so they cool themselves by moving into the shade or hiding in a hole. Some warm-blooded animals, however, maintain a steady body temperature of about 37°c at all but the greatest extremes of heat and cold they are likely to encounter. This gives them two big advantages. First, by remaining active—or at least capable of activity—at all times, they are less vulnerable to attack than creatures subject to periods of torpor. Secondly, their comparative independence of outside temperatures has enabled many of them to colonize a wide variety of climatic regions, so that they are less likely to become extinct if the climate in any one part of the world changes.

As most marine animals, including crustaceous zooplankton and fish, are cold-blooded, it is clear that those in cold seas have a lower body temperature than those in warm ones. We might therefore be tempted to think that cold seas would be somewhat unproductive. But although temperature does affect the seasonal distribution of many forms of plankton, cold-blooded marine organisms acclimatize remarkably well to very low temperatures, and several cold seas are, in fact, very productive. There are also other factors that affect productivity, such as upwelling, which mixes water from deep down with water near the surface, thus replenishing the upper water with nutrients. Such upwelling, which can more than offset the unfavorable effects of low temperature, occurs much less in warm tropical seas than elsewhere. Also, the solubility of oxygen and carbon dioxide decreases with a rise in temperature, and this also reduces production in warm surface water.

Different species of plants and animals can tolerate different ranges of temperature. What decides whether or not a particular species can live in a given area is not the average temperature throughout the year, but the fluctuations of temperature that the area experiences from season to season, *within* a season, and—equally important—within a single day. Figures of mean monthly temperatures, mean daily temperatures, and so on, therefore tell us little about the forms of life that are likely to exist at any particular place. For instance, the mean temperature of a stream for an entire month may have been 6°c, but this does not tell us whether or not the brook trout (*Salvelinus*) could have lived and bred there. The eggs of this species develop only between 0° and 12°c; if the incubation period included a spell of more than 15°c the eggs would be killed. But it is worth noting that even if brook trout could not have bred in the stream, frogs would probably have been able to do so, for their eggs develop between 0° and 30°C, and are therefore more tolerant of temperature fluctuations.

In general, the temperature range of each species is determined by its genetic make-up. Sometimes, however, a single species may be divided into geographical races, or strains, called *ecotypes*, each with a different temperature range. For instance, the northern race of the marine jellyfish *Aurelia* (which occurs off Halifax, Nova Scotia) is most lively at 14°C; yet at this temperature the tropical race (found off Venezuela), which prefers a temperature of 29°C, would be very inactive.

Just as light affects the seasonal behavior of organisms (*photoperiodism*), so does temperature (*thermoperiodism*). Winter grains, for instance, are sown in the fall and the young plants need the cold of the winter in order to flower the next summer. If sown in the spring they fail to flower. A period of cold is also essential before most biennial plants can flower. Bulbs, too, need a period of cold before they can start a new season's growth. This fact is exploited by growers to produce hyacinth and tulip bulbs that will flower at Christmas; this is done by chilling the bulbs for some weeks to 9°C during August and September—in effect subjecting them to a "false winter." When planted in October in a warm soil, in a greenhouse or an ordinary living-room, the bulbs soon start to grow, and flower in December.

Temperature also influences living things indirectly through its effects on soil and water. Alternate heating and cooling causes rocks to expand and contract repeatedly, often splitting them in the process, and water that has percolated into cracks and crannies exerts tremendous force if it freezes and consequently expands—breaking rocks still further. Gradually the rock surfaces break down into tiny particles that form the mineral part of the new soil. In fully formed soil, changes of temperature bring about changes in the rate of activity of microorganisms, and so affect the rate of breakdown of dead organic matter and the turnover of nutrients. The amount of water available to plants depends largely on temperature. As air becomes warmer, more water evaporates from leaves and soil, because warm air holds more water vapor than cold air. So plants often go short of water not because of low rainfall but because of high temperatures.

Rainfall and humidity

We have already noted the many essential roles that water plays in life. All land organisms are ultimately dependent on precipitation for their water supply, and the amount of precipitation depends on temperature, wind directions, atmospheric pressure, elevation of land, the nearness of mountain ranges, and so on. But in many parts of the world the types of flora and fauna depend not so much on the annual rainfall as on the severity of the dry and wet seasons, and how long each lasts. In those parts of the tropics and subtropics with a wet and dry season, rainfall is the main factor governing the seasonal behavior of organisms, especially their reproductive cycles. This contrasts with temperate climates, where seasonal behavior is governed mainly by light and temperature.

In determining what kinds of flora and fauna an area can support, the actual amount of rainfall is of less importance than the ratio between rainfall and *evapotranspiration* (evaporation from soil and plant surfaces plus plant transpiration). This explains many apparent discrepancies in plant distribution. Discrepancies arise because areas of equal rainfall may have very different rates of evapotranspiration; and this can make all the difference, for example, between grassland and desert, or between grassland and forest.

Humidity can be expressed either as an absolute quantity or as a relative quantity. Absolute humidity is the mass of water vapor in a given volume of air. Relative humidity—of more importance in ecology—is the percentage of water vapor in the air compared with the maximum it could hold (saturation) at the same temperature and pressure.

It is an oversimplification to talk about humidity on its own, because its effects on living things depend very much on other factors, especially temperature. Temperature affects many organisms most when humidity is very high or very low, and conversely the effects of humidity become more critical at extremes of temperature.

Evaporation is quicker when the air is relatively dry; less water is then available for plants, and animals may be in danger of dying from the effects

 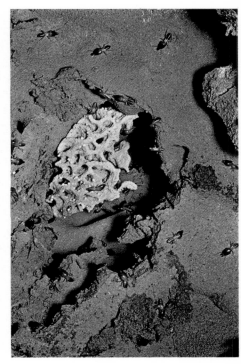

Above, a section of a termitarium, and right, termites cultivating fungus as a food crop. The fungus is not found anywhere else, because it needs the special degree of humidity that is found inside the termitarium.

117

of dehydration. Some animals, such as the woodlouse *Asellus*, seek out areas of high humidity, whereas others, such as snails, move about at night, when the humidity is usually at its highest. But very high humidities can also be dangerous to animals by slowing down the rate of evaporation too much and hence reducing the cooling effect; some organisms therefore overheat dangerously in very humid conditions. The boll weevil, for example, is unable to live at high temperatures when the humidity is also high. In one mosquito genus (*Aedes*), humidity determines both the distribution and the abundance of the various species, which may vary from night to night in the same locality.

When the air is extremely dry many plants transpire more water than they can absorb from the soil. The immediate result is that individual cells are no longer tightly filled with water, and the leaves, branches, and even the main stem become floppy—the plant wilts. A defensive mechanism comes into operation by which the stomata of the leaves (through which plants normally transpire) automatically shut. But the wilted plant cannot survive indefinitely, and if there is no rain for an abnormal length of time, any plant will die.

Wind and currents

Although wind is one of the least investigated factors of the physical environment, we know that its influence on living things is considerable. Strong winds spread the dispersal forms of many organisms (pollen, fruits, and the spores of certain lower animals) over wide areas. Gales and hurricanes blow down hundreds of forest trees, and their destruction may affect other parts of the forest for many years. In coastal and alpine areas especially, continual strong winds stunt the growth of trees, and dwarf pines are a common sight. Less spectacular but no less important is the effect of winds on transpiration. Transpiration adds moisture to the air around plants; by blowing this moist air away and replacing it with drier air, winds increase the rate of evaporation from the stomata of leaves.

The most important physical factor of aquatic environments is light, which enables photosynthesis to occur in the upper waters. Next in importance are temperature and the concentration of nutrient salts. In many sea areas, as we have already noticed, currents and upwellings of water due to winds bring nutrient salts from the unlit depths to the photosynthesis zone, greatly stimulating the growth of phytoplankton. In shallow freshwater areas, the turbulence produced by wind-generated waves and currents is particularly important. Waves, even small ones, increase the surface area of lakes and rivers, so increasing the amount of oxygen they can absorb—which for many organisms can make all the difference between life and death.

Topography

The contour of the land affects every aspect of climate and so exerts a tremendous influence on all forms of life. We shall here look at some of the main effects of topography.

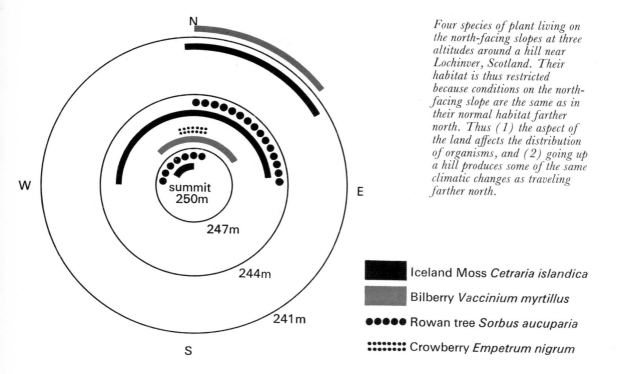

N
W
E
S

summit
250m

247m

244m

241m

Four species of plant living on the north-facing slopes at three altitudes around a hill near Lochinver, Scotland. Their habitat is thus restricted because conditions on the north-facing slope are the same as in their normal habitat farther north. Thus (1) the aspect of the land affects the distribution of organisms, and (2) going up a hill produces some of the same climatic changes as traveling farther north.

Iceland Moss *Cetraria islandica*

Bilberry *Vaccinium myrtillus*

●●●●● Rowan tree *Sorbus aucuparia*

:::::::: Crowberry *Empetrum nigrum*

Mountain ranges, even quite low ones, often have a higher rainfall than the low land around them. In Britain, for example, the annual rainfall at sea level averages about 90 cm. on the west coast, but at a height of 150 meters in the western uplands it averages 140 cm. Moreover, because temperature drops with increasing height—on average by about 1°c for every 160 meters—mountains show a vertical zonation of plants and animals. Indeed, in climbing a mountain 3000 meters high one may pass through as many different zones of flora and fauna as in traveling from south to north through 60 degrees of latitude. The considerable difference in light intensity between the north-facing and south-facing sides of a mountain also commonly results in a large difference in primary productivity; and the side sheltered from prevailing winds may support different forms of life from the unsheltered side.

Drainage, as well as climate, depends largely on topography. In some large flat plains—for instance, in Pakistan—drainage is so bad that water-logging inhibits plant growth. At the opposite extreme, water may flow over steeply sloping land so swiftly that the fertile soil is washed away.

Gravity

The trouble astronauts have in adjusting to conditions of weightlessness indicates that gravity's effects on living things are profound. Gravity is the main cause of the downward growth (*positive geotropism*) of most plant roots, into depths of soil where they can find sufficient anchorage and water. Together with negative geotropism and phototropism, it is also responsible for the vertical growth of the main stems of plants and for the horizontal growth of side branches, into the best positions to receive light and air.

Most animals, including man, rely on gravity for their sense of orientation. A rather amusing, though unkind, experiment can be done to show that this is so, for example, with a crayfish. This crustacean has a hollow organ, called a *statocyst* or gravity receptor, which it normally fills with particles of sand; as the animal changes its position, the sand presses against various sensitive hairs, so informing it which way up it is. If the sand is replaced with iron filings and a magnet placed above the crayfish, it swims upside down.

Fire

It is very easy to think of forest and grassland fires as disasters caused by man. Very often they are, but there have always been, and still are, fires caused by volcanoes and lightning. While, in most regions, these do not happen very often, their effects on an ecosystem may last for many years. Indeed, in the forests and grasslands of temperate regions, as well as in tropical regions with a severe dry season, fire—whether natural or man-produced—is a very important ecological factor and often a beneficial one. During long droughts,

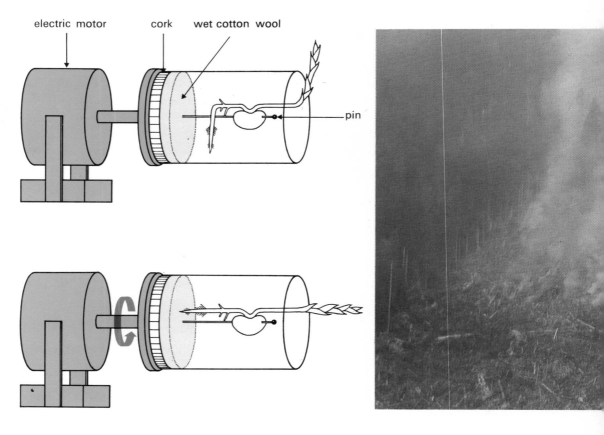

In a stationary klinostat the root of a seedling grows downward and the stem upward. When cylinder rotates, both grow horizontally because gravity pulls equally on all sides. Light is excluded to prove that gravity alone acts.

surface litter in grassland and forest becomes so dry that the decomposing bacteria and fungi are unable to break it down. Fire helps to free the nutrients locked up in the dry litter and return them to the soil, so it is often followed by a brief flush of vegetation. There is also evidence that mild surface fires increase the activity of nitrogen-fixing soil bacteria.

One example of forest fires that are useful to man occurs in the coastal plains of the southeastern United States. Here the long-leaf pine, one of the world's most valuable timber trees, is less vulnerable to fire than other plants in the region, because the tips of the delicate growing shoots are protected by circlets of fire-resistant pine needles. In the absence of fire, scrub hardwoods flourish among the long-leaf pines and tend to crowd them out. But an occasional fire destroys these competitors and enables the long-leaf pine to take over. This is an instance of *fire-controlled vegetation*, or a *fire-climax*.

Incidentally, in natural conditions a light fire burns up surface litter that, if left to accumulate, could encourage a very serious "crown" fire—that is, one that would destroy all or most of the vegetation. Some people think that the zealous prevention of light fires in parts of California, for example, has actually increased the severity of the occasional devastating crown fires that destroy not only timber but also homes.

The photographs above show the stages before and after control of vegetation with fire in an American forest. Primitive tribes and rich countries alike still make their mark on ecosystems by the deliberate firing of vegetation.

The Palm House of the Royal Botanic Gardens, Kew, England. Plants and trees from many different parts of the world flourish in the glasshouses, each of which is provided with its own carefully controlled environment.

7

Limiting Factors and Tolerance

Every living thing is affected by the combined action of many factors in its environment. Too much or too little of any single requirement may destroy it, or at least reduce its productivity. In the last two chapters we saw how rigorous such requirements can be, particularly in terms of light, heat, rainfall, humidity, and food. Partly because one must begin somewhere, and partly for the sake of clarity, we tried to explain how and why this happens by treating each factor more or less separately. Our approach, in fact, was similar to the neat and tidy approach of the German chemist Baron Justus von Liebig (1803–73), who was one of the first scientists to study the effects of inorganic nutrients in agriculture.

In 1840 Liebig pointed out that the growth of crops was often limited by whatever essential element happened to be most scarce in the soil; in other words, productivity was checked by the weakest link in the environmental chain—what we now call a *limiting factor*. But to begin with, scientists in the 19th century sometimes oversimplified their ideas on limiting factors, and mistakes were made. In one instance some scientists decided to improve the productivity of phytoplankton in an arm of the sea where the water did not change at every tide by adding nitrates and phosphates, which tend to be scarce. Nothing happened, and further investigations showed that a shortage of iron and silicon was also limiting growth. When these two elements were added, the phytoplankton were able to make use of the nitrogen and phosphorus available, and productivity rapidly increased.

For many years after this, biologists and pioneer ecologists took to their laboratories and systematically investigated the effects of *individual* environmental factors (both chemical and physical) on *individual* plants and animals. As far as possible, in each experiment they held all but one of the factors constant, so that they could assess how variations in the remaining one affected the organism they were studying. As a result of such experiments they could say, for instance, "All other factors remaining constant, such-and-such a plant grows best at such-and-such a light intensity."

This simple approach was perhaps the only way to begin to understand which factors in any given situation were encouraging productivity and which were limiting; but it had its dangers. For one thing, the early investigators found it all too easy to generalize from their findings in the laboratory. In other words, there was a strong temptation to say: "*Plants* grow best at such-and-such a light intensity," instead of: "This plant and that plant grow best at such-and-such a light intensity in certain conditions." It was also easy to forget the huge pitfall contained in the proviso "All other factors remaining constant." In fact, as we shall see later, it is not always possible to alter one factor while keeping all the others constant, even in a laboratory; and in a natural ecosystem it can never be done.

Nevertheless, early laboratory workers discovered much about limiting factors that was useful as a background to the study of natural ecosystems and agricultural productivity. For instance, when crops become diseased, or when yields begin to fall, a broad general knowledge of what limiting factors may be at work could well give an immediate answer to the cause of the trouble. The culprit—perhaps a deficiency of one particular micronutrient—can then often be found quite quickly by more careful tests.

The laboratory approach still plays a very important part in ecological studies, but it is now practiced at a far more sophisticated level, and with a healthy awareness of its own limitations. It is now realized, for instance, that we cannot always vary one factor without varying others. Changes in light intensity, for instance, inevitably affect the aperture of the minute stomata in the leaves and this change in turn affects the carbon dioxide and water relations in *and around* the plants. The close control of environmental conditions is far too expensive to practice in natural ecosystems or on any sizable field crop. But in recent years it has proved possible to construct enclosures in which a great variety of outdoor conditions can be very accurately simulated. The most famous example is the collection of greenhouses (or phytotron) at the California Institute of Technology, Pasadena. Here all the environmental conditions can be permutated in each of 54 separate chambers.

In outdoor agriculture, however, as in natural ecosystems, many conditions vary—often unpredictably—all the time. Agricultural research stations therefore approach the problem from the other end by growing plants in their natural environment, observing what happens when conditions change, and analyzing the results by statistical methods. This approach also yields valuable information about limiting factors, even though it may do so only with great difficulty. (It is not easy, for example, to tell whether nitrogen is limiting growth in a field if it is applied just before an unexpected drought that prevents the plants from making full use of the nitrogen.)

Interaction of factors

Nowadays neither laboratory investigators nor field investigators are given to making the kind of sweeping generalizations that were fairly common up to the 1930s. They know that such tags as "The more light, the more photosynthesis" and "The higher the temperature, the higher the productivity"

Above: crops in an outdoor experimental field at Rothamsted, England, growing in test plots, all subject to the same climate but differing in soil conditions.

Below: inside this Australian phytotron are numerous cabinets (below right), where temperature and other climatic factors can be accurately controlled by a computer.

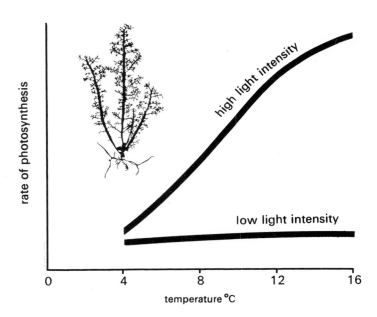

Every organism is affected by many environmental factors, but it is not easy to guess what the effect of a change in one factor will be on the life of an organism. For example, it need not follow that an increase in temperature will increase photosynthesis in a particular plant, although this may be true of many other plants. Thus in the alga Gigartina *(right) temperature did not influence photosynthesis at low light intensities but it did do so at slightly higher intensities.*

cannot be universally applied. Nature is far more subtle than this. First, changes in one environmental factor often modify the effects of others. Next, different organisms have different tolerances to temperature, light intensity, humidity, and so on, and therefore respond to them in different ways.

We have already seen that humidity, for example, modifies the effects of temperature, and vice versa. This happens to such an extent that even when we know that one or other is acting as a limiting factor, we cannot easily say which. Thus the boll weevil in the previous chapter can withstand high humidity or high temperature, but not both together. We also saw in Chapter 5 that plants may suffer from a shortage of mineral nutrients when those nutrients are actually present in the soil in sufficient quantity. This is often the result of antagonism between two minerals, each of which inhibits the absorption of the other, and it underlines the fact that there must be a correct *balance* of nutrients in the soil, as well as a sufficient *quantity* of each.

Another example of interaction between environmental factors concerns light and carbon dioxide. In general, the rate of photosynthesis increases *up to a point* with light intensity, and is little affected by the concentration of carbon dioxide in the air. However, the proviso "up to a point" is important. When the light intensity reaches a certain level, any further increase makes no difference to the rate at which a plant grows; but if both light intensity and carbon dioxide concentration are raised, the rate of growth increases. Thus light can be a limiting factor at one point, carbon dioxide at another.

It must also be borne in mind that organisms are often inhibited not by a deficiency but by an excess of certain factors in the environment, or even by a change of balance between them. A striking instance of this occurred some years ago in Great South Bay, Long Island Sound, New York, where the establishment of duck farms along the rivers leading into the bay brought considerable quantities of duck manure into its waters, and because the currents there are slow moving, the nutrient-rich manure accumulated instead of being washed out to sea. As a result, there was a great increase of phytoplankton. But because duck manure had a lower nitrogen-phosphorus ratio than was previously present, most of the former species of phytoplankton disappeared, and were replaced by a rare flagellate. The valuable blue-point oysters in the bay were unable to digest this new species, and so starved to death, their guts filled with flagellates.

Tolerance

The same environmental factor that is limiting for one organism is not necessarily so for another, because different species have different ranges of tolerance to various factors. For instance, the most important factor affecting the growth of tomatoes is night temperature, whereas a ragwort called cocklebur (*Xanthium*) is affected most by ratio of day length to darkness, and the lily *Veratrum* by seasonal variations in temperature. Some species are affected equally by more than one factor: the strawberry, for instance, grows well only when both the temperature and the daily exposure to light are just right. The tiger beetle is even more exacting. For its egg development it needs moist, warm, shaded soil on sloping ground, well-drained, porous, sandy, and with little humus.

Within a single species, tolerance to any given factor may vary considerably from one stage of the life cycle to another. Thus adult blue crabs can tolerate brackish water, but their larvae cannot; and, of course, many seedlings are far more vulnerable to lack of water than are the adult plants.

The study of limiting factors is complicated by yet another consideration: when one factor is not present at optimum concentration, tolerance to other factors may be reduced. For instance, when the temperature falls, certain animals living in estuaries become less tolerant to water with a low salinity. Indeed, in the Blackwater Estuary in southeast Ireland during the hard winter of 1963–4, there was heavy mortality among certain marine snails (*Urosalpinx*) that are destructive to oysters. They died because at low temperatures they could not tolerate the salinity of the waters they had always lived in. Similarly, on land, a deficiency of nitrogen reduces the ability of grass to withstand drought. Also, some plants need more of the trace element zinc in full sunlight than they do at lower light intensities, so with them zinc can be a limiting factor in full sunlight but not in the shade.

Finally, even if an organism does have the best possible physical and chemical environments, there is no guarantee that it will be able to take full advantage of them. The reason for this is that by the very act of growing, it may well bring about changes in its own environment, so that the optimum conditions no longer exist. For example, most plants, by producing new leaves, alter the temperature and humidity of the air immediately around them, and also change the intensity and wavelength-composition of the light filtering through to their lower leaves. Moreover, predators or parasites may prevent an organism from taking full advantage of ideal conditions; and other organisms that are neither predators nor parasites may do the same thing. Many microorganisms, for example, secrete into the soil substances that inhibit other organisms, and the roots of some plants also secrete inhibiting substances that prevent other plants from growing too close.

Success without optimum conditions

Few organisms ever have the luck to live in the best possible surroundings. But oddly enough there are some species of plants that actually flourish best in an environment where one or more factors are *not* optimum. Certain tropical orchids that grow naturally in shady situations grow better in full sunlight as long as they are kept cool. So in a cooled greenhouse it is possible to increase their growth by increasing the light intensity. We might be tempted to conclude that lack of light in the wild state limits their growth. But we should be wrong, because in natural outdoor conditions an increase in light intensity could be gained only by exposure to full sunlight, and this would mean a greater increase of temperature than the orchids could tolerate.

In natural ecosystems, evolution has seen to it that organisms are well adapted not to optimum conditions of every kind, but to *prevailing* conditions. If one factor limits growth to some extent, this may not matter much—partly because organisms have learned to put up with it, and partly because there may be advantages that compensate for it. For example, plants in the mountains of Colorado, where the light is fairly intense, reach their peak of photosynthesis at high light intensities; but in the Yukon Territory, about 2000 miles to the north, where intensities are lower, alpine plants reach their photosynthetic peak at lower intensities. So although it is true that the plants of Colorado have a higher rate of photosynthesis than those of the Yukon, it is also true that the plants of the Yukon would not benefit by having the higher light intensities of Colorado. So one cannot say that the flora of any one region is more successful than that of another.

In short, in natural ecosystems, evolution has minimized the adverse effects of environmental factors that are not as good as they might be. The study of limiting factors in relation to particular species is therefore most useful where man intends to alter existing conditions, either in agriculture or in ecosystems that have so far been natural ones. Perhaps the most useful application of all will be in up-to-date, controlled greenhouses.

Left: a river polluted by detergent and (far left) a canal that has been used as a dump for domestic trash. Inland, estuary, and coastal waters are some of the most productive parts of the biosphere, yet daily millions of tons of sewage and industrial effluent are emptied into them. This in turn leads to drastic changes in the underwater environment, so that often few organisms survive. Recent efforts by governments have had some success in the industrialized West in reversing the consequences of dumping, and life is becoming more abundant in our lakes and rivers (above).

A cliff of sedimentary rock. Weathering has created a loose "dust" on the surface, allowing seeds of pine (top left) and shrubs (three in center) to germinate. There was no prior preparation by lichens, mosses, or grasses.

8

The Growth of Ecosystems

Interaction between the nonliving and the living environment not only affects ecosystems; it also *produces* them. For example, the natural forests we see today did not appear on earth ready-made. At some time in the past the land where they now flourish was barren ground, such as rock, sand, or recently cooled lava. The first forms of life to find a foothold there were few and relatively simple; but gradually, and in increasing order of complexity, other communities of organisms took over until finally the forest emerged with all its multifarious life.

When one community replaces another, it is called *succession*, and all the stages together are called a *sere*. A sere (another word for series) is complete only when a community of great stability is achieved—one that is capable of continuing indefinitely without further basic change. This is known as the *climax*, and for the time being, at least, we can regard it as self-perpetuating.

Sometimes, all the stages of succession can be witnessed within the span of a single human life. Most climax ecosystems, however, take far longer to develop—maybe 1000 years or more. Yet we still know pretty well how they arise, because at any one time we can see young stages and older stages in different parts of a region. And by looking at any stage before the climax itself, an expert ecologist can predict the likely future stages.

Few parts of the world are now covered with climax vegetation, for two main reasons. First, many natural ecosystems have not yet reached the climax. Second, man has either destroyed many climaxes altogether, or has deliberately prevented one stage from giving place to its natural successor on the road toward the climax. In many parts of Britain, for instance, the climax vegetation is oak forest, yet only a few such forests now remain, in secluded parts of the northern uplands. Elsewhere, they were cleared centuries ago to make way for cultivated fields, or grassland maintained for grazing. If the fields were abandoned, succession would eventually cover many of them with oak forests once more.

What causes succession? We can say that the type of climax vegetation that eventually appears in a particular area depends on the regional climate. But we cannot say that the climate *causes* succession, because regional climatic

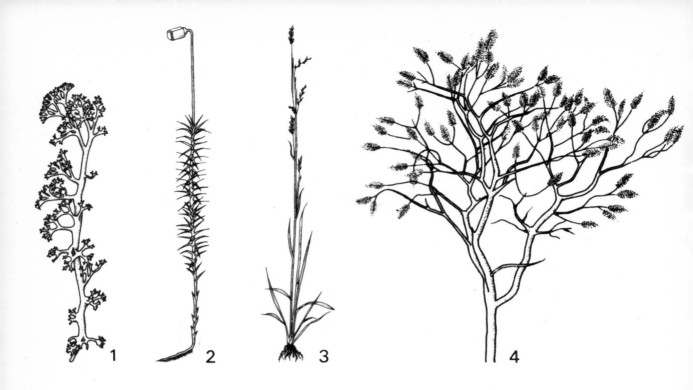

conditions remain roughly the same from start to finish. So we must conclude that succession is caused by living communities themselves, and that when the first organisms arrive they change the local environment in such a way that new organisms are able to move in, ousting the old.

In this chapter we examine the broad principle of succession without confusing matters by describing all the different species that exist at each stage. Following the custom of some ecologists, we divide the process into *primary* and *secondary* succession. Primary succession is the community changes that begin on a sterile area, or on sites not previously occupied by organisms: examples are newly exposed sand dunes, cooled lava flows, and newly exposed rocks. Secondary succession is the sequence of community changes that occur on sites favorable to life, or where there have previously been organisms: examples are new ponds, abandoned farmland, and burned-down forests.

Although these definitions are not inexact, when we scrutinize them it is not quite clear into which category we should put many examples of succession. Taking primary succession first, "sterile area" means purely inorganic. Thus, if life begins on bare rock or in fresh water containing no organic matter, this must be primary succession. But the phrase "not previously occupied by organisms" is confusing. Can we be sure that a newly exposed sand dune or a hill stripped bare by a glacier has not, in ages past when the climate was quite different, supported life? What we really mean, of course, is that living things have not *recently* occurred on such-and-such a site, but then, how recent is recent? We shall come back to this question later.

An abandoned field, a forest destroyed by fire, a stream heavily polluted with sewage, all contain many organic and inorganic nutrients, and so conditions here are favorable to life, and these are clearly cases of secondary

5

Five plants taking part in the primary colonization of a bare rock surface. Cladonia (1) is one of the lichens that are the first arrivals. They start to decompose the rock; windborne dust lodges between the plants. Then mosses, such as Polytrichum (2), move in and build up a real soil. Grasses, such as the broom sedge Andropogon (3), then arrive, followed by shrubs, for instance sumac Rhus (4). Finally trees, including oak (5), form the permanent and stable flora.

succession. Now let us look at a new pond or reservoir. Although it is a brand-new environment, the water still contains organic matter that quickly diffuses out of the flooded land vegetation that was there before; also, all surface fresh water contains some organic matter and organisms. These examples too, then, come under the heading of secondary succession. But what about the new concrete dam of the reservoir? This is, from the start, inorganic and it forms a special habitat for encrusting organisms not present in the reservoir itself; this would be an example of primary succession.

The best way to define primary and secondary succession is to say that primary succession always occurs in or on some inorganic medium and is therefore usually very slow to get started. All other cases are, quite simply, ones of secondary succession.

Primary succession

Perhaps the best way of describing primary succession is to start with a piece of land that has not previously supported life at all, and to trace in outline one sequence of events that could eventually cover it with climax flora and fauna. We start with a stretch of bare igneous rock that is exposed in a temperate climate. First, wind, rain, and frost weather the surface so that it becomes pitted with small crevices and irregularities. Water that collects in these hollows also dissolves some of the minerals out of the rock surface. The minerals are enough to allow simple plants—for example, lichens—to establish themselves on the rock surface; the lichens, in spore form, were wind-borne from elsewhere, and took hold wherever conditions favored them. Once established, some lichens (but not all) secrete acids that dissolve more of the rock surface, releasing even more nutrients. After a period, maybe months or maybe years, the rock hollow fills up with a mixture of inorganic wind-borne dust and fine soil from neighboring areas, and organic matter from dead lichens.

This first soil provides a livelihood for the soil microorganisms that start cycling the nutrients from the dead lichens so that the living lichens can multiply still further. At this stage the soil is no more than a shallow anchorage; but as the lichens grow and die, the soil and lichens build up enough to allow a few animals such as mites, ants, and spiders to move in. There are not many of these because only a few species can stand such harsh conditions. Nevertheless, their dead remains bring about a further enrichment of the scanty soil. This first combination of plants, microorganisms, and small animals forms what we call the *pioneer community*.

After a time other species—forming the *secondary communities*—begin to move in. Mosses may be the second plant immigrants, and soon there are fresh animal arrivals, including new species of mites and other groups, such as springtails (wingless jumping insects). Without plants, or with only a sparse growth of lichens, much of the new soil would sooner or later be washed off the rock by rain, or else be blown off by strong winds. But as the mat of moss becomes more extensive it holds the soil down firmly. Its depth, meanwhile, is further increased by the decay of plants, as well as by the decay of the animals. Thus the formation of the soil depends both on the physical environment and on organisms themselves.

At this stage the soil is very different from the mainly inorganic dust that was there before, and is far more like garden soil. It contains more minerals, because weathering and organic acids have continued to dissolve the underlying rock; it also contains organic matter, called *humus*, from decayed and decaying lichens, mosses, and soil organisms. Because humus has good water-retaining properties, the soil now remains damp for longer periods, so plants and other organisms can survive longer periods without rain. New soil microorganisms now arrive, their spores being blown in by winds or perhaps carried on the bodies of birds and insects, and these accelerate the breakdown of organic matter, so that still more nutrients are available for future organisms.

Soon various biennial grasses appear, and these are followed by perennial grasses. The grasses provide both cover and food for new animals, and nematodes (such as thread worms, eel worms, and round worms) join the existing animals in the deepening soil. Because the grasses are taller than the previous plants, the physical conditions just above the surface change. The grasses provide small windbreaks, which increase the local humidity by slowing down the dispersal of moisture by evapotranspiration; they also create shade, thus altering the temperature and light intensity just around them. In effect, the grasses have helped to make a new microclimate so that small shrubs, such as heaths, can and do establish themselves; finally, tree seeds can germinate, and perhaps, after several centuries, what was once bare rock becomes a climax forest. The sere is now complete and no further change is likely so long as the climate stays the same.

When the climax is reached, not only are there more species of plants and animals than in the early stages, but the physical environment within the ecosystem has also been modified more than at any previous stage. The light, temperature, humidity, and wind speed in the forest are now quite different from what they were before; they are also quite different from what they are outside the forest. All this has a marked effect on the germination and growth of, say, a tree seedling. The seedling may need cool weather and a reasonably moist soil. Outside the forest, if the weather happened to be very hot and windy, it could find neither, and would probably die. Inside the forest it can always find both, because as well as providing shade, the trees slow down the wind and reduce the rate of evaporation from the soil. Within its confines the forest thus changes the main regional climate by creating its own micro-climate, and in doing so, it buffers organisms against extremes of weather.

In our example of primary succession, the sequence of primary producers was lichens, mosses, grasses, shrubs, and finally trees. There are, however, other possible sequences: for example, grasses, shrubs, or even trees can in some cases establish themselves straight away after weathering and wind have built up a thin layer of soil to get the seedlings started. Also, the actual species that establish themselves at each stage of primary succession depend both on the regional climate and on their availability in that region. For instance, at the grass stage, the species of grass in Africa would not be the same as in Asia; and in prehistoric Australia, the biggest herbivores feeding on grass would be not bison, as in North America, but kangaroos.

But whatever the details of succession, all cases have certain things in common. There is an orderliness and predictability in the sequence of changes that occurs in any given case of succession. Thus in our original example there is only one period at which shrubs could be the main vegetation: after the grass stage and before the tree stage. The shrubs needed the prior preparation of the environment by grasses, including the formation of a deep enough seed-bed; but the shrubs could not continue indefinitely, because they paved the way for tree seedlings and in doing so sealed their own doom. The essence of succession, then, is that one stage changes the environment so that it becomes less favorable for itself and more favorable for others. Finally, there comes a time when no further change occurs—when the climax allows no further colonization by new species. All the existing species are then in equilibrium.

The areas of the earth where life starts from scratch are fairly numerous. Soil, rock, volcanic ash, sand dunes, and new volcanic islands emerged from the sea are forms of purely inorganic matter that are all ready to be colonized by organisms in the process of primary succession. Landslides in alpine regions might strip bare the sides of mountains, and life has then to start anew on the exposed surface. In aquatic environments the building of a dam provides a new habitat on which organisms can live. Over longer periods of geological time, of course, new habitats have been created periodically on a vastly greater scale, as when the ice-sheets retreated after each of the four glacial ages.

Secondary succession

In secondary succession, some drastic event, such as a change in climate, raging forest fires, long-term flooding, the abandonment of agricultural land, hurricanes, typhoons, or the pollution of streams by excessive sewage or industrial effluents, may destroy or severely alter the previous communities so that new ones begin to develop. Secondary succession differs from primary succession only in its early stages.

Because man has turned much of the earth's surface into farmland and then moved on, let us follow the main stages of secondary succession in an abandoned field in the southeastern part of the United States. We can safely assume that there is still a fairly thick layer of soil, containing nutrients, soil micro-organisms, and humus. The long, slow stages of primary succession are therefore skipped. At first, crabgrass and horseweed are the dominant plants (see figure below), and in the second year these are followed by wild aster and ragweed. In the third year, and for the next few years, broom sedge assumes dominance, but gradually pine seedlings appear between the tufts of sedge. The pines, having made their first appearance between 5 and 8 years after the field was abandoned, increase and multiply, until after about 16 years they are the only tall plants present in the ecosystem. Then certain hardwood (broad-leaf) species appear, such as dogwood, red gum, red maple, and black oak. Some of these eventually create shade that the smaller pines cannot tolerate. As the pines die one by one, they are replaced by hardwoods, especially oak and hickory. After about 150 years, the pines are entirely eliminated, and a climax oak-hickory deciduous forest becomes firmly established.

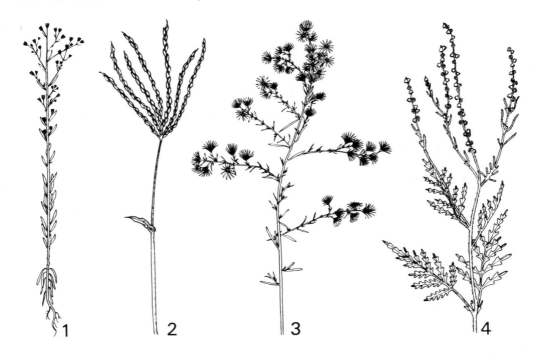

Above: plants in secondary succession in an abandoned field in Carolina, USA. First year: horseweed (1) and crabgrass (2) are dominant, with small wild aster. Second year: aster (3) is dominant, with ragweed (4) beneath its shade. Sedges, then pines, and finally hardwoods, take over.

The clearing shown above has been put back to earlier successional stages, and contrasts with the sparse undergrowth of the surrounding beech forest. Grasses, shrubs, and young trees compete, but beeches will eventually predominate.

In 1963 a new island— Surtsey—appeared. Already life is appearing on the lava, providing scientists with a good example of primary succession.

In describing both primary and secondary succession, we chose places where there was a reasonable amount of precipitation and soil moisture, neither too much nor too little. Such a succession is called a *mesarch* (*mesos* = middle, *arche* = beginning) type. But succession also occurs where there is abundant water— called a *hydrarch* (water-beginning) type—and in extremely dry regions we find the *xerarch* (dry-beginning) type.

Let us look first at hydrarch succession, starting with, say, a pond or a shallow lake that is already inhabited. The animals and plants may be either completely submerged or semiemergent, but they all depend on plentiful water. If such an area is left undisturbed for a long time, however, it often slowly dries up. Silting-up by soil erosion from the surrounding land, coupled with the gradual but continual accumulation of organic debris, slowly reduces the depth of water until what was formerly a lake or pond becomes a swamp. In other words, the physical environment (by way of silting) and living things (by death and decay) jointly change a small part of the face of the earth in such a way as to make it less suitable for its original plants and animals and more suitable for others. After the swamp develops, there is an uninterrupted succession, producing a marsh thicket with average soil moisture conditions. In a moist, temperate climate the marsh thicket might eventually be replaced by a forest; but in a slightly drier region, a climax prairie might finally develop.

At the other extreme, xerarch succession begins where moisture is severely limited, as in deserts. Nevertheless, over a long period of time, a few species of plants might establish themselves and gradually build up a soil with fairly good moisture-holding properties. Thereafter, succession can proceed normally, and it is possible that a fairly extensive plant cover will eventually appear. Often, then, both hydrarch and xerarch types of succession tend toward the intermediate mesarch type.

We have so far said little about animals, mainly because they depend on plants for food and shelter; and the succession of plants largely decides what kinds of animals are present. But this does not mean that animals play a subordinate role in ecosystems; on the contrary, they may affect the vegetation very much, especially when the plants are small. For example, most young oak seedlings fail to survive because they are nibbled away by herbivores. If the herbivores became too numerous, there would be no oaks left at all, but in a balanced climax forest there are always enough predators to keep the herbivores in check. Insects also play an essential part in ecosystems by pollinating many plants, and birds and rodents help to disperse the fruits of certain trees.

Dominance

We often refer to stages of succession by the names of only one or two of the most important species of plants, as when we talk of a grass stage or an oak-hickory climax. There are two good reasons for this. First, it would be extremely tedious (and often impossible through lack of detailed observation) to name or describe all the plants growing at each stage. Second, it is not really necessary

Hydrarch succession in a Michigan lake. The lake became blocked by silting, and the growth of grasses and sedges from the margin inward turned it into a bog, then into a bog thicket (view above), and finally into a climax forest.

138

to do so, because not all of them are equally important in determining the nature of the ecosystem. Only a few species exert a major influence, and those few we call *dominants*. The dominants control the environment and influence other organisms in the ecosystem far more than the other species, either because they are the most productive, or because they are more numerous, or for some other reason. We can, if we like, reverse the argument and say that removal of these dominants would radically alter the whole ecosystem. The names of the dominants therefore provide meaningful, as well as convenient, labels for various stages of succession, even though a shrub stage, say, can have scattered oak trees, and an oak-forest stage a lower layer of shrubs.

At all but the earliest stages of succession there are many different species present, often reaching many hundreds in the climax. Yet there are seldom more than a few dominants at any stage. If one or two of these happened to have a wide range of tolerance to changing environment factors, they might persist through more than one stage, but their importance would nevertheless decline. So species that are dominant early are unlikely to be so later on.

Both the total number of species and the number of dominants tend to be smallest where the climate is harsh or extreme, as in desert or tundra regions. In general, too, high-latitude communities have fewer dominants than low-latitude ones. In some northern forests, for example, only two species of trees account for 90 percent of the total standing crop. A tropical forest, on the other hand, usually has a dozen or more dominants. Indeed, that is why we do not call tropical forests by the names of their dominants—there are simply too many of them.

Climax, complexity, and stratification

We have already seen that the numbers of animal and plant species are much greater in the climax than in the early stages of succession. The total biomass is also greater, and the amount of dead organic matter in the soil, too. In fact the climax is the most complex stage of all, and it is also the most stable.

On the face of it this may seem extraordinary, because usually the more complex an organization is, the more things there are that can go wrong. Indeed, the reason why the climax combines complexity with stability is a subtle one. One way to explain it is to compare it with the mammals, which are both the most complex and the most successful of the vertebrates. Their success is partly due to superior "intelligence," but it is also because they possess delicate mechanisms for maintaining the constancy of their *internal environment*. For example, they have complicated mechanisms that keep their body temperature steady at the ideal level for internal chemical reactions in both cold and hot weather, and can therefore live in a wide variety of climates; because such mechanisms have evolved slowly, they rarely go wrong, except in old age. And just as the mammals' success in their external environment is largely due to their control over their internal environment, so the success of the complex climax is largely due to its ability to control its external environment.

Another reason why the climax supports a greater number and variety of species than the early stages of succession is that it offers both *more* food and more *varied* food; it also provides a greater variety of places for animals to live in. Let us imagine, for the sake of argument, a forest made up of only tall trees in which animals live only in the canopy. This would limit the number of organisms in two ways. First, only those animals that were able to eat the food available in the canopy would exist; second, the animal numbers would be limited by the living-space in the canopy. Now imagine a forest in which other plants are arranged beneath the tall trees. There would then be more plants *in the same area as before*, and hence there would be more food for animals. There would also be more space in which animals could live, and a greater variety of habitats, each of which could provide food for a special type of community.

The broad arrangement of an ecosystem into layers is called *stratification*, and this is especially noticeable in the climax. Forests show very marked stratification: as an example, we can take a climax maple (*Acer*) forest, which has four main layers. The tallest trees (the *overstory*) form a canopy; then come smaller trees (the *understory*) that prefer some shade; then come shrubs, and finally herbaceous, soft-stemmed plants, which are still more shade-tolerant. We could even extend our list to include several layers of organisms beneath the surface of the soil. Each layer has its own type of shelter, food, light, temperature, and humidity, so that each is different and each is suitable to harbor particular populations.

One might imagine that stratification would not prevent animals from moving from one layer to another, but in fact most animals, including many birds, live most of their lives in a single layer. Thus in the evergreen forests of New England the magnolia warbler occupies the low levels, the black-throated warbler the middle, and the Blackburnian warbler the high. The diagrams opposite show examples of stratification.

We can, however, take the idea of stratification further. Within each of the main layers, there are many quite different habitats—as different as the forest canopy is from the moss on the ground. Fallen logs, stones, the leaves or roots of particular flowers and trees, all provide more or less ideal conditions for species that could not live elsewhere. Anyone who tries collecting beetles, for example, soon learns that some species can be found only in fallen, rotting branches, or in the dung of a particular animal.

Succession and energy flow

We have seen that a supply of energy is essential for the growth of all organisms (Chapter 2) and we have also seen (Chapter 4) how the original energy is passed on from one stage of the food chain to another. Now we look at the flow of energy in a succession. We saw that part of the stored energy in the total protoplasm produced (the gross production) is lost, first by plant respiration and then by animal respiration. We also saw that animals lose more energy by respiration than plants. In the early stages of primary succession

24

18

12.5

7

1.5
meters

Typical forest (above) has five layers. In an oak-hickory forest, these would be: (1) overstory forms a canopy (oak and hickory); (2) below this is shade-tolerant understory, with smaller overstory trees (smaller oak, hickory, and maple); (3) transgressive stratum has even more shade-tolerant plants, growing overstory and understory trees (oak, hickory, maple, sourwood, dogwood, red cedar, and black gum); (4) seedling stratum (trees of other layers); (5) herbaceous stratum (wild ginger, etc.).

Layering, or stratification, also occurs in fresh-water lakes, such as Lake Windermere (below), England. In autumn, this lake has a uniform temperature of 4°C. In summer, it is stratified into the epilimnion, which is warm and therefore floats on top, the thermocline, where temperature rapidly changes with depth, and the hypolimnion, a layer of almost uniformly cold water. Plant life occurs solely in the warmer top layer; the bottom layer often lacks oxygen and supports only certain organisms that can do without it.

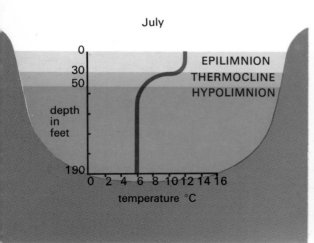

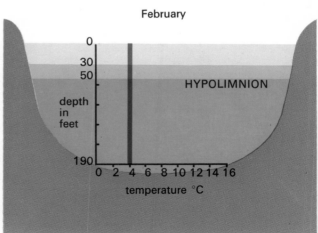

SPECIES STRUCTURE

SPECIES COMPOSITION	Changes rapidly at first, then more gradually
NUMBER OF SPECIES OF AUTOTROPHS	Increases in primary and often early in secondary succession; may decline in older stages
NUMBER OF SPECIES OF HETEROTROPHS	Increases until relatively late in succession
SPECIES DIVERSITY	Increases initially, then stabilizes or declines in older stages

ORGANIC STRUCTURE

TOTAL BIOMASS	Increases
NONLIVING ORGANIC MATTER	Increases

ENERGY FLOW

FOOD WEB RELATIONSHIPS	Becomes more complex
GROSS PRODUCTION	Increases during early phase of primary succession; little or no increase during secondary succession
NET PRODUCTION	Decreases
RESPIRATION OF WHOLE COMMUNITY	Increases
PRODUCTION/RESPIRATION RATIO (P/R)	P > R to P= R

there are few animals to eat the plants, and the loss of energy in the entire ecosystem by respiration is therefore low. So the gross production of the plants is considerably greater than the total loss through respiration. The plant biomass therefore increases, and the amount of organic matter in the soil accumulates. In the later stages of both primary and secondary succession, there are more animals present, so the loss of energy by respiration is greater. In the end, losses through respiration tend to draw level with gains from gross production, so that *net* productivity diminishes. Finally, in the climax the total biomass tends neither to decline nor to increase, and there is little or no further accumulation of living or dead matter in the ecosystem; in other words, the climax is stable. The table above shows a summary of the tentative conclusions of Howard T. Odum on the changes of energy flow in one kind of succession.

Time and stability

So far, the picture we have drawn is a simple one—succession is orderly and goes in only one direction. Now we must modify that picture a little in terms of the time taken to reach a climax and then of the stability of the climax itself.

In primary succession, the pioneer stages are slow to start, because the environmental conditions near the ground are severe; the ensuing stages are quicker, but the rate then slows down again as the climax is approached. On lava flows or sand dunes it may take at least 1000 years for the climax to appear. Secondary succession, with its head-start, is usually more rapid, and in moist temperate regions it would take about 200 years for a forest to return to abandoned farmland. The time needed to reach the climax in secondary succession obviously depends partly on what stage we start from, but it is also affected by other conditions, such as the regional climate. The grasslands in the severely overgrazed and eroded regions of Spain might take more time to reestablish themselves than would be taken for a forest, for example, to appear, by primary succession, on a newly emerged volcanic island in the moist climate of the South Pacific.

The harsher the environment, the more rapidly succession usually reaches its final stage. For instance, in the plains of west-central North America, there are few trees or dense shrubs to buffer the effects of regional climate, and secondary succession produces climax grassland in 20–40 years after abandonment; in desert and tundra regions, where climatic conditions are even more extreme, succession is usually completed in a much shorter time.

Succession is most rapid where organisms can scarcely modify their environment at all. In the sea, organisms are at the mercy of an enormous bulk of water moved by winds and currents, and can do very little to change their surroundings; they may affect the nutrient balance of the water but certainly not the

The sharp line between this stable climax New Zealand forest and the grassland is maintained by red deer grazing any young tree seedlings that happen to germinate in the pasture. Without the deer, the forest would extend its boundaries.

143

light intensity or the temperature. Marine environments thus produce a very brief plant and animal succession each season, or even several times a season. Since the latter stage lasts only a few weeks, it certainly cannot be regarded as stable in the usual sense; so if we continue to define a climax as the stable end-product of succession, we must say that in marine environment there is no such thing as a climax comparable to that on land.

Complications of a different kind arise where succession is very slow. So far we have assumed that succession proceeds one-way and inexorably toward the climax, but when a climax takes a very long time to develop, the process seldom goes smoothly. Somewhere along the line there will almost certainly be some setback, such as a fire, severe drought, or hurricane, that drastically changes the environment and pushes things back to an *earlier* stage. Indeed, such events may occur so frequently that the climax is never reached but postponed indefinitely. Of course, such catastrophes can also occur after the climax is reached, turning the clock back. For example, after a long drought, some mature climax grasslands revert to an earlier stage of succession, with more annuals and short-lived perennials. The climax quickly reappears, however, with the return of normal precipitation. But if severe droughts occur at frequent intervals, then the climax never has a chance to develop. In general, the later stages of succession, and especially the climax, are less liable to be pushed back than are earlier stages. This was dramatically shown during the exceptionally severe droughts of the early 1930s in the Great Plains of North America. Here mature climax grasslands survived and held down the soil; but in overgrazed regions (where overgrazing itself had already set things back to an earlier stage of succession) plants perished and severe erosion followed.

These examples show that the course of succession is not always one way or completely predictable. Nor can we claim that the climax appropriate to the regional climate—what we call the *climatic climax*—is bound, sooner or later, to appear or even that it must always be self-perpetuating if it does appear. All we can say is that while the main regional conditions remain the same there will always be a *tendency* toward a certain type of climax.

However, "main regional conditions" are not necessarily identical with "main regional climate." We must be careful, therefore, to qualify the word *climax*. For example, the long-leaved pine forests of the southeastern coastal plains of the United States are actually maintained in a stable state by periodic fires. Because fires seem likely to go on occurring for a long time to come, we might well suppose such forests to be the normal climax of that region. But some ecologists say that if fires could be prevented, broad-leaved forests would eventually take over, and that therefore the true climatic climax of that region should be called the broad-leaved forest. However, some of the fires are started by lightning, so it is possible to argue that fires are just as much a part of the physical environment as rainfall. This being so, the concept of a climatic climax in such a situation is useful only in theory. In practice, it is more realistic to call the pine forest a *fire climax*.

There is also another common reason why a stable climax can be maintained that differs from the climatic climax appropriate to the region. For instance, in part of Texas we find very stable desert communities of creosote bushes, mesquite, and cactus that are in fact maintained by constant overgrazing. The climatic climax in these areas is grassland, but an earlier stage of succession is kept going, and kept stable, by the powerful influence of a non-climatic factor—man's cattle.

From what we have just said it must by now be obvious that ecologists use certain terms in different senses. This is not because they necessarily disagree with the facts themselves, but ecology is a young science and there is not yet international agreement on the precise meanings of all ecological terms. But

American C-123 Providers spraying defoliating chemicals on a Vietnam jungle. By laying bare well over a million acres, man yet again asserts his dominance over nature. But what the long-term effects of such actions will be we do not as yet know.

145

this does not matter all that much, so long as we carefully qualify what we mean in each particular case. This leads us to the last point we must make concerning the use of the word *climax*. Some ecologists say that in any large region only one climax—the climatic climax—can be in perfect equilibrium with the environment, and that all ecosystems within the region must eventually reach this type. Yet, for example, in parts of the tropical rain forests of Nigeria that are as yet unaffected by man, we can find many small areas where there are stable communities quite unlike those of the surrounding forest. For instance, there are marshes and grass patches that result from the fact that the lie of the land creates distinctive forms of drainage. Here drainage plays a major role in deciding what plants can flourish, and in this case it is difficult to imagine how those organisms could ever change their environment sufficiently to make it less suitable for themselves and more suitable for others. Furthermore, the different types of rock that underlie the various parts of the forest affect the type of plants that can grow, and this is also something that organisms can do little to alter. Thus geological factors, as well as drainage, may produce local deviations from the climatic climax. Such local deviations within the climatic climax are called *edaphic* climaxes, from the Greek word *edaphos*, meaning the ground or base of anything.

Some ecologists have questioned whether any climax could persist unchanged forever, even if the climate remained uniform forever—which, of course, it does not. They claim that after a time the regeneration of nutrients lags behind the ecosystem, and that disease and other factors would ultimately kill off plants faster than they could be replaced. We must for the time being keep an open mind about suggestions such as this. We need to know much more about ecosystems before we can be certain that we understand them fully.

Succession and human welfare

The study of succession is interesting in its own right, but it is not just an academic exercise; it can make a real contribution to man's well-being. For instance, even though we cannot be sure that a climax will last forever, we do know that with its great diversity, large biomass, and balanced energy flow, it is more stable and better able to buffer the physical environment than a younger community. On the other hand, it is less productive than the stages just before the climax. So farmers who are rightly intent on high productivity tend to concentrate on these stages. Most game birds, many freshwater game fish, and many of the most valuable timber trees, for example, thrive best and are most productive in these earlier stages, so farmers and foresters deliberately prevent the ecosystems in which they flourish from reaching the climax. However, we must combine high productivity with environmental stability, and this can often be best achieved by the judicious mixture of young and climax stages.

A good example of this is to encourage young rich grassland on alluvial plains while preserving a climax forest on the neighboring hills. This gives the farmer the best of both worlds; the fact that the hill forest is climax means that it can be expected to last indefinitely and so protect the steep slopes from erosion. Also the climax forest, with its undergrowth and spongy soil, releases rain water gradually into the streams that feed the plains.

Any interference with the climax forest in this combination can be disastrous. Reckless felling of trees is one form of interference that has led to severe erosion of hillsides, when the topsoil was washed by rain on to the plains, blocking rivers and flooding pastures. Moreover such hillsides cannot become productive again until a new fertile topsoil forms, and this, as we have seen, can take a very long time. There is another common type of interference—the introduction of new animals into the climax—such as rabbits in Australia and deer in New Zealand (see pictures below).

Two animals that man introduced to a new environment—rabbits into Australia and deer into New Zealand. Both animals upset the stable ecosystems in these parts and caused severe over-grazing and erosion—clearly visible (above left).

Before the coming of agriculture, man lived by hunting animals and gathering wild plants. Even today, this Australian Aborigine catches fish with a three-pronged spear resembling those once used by early Neolithic peoples in Europe.

9
Man the Hunter

The opening section of this book set out to examine how living things of all kinds fit into nature's network. We tried not to place special emphasis on man, yet he has repeatedly managed to loom larger than any other species. In fact, had he not done so, our picture would have been quite false; for man, unlike any other creature, has left some mark on every biome, and has altered large areas of the earth beyond recognition. Indeed, so special is his place in the world, that we must distinguish between natural ecosystems and man-controlled ecosystems. In other words, we must make a distinction between man and nature, even though he is part of nature. The history of man is, in an important sense, a record of his increasingly successful attempts to develop tools and techniques by which he has been able to manipulate nature, even if he cannot control her.

That these attempts had humble beginnings is certain, and in this and succeeding chapters we trace how man's impact upon nature, initially very small, has become ever more drastic as his technology has developed. For most of his time on earth *Homo sapiens* has set his sights little higher than mere survival. Throughout the Paleolithic, or Old Stone Age (that is, until about 10,000 years ago), he was in most essentials little more than just another member of his ecosystem. During this period, as we shall see, he developed enormous skill at food-getting and acquired deep knowledge about the living organisms upon which he relied for food. The point to remember, however, is that such skill and knowledge are more akin to the instinctive hunting ability of, say, a lion or an eagle than to the scientific and technological apparatus that modern man brings to bear on the problems of survival.

Although archaeologists have discovered relics of Paleolithic peoples in many parts of the world, their finds tell us very little about the lives of these peoples in ecological terms. Fortunately, for our purposes, we do not have to delve into prehistory to get such information. Man has not advanced toward civilization on a united front, and even today there are surviving remnants of human groups whose way of life differs little from that of their Paleolithic ancestors. Such peoples have been found in many parts of the world during the last two centuries: as one would expect, they vary widely in "racial" type and in many aspects of their culture. But they all have one important thing in common: they obtain food by hunting wild animals and by gathering edible plants in due season. None of them attempts to herd livestock; none of them cultivates plant foods. Some undoubtedly know how to do either or both as a result of contact with more advanced neighboring peoples. That they have not done either is due to a variety of reasons, some of which we shall consider here. But the theme of this chapter is that the hunter-gatherer accepts nature on her own terms: if she is kind, he prospers; if she is hostile, he goes hungry.

In terms of material culture, two of the most backward peoples in the world today are the Bushmen of southern Africa and the Aborigines of central Australia. Although most of the surviving remnants of both these peoples are now in permanent touch with more advanced cultures, small bands of Bushmen and Aborigines continue to live their lives in total isolation from all outsiders (except, perhaps, for the occasional party of anthropologists). In both cases their isolation is due mainly to the fact that they inhabit regions of semidesert that have little attraction for potential competitors.

Bushmen hunters

The Bushmen live in the interior of the Kalahari Desert, which extends for about 800 miles from the Okavango River in the north almost to the Orange River in the south, and westward from central Botswana to the Nossob River

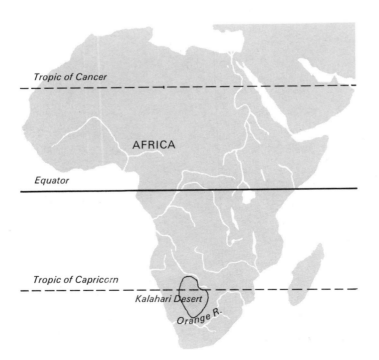

Map of Africa (left) showing the location of the Kalahari Desert, home of the nomadic Bushmen. The Bushman shelter (right), made from branches of trees, offers some protection from wind and sun by day as well as from wild animals at night.

Tropic of Cancer

AFRICA

Equator

Tropic of Capricorn

Kalahari Desert

Orange R.

in South-West Africa. Part of the great plateau of southern Africa, the desert lies at an altitude of 2500 to 3000 feet and in profile resembles a gigantic, shallow saucer. Although not true desert, the interior consists mainly of a seemingly endless succession of gently rolling plains of red or white sand. Every form of life in the desert is dominated by the climate. For nine months of the year (March to December) there is no rain. During the winter months of June and July, midday temperatures of 27° c are common, but at night the temperature drops below freezing point. For the rest of the dry period the heat of the day is oppressive to man and animals alike; the temperature soars to 46° c and above, and the air seems to undulate in great waves. The total annual rainfall, which varies between 6 and 10 inches, comes in brief but torrential thunderstorms between mid-December and March. Quite suddenly the thin patchy vegetation turns a richer green; the plants flower and fruit quickly and the water holes are briefly brimming.

For most of the year, however, the vegetation is meager, consisting mainly of low bushes and tufts of coarse grass interspersed with briar, thistles, and other hardy plants. In some areas there are groves of mangetti trees, which yield edible nuts. Occasionally, too, one comes across a baobab tree, which may attain a height of 200 feet or more, its roots striking deep into the soil and denying moisture to any other plant within 60 feet of its base. The mainly sandy soil of the Kalahari offers plants a better chance of survival than other, potentially more fertile soils. The reason for this is that less permeable soils

Left, above: four Bushmen hunting game. As is the case with most pre-farming peoples, it is the men who hunt animals and the women who gather plant foods. Left, below: distributing meat after a kill. The food is shared among every member of a family group.

Above: A Bushman woman digging for edible roots. She uses a digging-stick similar to that in the drawing. Typical Bushman plant foods shown on the right include two species of melon, a type of parsnip, and (between them) a small, onion-like root.

In parts of the Kalahari, sun and wind evaporate all moisture from the ground surface. Bushmen obtain water, often from as much as five feet below the surface, with the help of plant stems. They suck the water up through the stems and store it in ostrich egg shells. In many areas the Bushmen could not survive the dry season without this technique.

would retain most of the available moisture at or near the surface, where it would swiftly evaporate. As it is, the water sinks into the sandy soil and, in the rainy season at least, can support a considerable variety of vegetation. In particular, soil and climate put a premium on plants with large underground storage organs—roots, tubers, and bulbs—that can store enough water to keep them alive during the dry months. The existence of such plants is vital to the ecology of the Kalahari; without them the Bushmen could not survive.

Survival in the desert

The life of Bushmen revolves around the never-ending search for food, which, except in the rainy season, is always in short supply; by the end of the dry season most Bushmen are weak and emaciated. The hunter-gatherer way of life is forced upon them: the soil is too dry for too long to enable crops to grow even if the Bushmen knew how to farm, and reliable sources of water are too few and far between to permit the herding of animals. Because they are constantly on the move, the Bushmen have no permanent bases or villages, though each band, numbering up to 20 or so related men, women, and children, gathers plant food only in its own territory. The essence of this territoriality is not so much land acreage as such, as the number of places in an area where certain kinds of plants grow. Each band has undisputed ownership of a number of these places, on which other bands will not trespass even if starving. Similarly, a certain number of water holes are owned by each band (or, sometimes, shared between two neighboring bands); in some areas of several thousand square miles in the Kalahari there are only two or three places that yield water all the year round.

Survival of these people depends on their ability to find plant foods where, to even the most searching European gaze, none appears to exist. Even young Bushman children can spot the small, dry stalk, hidden in the grass, that is the only visible sign of an edible plant root, or the tiny cracks in the dry sand that betray the presence of truffles. Even more remarkable, Bushmen can remember the exact location of such a food source weeks or even months after they first noticed it on their daily walkabout. During the rainy seasons, food finding is comparatively easy, because certain species of melon and cucumber are available in small, isolated patches of the desert. The worst part of the year is from August to December, when the heat increases daily, most water holes dry up, and all surface-growing plant foods have long since withered and died. For this grim five-month season the Bushmen rely on roots and tubers not only for their scanty nourishment but also for their main water supply.

Bushmen eat a variety of small animals, including tortoises, snakes, lizards, and grasshoppers; they especially relish termites and ants, which swarm in vast clouds in their mating flights and are easily caught. Other small animals include porcupines, hares, ostriches, and game birds such as guinea fowl and partridge. None of these animals is abundant in the Kalahari, however; they are the lucky bonus that may occasionally occur.

Catching such animals, like gathering and cooking plant foods, is mainly the work of women. The principal role of the men is hunting a variety of species of larger game, including springbok, gemsbok, wildebeest, hartebeest, and eland. All these antelope are extremely difficult to hunt. They are not at all numerous and, because they can go without water for long periods, they do not need to graze within easy reach of a water hole. Moreover, the desert offers little natural cover for the hunters, who need exceptional skill to get within range of a herd of these fleet-footed, keen-scented animals.

The Bushmen hunt with bows and arrows. The arrows have tips of bone or metal (the latter obtained through barter with neighboring peoples). The tips are smeared with poisons extracted from plants. The hunters can instantly recognize the age of an antelope spoor, and so determine whether or not it is worth following. Once the animals are sighted the hunters approach them silently and with enormous patience, until they are within range. Many of the hunters are superb marksmen and can put an arrow into any part of an antelope's anatomy at a range of 100 yards. Even then, however, the hunt is not over. Once it has been hit in a vulnerable spot, the antelope is doomed— but it may not succumb to the poison for two or three days, by which time a mature buck may have struggled on for a hundred miles or more unless it can be overtaken. The Bushmen, though not especially fast runners, have extraordinary stamina; many are capable of running almost at top speed for 20 miles in pursuit of game, their keen vision unerringly distinguishing the spoor of the wounded animal from those of the rest of the herd. When the game is caught and killed, it is divided between the hunters, who in turn share it out among all members of their band. Very little of an animal is wasted.

Deserts of ice

We tend to limit our use of the word "desert" to the hot dry wastes of sand and scrub. In fact, it applies equally well to the polar regions, which provide just as harsh and unrelenting an environment for all forms of life. For thousands of years before the coming of European settlers, groups of Mongoloid peoples carried on a Stone Age existence in regions beyond the Arctic Circle in the far north of Canada, Greenland, and Siberia.

Canada's north coast is studded with large and small islands. For nine months of the year, however, during the Arctic winter, the land and sea are almost indistinguishable: both are frozen and covered with thick snow. The average temperature is about $-35°$c, and for a couple of months during the dead of winter the sun hardly rises above the horizon. Nothing can grow on land during this period, for the soil, even if cleared of snow, is frozen to a depth of several feet. With the spring thaw, some of the land and sea ice melts; later, the summer sun briefly thaws out the surface of the soil and for a little time the land is ablaze with shallow-rooting flowers and shrubs.

For man and other omnivorous animals it is the sea, not the land, that provides food in this region. For the Eskimos of Canada, life is exceptionally hard, but, with skill and patience, they can wrest a living all the year round from the sea, even when it is covered with ice. During the winter some groups build igloos on sea ice, carefully constructing these dome-shaped dwellings from blocks of snow. The snow, which is filled with innumerable tiny pockets of air, is an excellent insulating material.

The Eskimos' main winter food is the seal, which provides them also with skins for clothing, sinew for fishing lines and bow strings, and blubber for fuel. Because the seal is an air-breathing animal it makes blowholes in the ice as soon as the sea begins to freeze. Usually the blowholes become covered with snow, but the Eskimos rely on the sense of smell of their husky dogs to find them. When a seal comes up for air, the Eskimo stabs a barbed ivory-tipped harpoon in the blowhole and then plays the seal, like a game fish, on a line attached to the harpoon until it dies or becomes exhausted; then he hauls it up through the blowhole.

In the spring, when the beginning of the thaw is heralded by small puddles of water on the surface of the ice, the Eskimos move to the shore, where they build conical shelters made from wooden poles covered with skins. Most of the timber used by the Eskimos is driftwood carried downstream from riverside forests often hundreds of miles to the south. The wood is used not only for summer shelters and for harpoon shafts, but also as a framework for the seal-skin hulls of the one-man canoe, or *kayak*, for the larger *umiak*, and for sleds.

During the summer Eskimos spend much of the time in their boats, hunting seals and the occasional walrus, whose tusks provide the ivory used in their tools and weapons. They also use kayaks for hunting caribou. Almost every summer herds of caribou migrate northward into Eskimo territory to feed on the short-lived vegetation. The Eskimos either shoot them down with

Right: map shows the distribution (brown) of Eskimos in Alaska, Canada, and Greenland. Their combined population in the three areas is around 50,000.

Below: Eskimos in the Canadian far north still catch fish in tidal rivers by the ancient but effective method of constructing pools or (as here) weirs from rocks. As the tide falls, the rocks rise above the water; the fish are trapped behind them and fall prey to the Eskimos' spears.

bow and arrow, or drive them into the sea where, slowed down by the water, the caribou are swiftly overtaken and speared by Eskimos in kayaks.

For the Eskimos, like the Bushmen, survival hinges upon the exceptional skill of the hunter. But sometimes, in the depth of winter, there is too little food to go round: the fish will not bite, the seals have moved elsewhere, and the group's stock of dried meat is exhausted. Today, even the remotest Eskimos can rely on Europeans for help. But, until quite recently, in such desperate circumstances the claims of the community as a whole took precedence over any individual, and the old or sick whose useful (that is, food-getting or child-bearing) life was over might be left to starve or freeze to death.

In the case of both the Bushmen and the Eskimos, climate and habitat conspire to produce an environment in which the people have little chance of breaking out of the hunter-gatherer mode of existence, even if they knew how. Nonetheless, hunter-gatherer peoples have persisted until quite recent times in more favorable regions of the world. In such cases, the fact that they did not break through into technically more advanced ways of life is due, at least partly, to the very fact that the environment offered them a comparatively easy life, free of the pressures to improve their food-getting and other techniques.

These are just examples of peoples whose ecological impact was minimal. They were almost completely subservient to their natural environment. Before man could begin to overcome that subservience he had to break away from his hunter-gatherer existence.

A seal (below left), harpooned at its blowhole in the ice, is "played" like a fish until it is exhausted. Then it is hauled to the hole and killed with another harpoon. In remoter areas the Eskimos' basic way of life has been little changed by acquisition of modern weapons such as the rifle.

Below and right : Eskimo hunting tools. From left : fish spear ; bird harpoon (with two throwing sticks) ; fish harpoon ; two fish hooks. Tools of this kind are made from local materials : driftwood and bone.

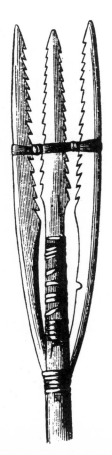

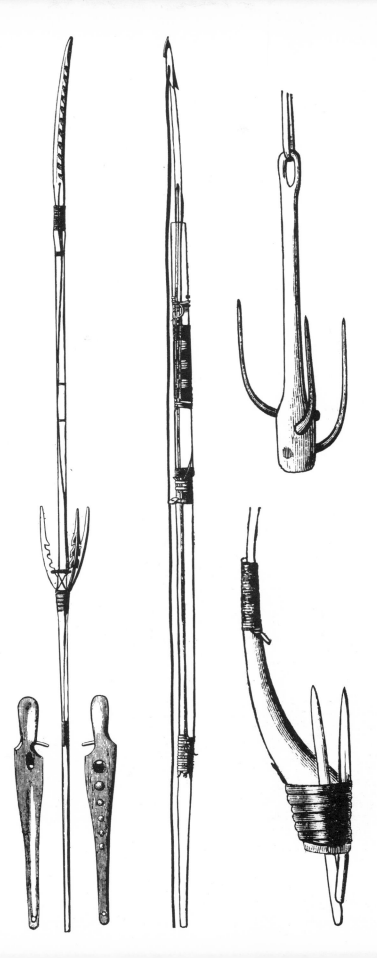

Rock painting made by African herdsmen about 5000 years ago at Tassili in the central Sahara. The pastoralists may have come originally from the Nile valley, 1500 miles away, at a time when parts of the Sahara were grassland.

10
Man the Herdsman

The lives of most hunter-gatherers center on the almost constant search for food. But not all Paleolithic (Old Stone Age) peoples were forced into this harsh existence. In some regions of the world, plant and animal foods were plentiful, at least at some seasons of the year, and the hunters could enjoy periods of leisure. That many peoples were able to use such leisure creatively is demonstrated by the magnificent cave paintings at Lascaux, Altamira, Tassili (in the Sahara), and elsewhere. Relative abundance of food animals in certain regions was undoubtedly one of the factors that first gave man the idea of permanently securing his food supplies by herding the animals instead of merely killing according to need.

The development of agriculture is justly called the Neolithic (New Stone Age) Revolution: certainly none of man's discoveries before or since has had more revolutionary or enduring consequences for life on this planet. Rightly, perhaps, we attach more significance to man's discovery of how to cultivate food plants (Chapter 11), but the herding and domestication of animals, which began at about the same time and in the same regions, also had far-reaching effects.

The agricultural breakthrough began about 10,000 years ago. The original center was probably the Near and Middle East—the region now occupied by Egypt, Israel, Lebanon, Jordan, Syria, Iraq, and Persia. At that time many of the areas that now consist of scrub and semidesert were covered by grassland and forests. It is impossible to determine which animals were the first to be herded, though it is reasonable to suppose that they were the ancestors of modern sheep and cattle. Excavations at Jarmo, in Iraq, which was a flourishing community about 9000 years ago, have yielded the bones of sheep (or possibly goats), oxen, hogs, and dogs, all of which are believed to have been domesticated by then.

There is an important distinction to be made between domestication and simple herding. The latter means following or driving the animals about their pasture, using them for their milk, fur, or wool, and eventually killing them for their meat, bones, and so on. Domestication embraces all these things but also includes the continuous process of breeding the animals in captivity. Later, as the art developed, breeding involved the preferential selection of certain individuals in flock or herd so as to perpetuate qualities that were desirable to the herdsman but were not necessarily advantageous to the survival of the animals in the wild. In this and other ways did such animals, individually and collectively, come to depend more and more on their association with man.

Domestication of animals

Almost certainly the first animal to be domesticated was the dog. Probably Paleolithic tribes had to contend with packs of wild dogs attacking their camps and attempting to steal their supplies of meat. Doubtless the hunters would sometimes placate them by throwing them unwanted bones or scraps. It may be that, in time, individuals or small groups of dogs became attached as scavengers to tribal groups, and that their puppies were adopted as pets and, later, trained to hunt and to help in herding. Certainly many hunters of the late Paleolithic used dogs for hunting. We have no idea of the species from which modern domesticated dogs are descended; even today some dogs will interbreed with wild canines such as the coyote and timber wolf. All we can be certain of is that modern breeds contain elements of many different species of wild canines, including some that are now extinct.

The wild ancestors of sheep can still be found in various areas in an enormous belt extending from the Mediterranean, through central Asia, to the Bering Strait. The three most important wild species are the mouflon (*Ovis musimon*), a native of Asia Minor, Iraq, and Persia; the urial (*Ovis vignei*), found in Afghanistan, Pakistan, and Turkistan; and the argali (*Ovis ammon*), found in the Pamir, the Altai Mountains, and eastward to the Pacific. The first sheep to be domesticated were used exclusively for meat and milk. All had coats of fur; it was only during the fifth or fourth millennium B.C. that woolly fleeces appeared, showing that herders were breeding sheep for wool as well as food.

The line of descent of modern cattle is obscure. It is likely, however, that their two principal ancestors were a short-horned species, *Bos brachyceros*, remains of which have been discovered in the 5000-year-old settlement at Anau, in Turkistan, and the long-horned wild ox, or urus (*Bos primigenius*), which survived in the forests of southern Germany until the Middle Ages. Hogs were developed from the wild boar, native to many parts of Europe, Asia, and North Africa, and still found in the forests of central Europe.

When tribal groups began to herd livestock, vast areas of the great northern continents were covered with dense forest. Indeed, even in Julius Caesar's day, Roman legionaries took several weeks to march through the huge forest of southern Germany. Even the countries of the Near East were heavily wooded in parts, notably with cedar forests. There were, however, considerable grassland areas. In many parts of the temperate zone, grassland disputes the title of climax vegetation with deciduous forest.

We have already noted that agricultural man is the greatest destroyer of forests. Many of the pioneer farmers who spread out from the Near East practiced mixed farming. The Danubians, for instance, who migrated up the Danube valley from about 5000 B.C. onward, cleared the forest for both cultivation and pasture. Later, the Battle-ax peoples invaded Europe from southern Russia, cultivating the soil and tending their flocks of sheep. These and other migrant farmers were responsible for early forest clearance in Europe and laid the foundations for the present-day grasslands and heaths of northwestern Europe and the Hungarian Plain. A similar pattern of events occurred in central Asia and in parts of Africa and North America.

Cultivator societies

These early farmers brought the practice of mixed farming from their homelands in the Near East and central Asia—where, however, the practice was doomed to extinction. The reasons for this we shall consider in the next chapter. The point we have to make now is that these regions are part of an enormous *dry belt* that extends from the Mediterranean through the Near and Middle East, then north of the Pamir and Himalayas through Outer and Inner Mongolia to the plains of northern China. At both the western and eastern ends of this belt mixed farming died out with the development of cultivator societies, and later empires, based on irrigation techniques. Over the course of several thousand years from about 4000 B.C. onward these powerful states gradually secured most of the agricultural land in the Middle and Near East and in the region near the present border between Inner Mongolia and China.

Two wild ancestors of modern sheep—the goat-like mouflon (left) and the argali (above). These and other fur-coated animals were first domesticated only for their meat, milk, and hides; selective breeding for wool was a much later development.

The pastoralists

As the cultivator states expanded, many peasant farmers refused to submit to the life of semislavery and took to herding, so swelling the numbers of the traditional pastoralists. But the problem of grazing land was acute. Throughout the dry belt, arable land—together with large areas brought profitably into cultivation by irrigation techniques—was surrounded and interpenetrated in many regions by desert or scrubland. More and more, the herders were pushed out of their traditional pasture lands toward the edge of the deserts. And though their flocks and herds could scrape a living from the scrubland during the rainy season, when the soil was temporarily carpeted with green plants, during the dry season the grasses withered and there was insufficient water for the animals. The herders had to move elsewhere. And so it was that they adopted the practice of *transhumance*: they grazed their herds or flocks on the outskirts of cultivated areas during the rainy season, and in wetter uplands and river valleys for the rest of the year.

Many of these dry-season pastures abutted on the cultivator societies, who were using them or had earmarked them for crops. Consequently there were constant border disputes between the cultivators and herders; many such disputes flared into long, bitter, and immensely destructive conflicts. In the Bible, this basic, deep-rooted antagonism is exemplified in the relationship between Cain, the cultivator and Abel, the herdsman.

At first the small, loosely knit bands of herdsmen had little chance of winning a fair share of the soil from the hierarchic, well-armed, and extremely powerful cultivator states. But the situation was radically altered by the domestication of the horse and the camel. Wild horses once roamed over much of northern and central Asia and Europe. For long the horse and its relative, the ass (a native of Africa), were used mainly as pack animals and secondarily as providers of milk and meat. During the second millennium B.C., however, herders on the Russian steppe began to use horses for riding; a little later, Arab Bedouins began to exploit camels for the same purpose. As a result, not only could the herders range much farther than ever before in search of pastures, but they also became more mobile and formidable adversaries of the cultivators with whom they competed for land.

The nomads

Transhumance gave way, in the case of many mounted herders, to true nomadism. The Mongol herders, for instance, spread all over central and eastern Asia from 1000 B.C. onward; in the west, Turkish herders held sway. Ranging far and wide, they absorbed innumerable groups of herdsmen, by conquest or by persuasion. Over a period of 2000 years these herder hordes grew large and powerful. By the 12th century A.D. their greatest leader, Genghis Khan (1162-1227), controlled an empire stretching from the Pacific almost to the Mediterranean and commanded a mounted army numbered in hundreds of thousands. Genghis captured huge areas of northern China and

Competing climax vegetation. Savanna, such as in Serengeti, Tanzania (right), comprises grassland interspersed with isolated clumps of trees with very wide-spreading root systems. The determining factor is the length and severity of the annual dry season.

most of Afghanistan, putting the cultivators to the sword and turning their land into pastures for livestock.

Throughout history, pastoralists have tended to destroy the very thing on which they depend: pasture. Pastures usually consist of a variety of perennial grasses and legumes. Just as regular cutting during the growing season is good for a garden lawn, so controlled grazing is good for pasture: it stimulates the plants to produce more stems and leaves. As long as the number of stock in a given area of pasture is just right (even in the temperate zone a single cow needs about two acres), and as long as the stock are not permitted to graze the same area continuously for too long, the pasture may endure indefinitely; indeed, in favorable areas it may even increase its annual output of food.

But overgrazing upsets this balance between stock and plants. Destruction of the pasture proceeds in stages. First of all, the perennials die out because their stems and leaves are continuously being bitten back to the surface of the soil. They are usually replaced by annual grasses, which are hardier but less nutritious than the perennials. At the end of each growing season the annuals die, the next season's crops normally developing from their seeds. Overgrazing, however, means that most of the annuals are eaten *before* they can set seed. The pasture may then be invaded by shrubs, many of which are unpalatable; moreover, they do not form a dense, continuous turf but grow in scattered

Overgrazing of this pastureland in Arizona
(above) has destroyed the vegetation and exposed
the soil to erosion. Rain-water runoff has already
cut deep channels in the ground.

Below: the field on the right (also in Arizona)
has been carefully managed after being overgrazed.
In the field on the left, overgrazing has almost
entirely denuded the soil.

clumps, leaving much of the soil surface bare and vulnerable to erosion by water and wind. At each stage in this cycle of deterioration the pasture offers less and less nourishment per acre, and so a given number of animals will require progressively larger areas of pasture. Once started, then, overgrazing becomes a vicious circle. In the dry belt, drought aggravates the effects of overgrazing. In combination, overgrazing and shortage of seasonal rain water were responsible for much of the desert wastes of the Near and Middle East and central Asia that once supported vast herds of sheep and other livestock. Similar destruction, though on a smaller scale, was the result of overgrazing cattle in parts of northern and central Africa, Spain, Italy, Greece, and (as late as the present century) in many of the dry western states of America. The scars remain on the land to this day. One example of such damaged land can be seen in the top photograph opposite.

Herdsmen of the temperate zone

Today the cattle- and sheep-farmers who produce the bulk of the world's meat, milk, and wool are in the temperate zone. Beef cattle and sheep, in particular, are products of open-range farming on the drier grasslands, such as the Argentinian pampas, a wide belt in the United States stretching from Mexico to the Canadian frontier, the steppe of the southern Soviet Union, and in the Australian states of Queensland, New South Wales, and Northern Territory. In a sense, these herds (though not their owners) pursue a nomadic existence, for they range at will over their enormous pastures for most of the year, tended by small teams of cowboys or sheepmen.

It is in the temperate zone, too, that the breeding of stock has made the most striking advances. Since the mid-18th century, when Robert Bakewell's pioneering work in selective breeding began to attract attention, the development of particular breeds of cattle and sheep has been firmly based on scientific principles. Among the many breeds of domestic cattle, about 30 are of world-wide importance, and a high proportion of these were originally bred from British stock. Whereas in ancient times cattle were general-purpose animals used for drawing vehicles and cultivators and for providing meat, milk, hides, and other products, modern cattle are highly specialized. In the temperate zone, at least, they are bred almost exclusively either for meat or for milk (though high-quality dual-purpose breeds are beginning to make headway). The use of artificial insemination has enabled breeders to exploit the best sires on a greater scale than ever before, and is rapidly improving the quality of herds all over the world. The story of sheep-breeding has followed a similar pattern. As we have seen, the wild ancestors of sheep bore fur instead of wool. Fur-bearing sheep are still raised in several parts of the world, notably in central Asia; other breeds, in North Africa, Arabia, southeastern Europe, and elsewhere, are used primarily as milk-producers. But in the temperate zone sheep are bred primarily for one of two products: wool or meat—the wool being generally the more profitable of the two.

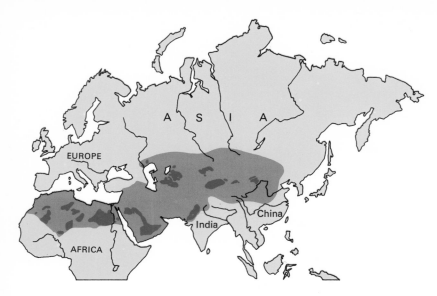

Left: map of the dry belt (red) that runs eastward from northern Africa, throughout the Levant and Central Asia, to the Far East. The fertility of much of this region was ruined in the long struggles for supremacy between nomadic herding people and cultivators. Areas reduced to complete desert are in brown.

Right: map of the sheepwalks and winter pastures (shown in green) of the Mesta in Spain during the 13th century.

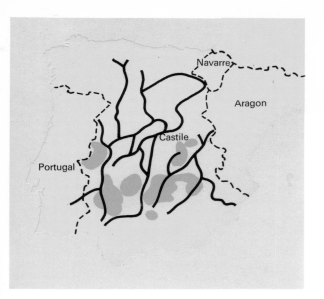

Above: Arab warrior-herdsmen of the seventh century B.C. The camel in the Near East, like the horse elsewhere, greatly increased the power and mobility of nomadic pastoralists from 1500 B.C. onward.

Nomadism is still practiced by the pastoralists of Inner Mongolia, who raise goats, sheep, cattle, and yaks on vast upland pastures, watered by the springtime rains, to the south of the Gobi Desert.

Left: Bedouin shepherd in modern Jordan. The decline of pastures into desert, common all over the fertile crescent, owed more to centuries of overgrazing by animals than to drought.

Stockfarming

In complete contrast to free-range grazing is the method of *static husbandry*, in which the stock graze in comparatively small, enclosed pastures, This method is very common in Europe, which lacks vast grasslands such as prairie or pampas. But there is a positive value in static husbandry, for it permits the farmer much stricter control than any other method. It has, in fact, enabled the farmer to domesticate his pasture as well as his livestock. Many of the pastures used by such farmers are *leys*—that is, temporary pastures specially sown by the farmer, grazed for a year or more, and then plowed up again and re-seeded, either with pasture plants or with rotation crops such as cereals or roots. The farmer can also exercise control over his permanent pastures by over-sowing them with desirable species of grasses or clovers.

Overgrazing: the lesson learned

In recent years plant-breeders have developed a large number of new varieties of grass and clover species. They do not merely offer improved overall yields: some, for instance, have been bred to provide good yields at certain times in spring, summer, or fall, when most other pasture plants are resting; some have been developed especially for pastures, others for cutting as hay. Breeders are also concerned with compatibility between different species of pasture plants; in permanent pastures or leys there may be six or more important species competing for light and soil, as well as many other species of minor food value. Ideally, a pasture consists of a plant community that provides a thick, lush growth of stems and leaves throughout the growing season; and this is possible only by using various species in combination. In order to get the best out of such mixed pastures, breeders have done much to reduce the natural aggressiveness of certain species that normally tend to crowd out other useful grasses and legumes. One of the most striking partnerships is that between perennial ryegrass and white clover. When they are sown together the pasture is twice as productive as one consisting of ryegrass alone; moreover, the yield of ryegrass itself is half as much again in the mixed pasture. The grass benefits because atmospheric nitrogen, passing into the soil, is converted into nitrates by bacteria living in nodules on the clover roots.

At any given moment a pasture or hay meadow contains a finite quantity of foodstuffs—carbohydrates, fats, proteins, vitamins, and minerals—needed by the stock for nourishment and growth. A proportion of the minerals is present in the soil; the rest is taken up by the plants through their roots. In a pasture, some of the minerals absorbed by the plants and eaten by the animals is returned to the soil in the form of excretions. But in a hay meadow, as in a wheat field, this source of replenishment is missing, and the balance must be restored by the farmer in the shape of fertilizers, lime, and other minerals. In pastures, a deficiency of nitrogen may result in the best grass species being replaced by hardier but inferior species. Similarly, soil that lacks phosphorus or lime will produce a poor clover crop.

Controlled grazing (right), with the use of movable electric fencing enclosing comparatively small pastures, is typical of European stock-farming. It enables the farmer to control closely both his animals and the plants on which they feed.

Most stock-farmers engaged in static husbandry, and particularly those specializing in dairy cattle (which require a particularly high standard of nutrition), supplement the diet of their stock in numerous ways. These include proprietary goods such as cattle cake, silage for winter feed, and various root crops. Many dairy farmers practice mixed farming, growing a variety of crops in rotation in years when the leys are not being used for pasture or hay production. Several such crops are used, at least in part, as supplementary foods for the stock.

The modern stock-farmer, then, has made a profound impact on the animals he values for food and other products. He has not only changed the animals physically but, at a local level at least, has transformed the ecology of the plant communities on which his animals depend for food. Indeed, so complete is the reliance of these domesticated animals on man that it is doubtful if they could survive in the wild. But this present success rests on a past in which man has learned—often at appalling cost to himself and to the land that supports him—at least some of the dangers of attempting to take more from nature than it is able to give. Modern stock-breeding and grassland management must collaborate with nature, not try to outwit it.

11

Man the Cultivator

Man began to cultivate the soil and raise plants from seed at the onset of the Neolithic (New Stone Age), about 10,000 years ago. The focus from which the knowledge and practice of cultivation was eventually to spread was probably the *fertile crescent*—a region that includes the Nile valley, Israel, northwestern Jordan, western Syria, Lebanon, southeastern Turkey, the Tigris and Euphrates valleys, and western Persia. It has also been suggested that, at about the same time, peoples in northeastern India and Burma began to cultivate root crops and to domesticate poultry and pigs.

We can only guess why hunter-gatherers in the fertile crescent became cultivators. According to one view, the increasingly dry climate of the Near and Middle East during the early Neolithic forced the peoples in this region to seek new sources of food. In earlier times and in other regions, shortage of food usually resulted in migration. But the peoples of the fertile crescent had been living in the region for so long that they had acquired a deep knowledge of the local vegetation, which was rich and diverse after the winter and spring rains, and had grasped the connection between the development of the seed head in, say, annual grasses in the fall and the growth of new plants in the following spring. The idea of deliberately planting seeds was a revolutionary one, certainly, but it may not have seemed so to peoples who were accustomed to harvesting wild-growing food plants, including wheat and barley, in due season. In caves on Mount Carmel (Israel) archaeologists have discovered sickles with flint blades set in bone handles; these sickles were used to harvest wheat or barley and were made, moreover, by hunter-gatherers, not cultivators. Here and at other hunter-gatherer sites in Kurdistan (on the Iraq/Persia border) archaeologists have also found shaped stones that were used for pounding and milling grain. It seems evident that all these peoples lived in semipermanent communities: they had no need to wander far and wide in search of food, because both animals and food plants were plentiful.

All these settlements, then, seem to have been on the threshold of a breakthrough to cultivation. Most of them were in the uplands that form the outer perimeter of the fertile crescent. Even today these highlands are the most abundant source—and were, perhaps, the original home—of wild wheat;

Once, vast areas of the world were covered with forest, their floors carpeted with rich, fertile soil. Man has steadily drawn on this valuable resource, felling the forests for building materials and fuel, and using the soil for his crops.

173

wild barley, also found here, is more widely distributed over western and central Asia. Significantly, the settlements that have yielded the earliest evidence of cultivation so far discovered are also in these highlands. The best-known of these are Jarmo (eastern Iraq) and Tepe Sarab (120 miles to the southeast, in western Persia); both these early farming villages are near the sites of older, pre-farming settlements; both are almost 9000 years old.

As the practice of cultivation spread, farmers came down out of the highlands onto more level ground, where larger fields could be planted and cultivation was easier. Most of them established farming communities on or near rivers in order to be certain of permanent supplies of water. At the eastern end of the fertile crescent many villages developed in the valleys of the Tigris and its eastern tributaries. An example of these was Hassuna, on the Tigris just south of the present city of Mosul.

Over a period of several thousand years, farming spread throughout the fertile crescent and beyond. According to one legend, Osiris, the mythical god-king of the ancient Egyptians, introduced the arts of cultivation into Egypt from his homeland in Syria or Lebanon. Cautious as we should be in accepting the evidence of legend, this does seem to indicate that farming was imported into Egypt and did not develop there spontaneously. The earliest Egyptian cultivators—the Tasians of the Middle Nile, the people of the Fayum depression (then filled by a lake), and the Merimdians of the delta—were all later than the earliest hill farmers of Iraq. Like them, the Fayumis and Merimdians domesticated cattle and sheep and hunted wild animals; and

Five wheat species: (1) wild einkorn, (2) emmer, (3) einkorn, (4) durum, (5) common or bread wheat. These and the other main species of wheat have evolved by hybridization in the wild, but man has improved their properties by scientific breeding.

their harvest of cultivated crops was supplemented by wild-growing plants. A little later, cultivation began in many regions to the north and east of the fertile crescent—at such places as Tepe Sialk, near Kashan (Persia), Anau (east of the Caspian Sea), and the valley of the Indus River (Pakistan).

The evolution of wheat

In all these regions, the principle cultivated plants were wheat and barley. Wheat, as we have mentioned already, is indigenous to the highlands fringing the fertile crescent. In order to perpetuate the species, the wild-growing plant must scatter its seeds over as wide an area as possible, so that they do not compete with each other for soil and moisture. As an aid to this, the central stem, or spike, of the head of wild wheat becomes brittle when the seeds are ripe and breaks off easily, so that the seeds of each ear are dispersed by wind. Valuable as it is to the wild plant, this property is disastrous to the farmer because it means that a high proportion of his crop will scatter its seeds when harvested (whether by primitive sickle or by combine harvester). Modern cultivated wheats have much stronger spikes that do not snap off; moreover, it is evident that even the earliest cultivators in the fertile crescent grew wheat with strong spikes. At first sight this seems to imply that these cultivators deliberately bred varieties of wheat suitable for cultivation. It is not quite as simple as that. In any large patch of wild wheat (and wild barley) there are usually a few individual plants with strong spikes. This is not a matter of luck: it is due to a recessive gene that is common to these individuals and is a property that is passed on to future generations. Now, although such individuals are comparatively rare, their seeds are likely to form a significant fraction of any harvest of wild wheat reaped by sickle, because a proportion of seeds from brittle spikes will have been scattered by the wind before harvesting. It follows, then, that in each successive season seeds from strong spikes will form an ever-greater proportion of the stock retained for planting the following season.

Wheat is a member of the Gramineae family of grasses. There are between 11 and 17 species of wheat (according to different methods of classification), but all belong to the genus *Triticum*. Each species belongs to one of three groups, depending on the number of pairs of chromosomes in its cells. The two 7-chromosome wheats, which are presumably the oldest and from which all other wheats are directly or indirectly descended, are einkorn and wild einkorn. Both normally have fairly brittle spikes, and einkorn was one of the earliest to be domesticated. Of the eight species of 14-chromosome wheats, two are wild-growing—wild emmer and timopheevi wheat. The domesticated species of emmer is the principal wheat discovered at Jarmo and other ancient sites, and is still cultivated in Egypt and elsewhere in the Near East. Today, the most important member of this group is durum wheat, which is used in the manufacture of spaghetti and macaroni. Two of the 21-chromosome wheats, common (or broad) wheat and club wheat, are grown widely; common wheat provides almost all the grain from which the world's bread is made.

All these wheats evolved by natural means—by selection, hybridization, or both—in which man played no part. Common wheat, for instance, was the result of a 14-chromosome wheat (probably wild emmer) hybridizing with a wild grass, perhaps of the closely related genus *Aegilops*. But although man has created no new species, he has wrought tremendous changes in the qualities of those he cultivates, especially common wheat, since he began to understand the principles of heredity and genetics. By producing hybrids from different varieties of the same species, he has enormously improved the quality of common wheat. During the last 80 years or so, plant breeders have increased the yield of wheat enormously. More than that, they have greatly improved the milling and baking qualities of the grain, and have increased the resistance of varieties to the many diseases that can ruin a crop.

We speak a little glibly of man domesticating the great grain plants—wheat, rice, and maize—and we tend to ignore the fact that, in a sense, such plants returned the compliment and domesticated man. Once he began to cultivate plants, man committed himself to settling down in one place: to cultivating the soil, to sowing and tending the plants, to harvesting their seeds.

But it was a price worth paying, for the cultivated grains were both more reliable and much more productive than their wild relatives, and provided the cultivators with a nutritious basic food in abundance. Arguably, indeed, the people of ancient Jarmo enjoyed a healthier diet than many peasants living in the same area today. Their diet of grains was supplemented with meat and with fruit, vegetables, and nuts that they gathered in due season.

Migration and cultivation

Man's capacity to produce more food than ever before led to steadily rising populations in all the centers of cultivation; this in turn had two main consequences. The first, which we shall consider in a moment, led to the development of the first urban communities in the great river valleys. The second was the spread of farming communities in all directions from their original focus in the Near and Middle East.

This spreading of the techniques of cultivation entailed not only the diffusion of knowledge but also the migrations of land-hungry peoples. The coasts and islands of the eastern Mediterranean were the first region to be hit by the migratory wave, which began around 5000 B.C. A little later, other migrants passed through Asia Minor and began to farm the lower Danube valley. These Danubians, as they are called, like other pioneer farmers had to clear the forests before they could begin to farm the soil. Yet, in one respect, they were luckier than most. The soil of much of the valley of the Danube (and of the Rhine, where they eventually spread) is *loess*—a form of loam consisting of clay that was ground to a fine, powdery consistency by the retreating glaciers of the ice age, and was later wind-blown into the valleys. These soils were less densely wooded than most of the rest of Europe; they were light and easy to till with hand hoes, excellently drained, and exceptionally fertile.

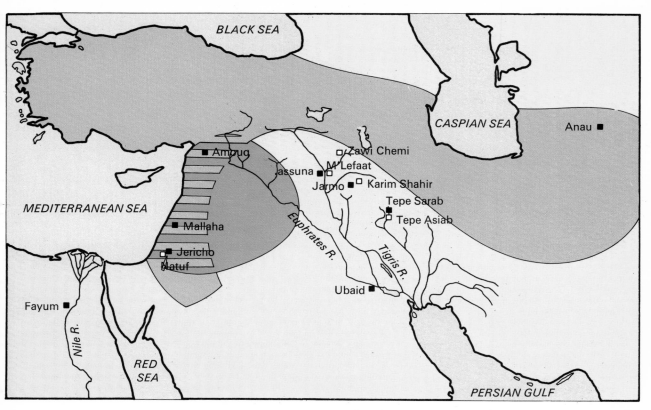

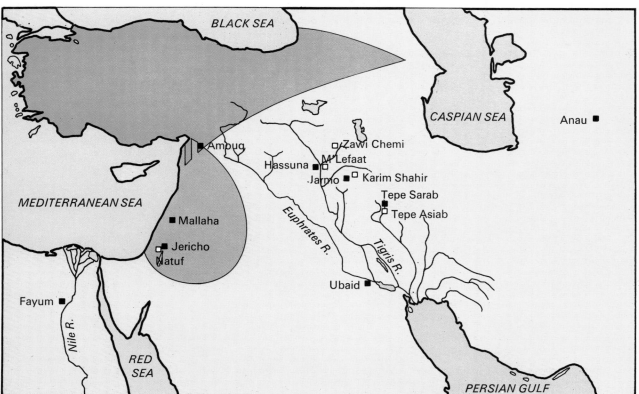

Maps show distribution of wild wheats and barleys in and around the fertile crescent. The Neolithic revolution began in this region about 10,000 years ago, when man began to cultivate the soil and plant the seeds of these and other wild-growing species.

Lower map: wild einkorn (ocher), probably the most ancient wheat species, and wild emmer (purple); location of pre-farming communities (open squares) and early farm settlements (black squares). Upper map: two-row barley (green) and six-row barley (red).

The Danubian method was to clear a small area of forest, using both fire and stone axes, plant most of the cleared area with wheat or barley (but reserving some as pasture for cattle and sheep), and farm the area every year until the soil was exhausted. Then they moved on to a new area and started the process all over again; their abandoned fields gradually recovered and eventually returned to forest. Archaeological maps of several areas of the loess "corridor" through central Europe suggest quite a high density of population among the Danubian farmers. In fact, such maps can be misleading. Evidently each Danubian group tended to settle in one area (as is suggested by the size and durability of their dwellings), divide it into sections, and farm each section in turn—rather like a very long-term version of the modern method of rotation. In other words, each Danubian settlement had several plots of land under its control: in any year, some would be in cultivation, some would be earmarked for future clearance, and some would be in process of recovery after being under crops for 10 or 15 years.

This method works if the population of the group is static. But this was rarely the case. The prosperous Danubians increased and multiplied, until the time would come when a settlement's population was too large to be supported by the land under cultivation, and the younger generations within the settlement would have to split off to found new communities. In this way, the Danubians spread westward and northwestward from the Balkans until, by 2500 B.C., they were cultivating the loess soil all the way from the mouth of the Danube to the Low Countries, northeastern France, and northwestern Germany.

At about this time, two other groups of farmers invaded Europe—the Beaker people, who sailed up the Atlantic coast from the Mediterranean; and the Battle-Ax people, who spread westward from southern Russia. The latter were pastoral farmers: though they cultivated the soil like the Danubians, they were essentially stock-farmers, herding horses, oxen, pigs, and especially sheep. Whereas the Danubian method of farming allowed the natural regeneration of the forest, that of the Battle-Ax peoples converted much of the cleared forest into grassland and heath; for their stock grazed the grasses and clovers and trampled or ate the shoots of trees as soon as they appeared.

Meanwhile, among the cultivators steadily pushing into northwestern Europe, another important, though accidental, development was taking place. We have seen that their principal grain crops were wheat and barley. In the Near East, whence these farmers had come, two other grain grasses commonly grew then, as now, as weeds of cultivation: wild oats and wild rye. The seeds of these wild species were harvested with the cultivated ones, and so were imported into Europe along with wheat and barley. Now, as the farmers gradually spread northward, the climate became cooler and damper; as a result, their yields of wheat and barley diminished, while those of the hardier oats and rye increased. Finally, both oats and rye were domesticated and, in many parts of northern Europe, were cultivated for their own sake.

Hydraulic societies

By about 5000 B.C., agriculture in parts of the Near East was undergoing a revolution that was to transform man's relationship both with his fellows and with the land.

Every year the silt-rich waters of the Tigris and Euphrates inundated the flood plain at least once, and often more than once. In other words, the farmers acquired a new layer of fertile, easily tilled topsoil during each rainy season. As long as the farmers could safely retreat to higher ground during the floods, they could cultivate the land, year in, year out, without exhausting the soil. Yet there were problems and drawbacks. This region is in the dry belt. Although the brief rainy season swelled the two great rivers and flooded the coastal plain, at other times of the year the plain was as dry as anywhere else in the belt; moreover, too drastic an inundation of the land could prove as damaging to agriculture as drought. Another problem was the steady rise of population in and around the flood plain as the area prospered. What was needed, then, was a method of conserving and directing the water in times of flood and drought alike. That method was irrigation.

Irrigation, like agriculture itself, did not suddenly emerge as a developed technology. In its crudest form—the digging of shallow ditches to divert water from streams into nearby plots—it had probably been practiced by the earliest dry belt farmers. But irrigation involves much more than this. It enables the farmer to control both the timing and the volume of water flowing into each field or plot; to drain water away from the land when necessary; and to prevent the destructive effects of flooding either by direct rainfall or by the river from which he channels his water. At its most sophisticated it includes complex systems of waterways, aqueducts, dams, sluices, and embankments.

In a large, flat area of land such as the Tigris and Euphrates flood plain, the nature and extent of flooding during the rainy season meant that irrigation had to be carried out on a large scale or not at all. Neither the individual farmer, nor even small groups of collectives, could hope to deal with the problem. The sheer size of the engineering works, and the quantity of labor needed to build and maintain the waterways and embankments, could be dealt with only by voluntary or enforced cooperation of all the able-bodied men in all the farming communities. In the event, it led to the planning and direction of these operations by a central authority. It was here, in fact, that the problem of social organization on a mass scale was first resolved by the development of a governing class—a group of leaders whose role was not directly productive but was concerned with enforcing the productive labor of others.

In most of the hydraulic societies, as they are called, mixed farming disappeared. Stock-farming was forbidden; and this, as explained in the last chapter, was one reason for the rise of the nomadic pastoralist hordes. It also had another important consequence. In the mixed farming communities, the animals provided the farmers with their main source of protein. Denied this source, the hydraulic societies had to look elsewhere, and they turned to the cultivation of leguminous plants—peas, beans, and others—that are a rich source of the nitrogen the body uses in the synthesis of protein. In the fertile crescent, the main cultivated legume was probably lentils, which are native to the Near and Middle East; in China and other oriental countries the soybean has been cultivated for at least 5000 years; in the Americas members of the bean family were cultivated.

We have said that civilization had its roots in these early hydraulic societies. The point to note is that the extremely diverse and complex social organization characteristic of civilization arises not from agriculture as such but from agricultural *surplus*. It was only when the hydraulic societies began to produce more food than they actually required, and so could exchange the products of agriculture for other commodities, that the multiple activities typical of urban society could begin to develop. In Sumer and Egypt two factors hastened these developments. First was the discovery of how to extract and work metals—first copper and later iron. Second was their need to trade with other communities in order to obtain these and other commodities (such as timber), which both Egypt and Sumer lacked. Important as this trade in goods undoubtedly was, it may be that its incidental benefits were even more significant in the long run. In particular, the exchange of ideas and techniques—new farming, metal-working, textile, and pottery techniques; the science of mathematics; the development of wheeled transport; the art of writing—all these spread throughout the fertile crescent as a consequence of trade. Moreover, trade spread the arts of irrigation far beyond the confines of the Middle East until it was practiced throughout the dry belt.

New World agriculture

Agriculture in the New World, which was isolated from the Old, was clearly a spontaneous development, as was the later development of hydraulic techniques. It was in Central and South America that the breakthrough began, based on corn (maize). The origins of this grass (an annual, like wheat, rice, and the others) are still not known for certain. The earliest examples of corn so far discovered were found in a cave in Mexico and are at least 4000 years old. They look rather like stunted examples of present-day pod corn, little larger than an ear of wheat and with seeds enclosed in husks. In size and shape they have little in common with the massive cob of the present day, with its

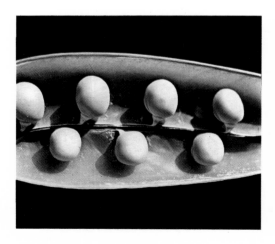

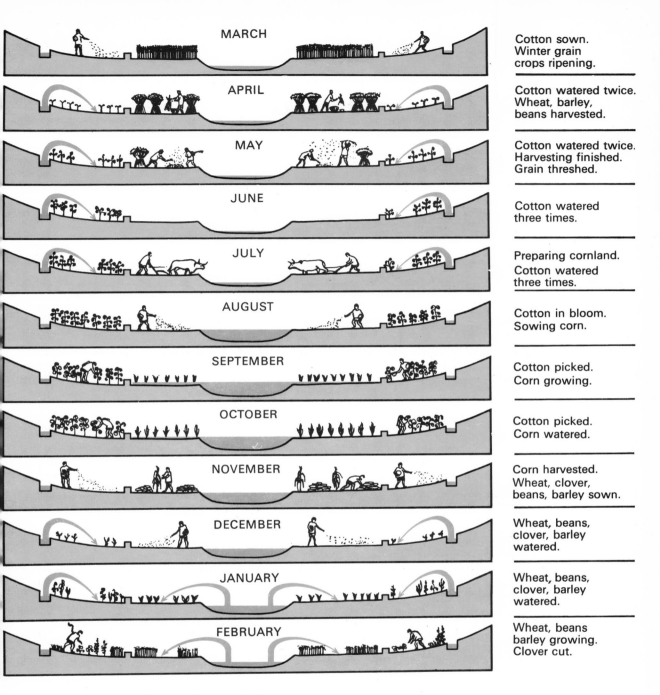

MARCH	Cotton sown. Winter grain crops ripening.
APRIL	Cotton watered twice. Wheat, barley, beans harvested.
MAY	Cotton watered twice. Harvesting finished. Grain threshed.
JUNE	Cotton watered three times.
JULY	Preparing cornland. Cotton watered three times.
AUGUST	Cotton in bloom. Sowing corn.
SEPTEMBER	Cotton picked. Corn growing.
OCTOBER	Cotton picked. Corn watered.
NOVEMBER	Corn harvested. Wheat, clover, beans, barley sown.
DECEMBER	Wheat, beans, clover, barley watered.
JANUARY	Wheat, beans, clover, barley watered.
FEBRUARY	Wheat, beans barley growing. Clover cut.

Above: chart showing monthly water-levels and present-day crop systems in the Upper Nile valley in Egypt. Dams divert water into irrigation canals during the dry season. The farmers of ancient Egypt had to rely on summer flooding, and planted wheat and barley in early fall. The river's silt comes mainly from the sluggish White Nile, which is driven north by the clearer, swift-flowing Blue Nile from Ethiopia.

Opposite: peas (left) and lentils. Since the earliest days of farming, these and other protein-rich legumes have been raised as an alternative source of protein to meat, which often has not been available to cultivator societies.

181

hundreds of large, tightly adherent, huskless fruit. But there seems little doubt that they were the ancestors of modern sweet corn, flint corn, dent corn, and the rest. Modern corn is a classic example of a truly domesticated grain, dependent entirely on man for its survival. Its fruits are attached too firmly to the cob to disperse naturally when ripe. If the entire cob fell to the ground, the massive fruits would have little chance of surviving, for they would have to compete in too small an area to obtain sufficient moisture or nutrients from the soil.

The earliest farmers of the New World used a tropical variant of shifting cultivation. The method is still used today in many parts of Central and South America, and we shall consider it in a moment. First, however, we must look at a unique method of hydraulic agriculture developed in the area of Mexico City during the late 12th or early 13th century A.D. Today the city stands on a dry plateau. At one time, however, the site of the future city lay beneath the waters of Lake Texcoco. And it was here that a group of Aztecs, fleeing from the armies of a neighboring chieftain, developed their "floating gardens," or *chinampas*. Their method was to create artificial islands out of mud from the lake shore and hold it in place with dikes made from reeds; trees planted as soon as the islands were completed threw down roots and bound the soil securely. The islands were constructed in groups intersected by canals. To ensure the soil was fertile, the farmers merely dumped fresh mud from the lake bottom onto the fields before planting their crops. Thus was built the city of Tenochtitlán, later sacked by Cortes and replaced by Mexico City.

The greatest hydraulic civilization of the Americas was that of the Incas, who at the height of their power in the 16th century controlled most of the Andean plateau and adjoining coastal strip from Ecuador to Chile. The civilization had its roots in central Peru where, sometime before the 10th century, groups of immigrant peoples established hydraulic city-states in the narrow mountain valleys. The rise of the Incas began in the 12th century when, from their base along Lake Titicaca, they began to establish hegemony over neighboring city-states. Like other hydraulic peoples, the Incas were superb irrigation engineers, and they brought the art of constructing terraces on steep hillsides to a pitch rarely equaled elsewhere. In many places they tunneled through mountains in order to bring water to their crops along stone-paved aqueducts. The coordination of agricultural output in a country consisting mainly of narrow valleys cut off from one another by high mountains required an all-powerful central authority and total submission by the peasantry. Though enjoying greater security and, probably, a healthier diet than at any period since, the Incas were essentially slaves. The capacity of the Inca emperor to impose his will on all his widely scattered peoples depended on good communications; and as road builders the Incas, who had neither horses nor the wheel, rivaled the Romans.

Old as it is, hydraulic agriculture has never been surpassed in its capacity to feed large, densely populated communities. Even during the first millennium A.D. in China, irrigation techniques were providing surpluses of food in vast

Right : map of the Indus River valley, showing positions of Harappa and Mohenjo-Daro. Both these cities—probably the twin capitals of a large hydraulic civilization—traded with the ancient city-states of Sumer, 1500 miles to the west.

Below : restoration of Mohenjo-Daro's ancient walls is carried out with locally made bricks almost identical to those originally used.

regions in which the population per square mile was considerably greater than it is in some European countries today; and this was achieved with few of the mechanical aids available to the modern farmer. In a purely technical sense (and given freedom from destructive interventions by nature, such as changes in climate) there is no reason why a hydraulic society may not endure forever. In China, after all, it is already 3500 years old; in Peru, in Mohenjo-Daro and Harappa, and elsewhere, it has been destroyed only by invaders.

The fertile crescent, and specifically the Tigris-Euphrates valley, however, is a special case in which the destructive effects of man have been conclusively aided by natural forces. We saw in the previous chapter that the farmers of this region were constantly at war with nomad stock-farmers. In addition, the hydraulic societies were constantly fighting among themselves. The power struggle in this region inevitably hinged on the control of water supplies. Large-scale irrigation works and dams at a given point on either of the two rivers inevitably meant that cities farther down the river had to do with less water. In a number of cases the actual courses of the rivers were changed by city-state engineers: at various times the cities of Ur, Eridu, and Lagash, all

Terraces on Luzon in the Philippines, where rice has been grown on steep-sloping hills for 2500 years. Many of the terraces are only a few yards wide; water from carefully dammed hill streams is led to each plot along bamboo pipes or troughs.

Above left: drawing shows how chinampa plots were kept in place by posts or trees. Fertility of soil was maintained with mud dredged from the canals. Aerial photograph (right) shows chinampas intersected by canals at Xochimilco, Mexico.

Below: terraces at Pilac, Peru, built originally by the Incas and still used today. Construction of complex engineering works of this kind, typical of the old hydraulic civilizations, depended on large concentrations of semi-slave labor.

on the Euphrates, were left high and dry because of this, and sank into oblivion through lack of water. Moreover, because the canals and waterworks were the key to wealth and power in this region, they were the prime targets for destruction in wars between competing city-states.

Even so, the hydraulic societies of Mesopotamia might have survived if the destructive effects of man had not been accompanied by natural forces equally, though more gradually, destructive of the foundations of their agriculture. One of the problems was salinization of the soil. Mineral salts are dissolved by rivers out of the rocks at their headwaters and are carried, like the silt, toward the sea. The waterworks of the hydraulic societies were designed to spread water over the cultivated land to a depth of only a few inches. Most of this water, bearing both silt and salts, either sank into the soil or was later drained away. But a proportion of it evaporated, leaving behind salts at or near the soil surface. Many of these salts are, of course, necessary to plant growth. Others, however, are harmful to the soil. One of these is sodium chloride (common salt), which, unless it drains in solution down to the water table below ground level, turns the clay particles in silt into a hard crust that cannot be penetrated by plant roots. This happened in many parts of the Tigris-Euphrates basin. Moreover, the seasonal flooding of the fields tended to raise the level of the water table, bringing ever-greater concentrations of salt to the surface. In parts of Sumer the increasing salinization of the soil has been plotted in detail over a 700-year period from 2400 B.C. onward. The evidence is partly direct, in the form of actual records of increasing salinization of specific plots of land made by contemporary temple surveyors. But there are also two persuasive items of indirect evidence. First is the gradual replacement in this region of salt-sensitive wheat by the more salt-tolerant barley as the main crop. Second is the decline of crop yields by almost two thirds during this period. Such a decline would have been serious enough in an ordinary community of peasant farmers. In the city-states, which required substantial

crop surpluses not only for external trade but also to support many "non-productive" citizens, the effects were disastrous.

Ironically, silt, the very commodity the farmers relied upon to maintain the fertility of their croplands, also helped to undermine the Mesopotamian civilizations. Every season, a certain proportion of the silt carried by the rivers was deposited on the beds of the irrigation canals and had to be dredged and piled onto the embankments. Unless the depth of the canals was maintained, the water would flow ever more slowly and, thus, deposit progressively more silt before it reached the fields. But the silt piled onto the embankments posed its own problems, for when it dried it turned into a fine dust that was carried by the wind into adjacent canals. Clearly, then, the large and complex waterworks of the hydraulic civilizations needed constant, well-planned maintenance if they were not to become choked. It seems probable that disruption of this work as a result of war, combined with the effects of salinization and of occasional disastrous floods, was enough to bring many of the city-states to their knees. The later invasions of the region by nomad hordes served to finish them off.

The irrigation systems of the Egyptian Nile escaped these disasters, for a variety of reasons. From the beginning of the dynastic period, Egypt was essentially united. True, its provinces, or *nomes*, were administered locally; but their rulers acknowledged the supreme authority of the pharaoh, who was both god and king. Egypt had none of the intensely competitive city-states so characteristic of Mesopotamian history. Civil war and invasion from outside were alike almost unknown during this period: hence the contrast between the heavily fortified Mesopotamian cities and the smaller, unwalled towns of the Nile valley. Living in peace, the Egyptian farmers could devote their lives to farming and to maintenance of their irrigation systems. Moreover, the Nile was a less exacting taskmaster than the Tigris-Euphrates. Its annual flooding was more gradual and more predictable, and the shape of its valley made it

Opposite: the soil of much of the Rechna Doab, in north-eastern Pakistan, is carpeted with a layer of salt. The cause of salinization here, as in many of the ancient city-states of the dry belt, is mainly inefficient drainage of irrigation systems.

Right: slash-and-burn farming on Mindanao, Philippines. Each plot, hacked out of the jungle, is used for cultivation of corn, bananas, and other crops for a few years. Then, when the soil is exhausted, the farmers clear a new plot, allowing the old one to return, eventually, to jungle.

easier to drain; in addition, it contained little salt and had less than half the concentration of silt carried by the Mesopotamian rivers.

It might be thought that the wet tropics, enjoying an abundance of rainfall and supporting the most luxuriant vegetation in the world, are ideal for intensive farming. In fact, they are not, for a number of reasons. Soil, in any region, consists mainly of inorganic matter—sand, silt, and clay, which contain minerals—and organic matter consisting of the remains of dead plants and animals that have been broken down by millions of living organisms in the soil. Humus is a kind of half-way stage between the breaking down of dead plant tissues and the releasing of their salts, which are needed by the next generation of plants; humus also acts as a moisture-retaining sponge. The heat and heavy rainfall in the tropics accelerate the processes of growth and decay; in addition, the heavy rains tend to remove the humus-containing topsoil and to wash the minerals to a much greater depth than in cooler, drier regions. This does not impede the dense growth of trees, which have roots deep enough to take advantage of the minerals and organic matter at greater depth. But it means that shallow-rooting plants—and this includes most of man's important food crops—swiftly exhaust the fertility of the topsoil.

Swidden farming

Farmers in the wet tropics have overcome this problem since the earliest times by adopting a method of shifting cultivation somewhat akin to that of the Danubians in Europe. The method is known variously as "slash-and-burn" or *swidden farming* (from an Old Norse word meaning "to burn"). The method varies from region to region but consists essentially of killing or felling the trees in a small area of forest, allowing them to dry out, and then burning them just before the onset of seasonal rains. The soil is then weeded and cultivated, and the seeds are planted in the fertilizing ash. Swidden farming is practiced throughout the wet tropics: in Central America (where it is known as *milpa*), in South America (*conuco*), in Uganda (*chitemene*), in Sri Lanka (*chena*), in Malaysia (*ladang*), in the Philippines (*kaingin*), and in many other regions. In the New World the main swidden crops are corn and beans; in Africa, millet or sorghum; in Southeast Asia, upland rice and yams. But at its most productive, swidden farming often involves the cultivation of many different plants in the same plot. In the milpa system, for instance, one of the commonest combinations is corn, beans, and squash; this has several advantages in addition to the obvious one of providing a choice of foods. For instance, the bean plants, which normally require to be supported by sticks, climb up the sturdy corn stems. The squash plants, with their vinelike stems, spread over most of the available space between the corn and bean plants, so offering the soil a measure of protection against the heavy rains, which otherwise might cause serious erosion on hilly ground. In the conuco system of South America an astonishing variety of plants may be grown; root crops such as manioc (known elsewhere as cassava); corn and beans; salad plants;

Broken idol at Copan, once a great Mayan city,
in the highlands east of Guatemala City.
Exhaustion of the soil, caused by overpopulation,
forced the Mayans to the disease-ridden
swamps of Yucatán, where their empire collapsed.

herbs and flowers. Cultivation is planned so that harvesting a proportion of the crops, and subsequent reseeding, can take place every week throughout the year.

The difficulty of swidden farming, of course, is that the soil is rapidly exhausted because it is deprived of the almost constant replenishment of organic matter that occurs in the rapid growth-and-decay cycle of the tropical forest. In most parts of the wet tropics the farmers cultivate the soil for a maximum of three or four years before moving on to newly cleared plots. In some areas it is impossible to raise crops for more than two successive years because the soil is invaded by weeds (mainly grasses) that compete with the crops, and it would take almost as long to clear and keep them in check as it would to prepare a new plot for planting. The abandoned plot may be left for anything between 10 and 20 years. At the end of this period it will have returned to forest: the soil's fertility will have been restored and the grasses and other weeds, denied light by the dense canopy of trees, will have disappeared. Now the cycle of felling, burning, and cultivation can begin again.

Nowadays we tend to associate swidden farming with primitive cultures, forgetting that the first of the great New World civilizations—that of the Mayas of Central America—was based on this system of agriculture.

Contour plowing, as here in Bell County, Texas, helps to maintain soil structure, conserve water, and minimize erosion. Like terracing, it represents a creative partnership between the farmer and his most valuable resource—the soil.

12

Land Use and Misuse

Man's earliest tillage instruments were digging-sticks little different from those used by hunter-gatherers. Often weighted with a stone, the sticks were used to dibble holes in the soil, into which seeds were dropped. The most primitive hoes were simply forked branches of wood. Early cultivators in Egypt and elsewhere manufactured hoes by joining two pieces of wood together at an acute angle, and it was probably from this type of hoe that the first plows evolved. When one of the pieces of wood was lengthened into a pole, and an upright handle attached at the junction of the two pieces, the implement could be dragged along by one or more men (or women) while another guided it through the soil with the help of the upright. Later, oxen replaced the traction power of man.

We do not know with any certainty when or where the first plows were developed. The earliest Egyptian paintings in which they appear date from a little before 3000 B.C. Soon after this date, the plow spread around the Mediterranean coast. By 1500 B.C. it had reached northwestern Europe, and a few hundred years later it was introduced into India and China.

This type of plow, together with another that seems to have been developed from the digging-stick, is known as a *scratch plow*. From earliest times, both types were progressively improved and adapted. The varying properties of different kinds of wood were exploited for different parts of the plow: oak, for instance, was often used in the construction of the share, because it is hard and less prone to rot under continuous use in damp soils. In Egypt during the early dynastic period, farmers often attached flint to the leading edge of the share to minimize wear. From the beginning of the first millennium B.C. onward, iron shares were being used throughout the fertile crescent, and by 100 B.C. they had spread as far as Britain.

As their name implies, dry belt plows do little more than scratch a shallow groove in the surface of the soil. Their job is to produce a fine, crumblike soil surface while keeping loss of moisture to a minimum. The usual practice today, as in times past, is to plow the ground twice, the second time at right angles to the first. Consequently, most dry belt fields tend to be small and square. Even today square plots, rarely more than two acres in area, are common in Mediterranean countries; and on some of the downlands of northern Europe (in the temperate zone) the original shapes of such plots are apparent in aerial photographs.

During Roman times, plows appeared that contained both an adjustable *share*, which cuts horizontally at the base of the furrow, and a *coulter*, which cuts vertically; some also had an attachment that pushed the earth to one side. But it was only during the sixth century A.D. that a plow was developed capable of dealing with heavy, damp clay soils. Probably a development of Roman models, it appeared first in eastern Europe. The most significant component of this plow was the *mold-board*, which turned the sliced earth over onto its side. The importance of the mold-board lay in the fact that, by lifting and turning the soil to a considerable depth, it helped to drain the soil, brought its nutrients nearer to the surface, and buried weeds. The first crudely shaped mold-boards underwent considerable development during medieval times; indeed it was not until the late 18th century that the correct shape of the board was scientifically determined.

The introduction of the mold-board plow ushered in the greatest period of forest clearance in northern Europe: it lasted, with interruptions, from the 8th to the 19th century, when the shortage of timber halted the process and new forests began to be planted.

Until the 12th century most farmers in northwestern Europe, as elsewhere, used oxen to draw their plows, though occasionally a horse was used to supplement the work of the ox team. On heavy soils, at least four oxen were needed for each plow. Draft horses were known to be both stronger and faster than oxen, but their use was held back by the design of the harness. During classical and early post-classical times the breast-bands and yoke of the harness were designed in such a way that, if the horse pulled hard on them, the breast-band slipped upward and risked throttling the animal. During the eighth century, however, the rigid horse-collar was imported into Europe from Central Asia. By transferring the load from neck to shoulders, it enabled a horse to pull four times the weight possible with the old breast-band and yoke. Although the horse-collar did not pass into general use in northwestern Europe until the 12th century, it was eventually to revolutionize not only plowing but land transport as well.

Rotation farming

Parallel with these advances in technology came equally important developments in the use of land in this part of Europe. Farmers on the light loess and upland soils used the simple *two-year rotation* introduced by the Romans. It entailed planting wheat or barley in the fall, harvesting it the following spring, and then allowing the land to lie fallow until the next fall but one. (The fall planting, incidentally, was due to the fact that in Mediterranean lands, where the Romans developed the technique, there is too little rainfall for planting in

Right : Egyptian tomb painting of about 1600 B.C. shows a primitive scratch-plow that simply cuts a groove in the soil. Below : diagram shows how a mold-board plow turns over the soil, so improving drainage and burying weeds and stubble. The mold-board enabled farmers to cultivate the heavy soils of northwestern Europe.

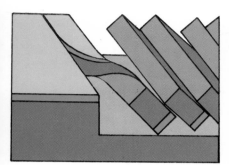

Right : 15th-century French painting showing mold-board plow in use. Below : 14th-century English painting of a horse-drawn harvest cart. Use of the rigid horse-collar greatly increased plowing capacity and revolutionized land transport.

the spring.) Farmers usually divided their arable land into two halves so that one or other half was planted each fall; even so, it meant that only half the land was productive each year. In Scandinavia, where the winters are too harsh for fall planting, the farmers used a two-year rotation based on spring sowing, with the crops harvested in the fall and the land then lying fallow for more than a year.

During the second half of the eighth century some enterprising farmers (probably monks, who were leaders in agricultural innovation during the medieval period) realized that the two rotations could be combined, at least in warmer parts of the temperate zone. The result was a *three-year rotation*: wheat or rye planted in fall and harvested in late spring; then, the following year, barley, oats, or legumes planted in early spring and harvested in fall; then a fallow for more than a year. For this system, the arable land was usually divided into three large fields. Each was at a different stage in the rotation at any one time, so that two thirds of the total land was constantly in use.

This advantage of higher productivity, together with the vast new areas of fertile soil brought into use by forest clearance, led to a steep rise in population between the 8th and 14th centuries (though between 1347 and 1351 the Black Death killed more than one third of the inhabitants of Europe). The three-year rotation also had another important advantage: it made possible the cultivation of oats on a large scale, which was vital if enough horses were to be bred to exploit the new advances in plowing and transport.

From early medieval times the farmers of northwestern Europe practiced mixed farming—sowing their crops on arable land and allowing their cattle, sheep, and horses to graze both on permanent pastures and on the grassy spaces between fields. The most critically important factor in mixed farming is that the animals provide an abundant and permanent source of manure with which the farmer can maintain the fertility of his arable soil. Mixed farming was indirectly responsible for the development of new, multistage rotations. One of these, the *Norfolk rotation*, was adapted from a Belgian system developed in the 17th century. It had four stages—wheat, root crops, barley, and clover—and eventually it enabled the fallow periods, which were brief and widely separated, to be discontinued. From now on, at least on the fertile heavy soils, the land was in constant food production. The use of clover as one of the four stages in the rotation was an important development. Hitherto, pasture plants had rarely featured in rotations. The introduction of leys, or temporary pastures, into the rotation made for more flexible use of the land; moreover, the nitrogen-fixing properties of clovers helped both to maintain the fertility of the soil and to bring extra protein into the food chain.

The development of farming in northwestern Europe is, on the whole, a story of an outstandingly successful relationship between man and soil, and one that contrasts strikingly with the fate of European migrants who have attempted to impose the same basic techniques on land in other parts of the world. The success in Europe has depended upon a constructive alliance between climate, soil, and cultivation techniques that have evolved over a period of more than a thousand years. It is important to stress the word "alliance." At first sight, the clearing of dense forest and the annual plowing of the topsoil to suppress the natural vegetation both imply the maximum possible interference with nature. So, too, does the elimination of the fallow period, which elsewhere has been vitally necessary to the continuing fertility of the soil. Yet, in northwestern Europe, the heavy clay soils have not only remained fertile, they have, if anything, increased in fertility, and today are among the most intensively farmed and highly productive in the world.

This alliance has often been disregarded by European colonists in other regions, who have used farming methods that both clashed with the dictates of climate and were careless of the needs of unfamiliar soils. In parts of tropical Africa, for instance, Europeans have suppressed the age-old native methods of shifting cultivation and have tried to import unsuitable techniques.

One of the best-known examples of this was the peanut scheme launched by the British government in Tanganyika (now Tanzania) shortly after World War II. At that time European agriculture was disrupted. Like many other countries, Britain was desperately short of fats, and it was hoped that peanuts, one of the richest sources of vegetable fat, would make good the shortage. The basic idea was a good one, but it was doomed from the outset by inadequate planning. Land surveys and research into soil and climate at the chosen sites would all have revealed in advance the almost insuperable difficulties facing

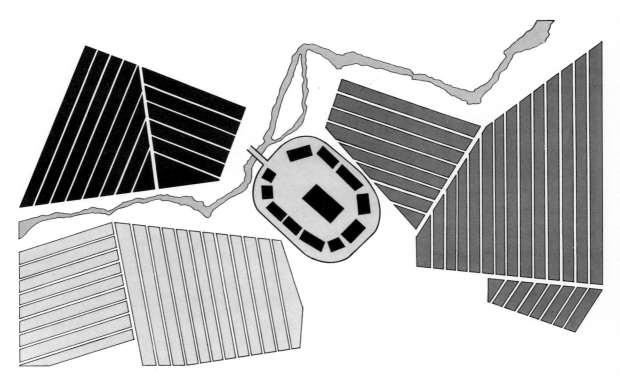

Three-year rotation system in a medieval village. Each group of fields was divided into strips farmed by individual peasants. In any year, one group was left fallow, another was planted in the spring, and the third was planted in the fall.

195

the scheme in the particular areas chosen. The main area, at Kongwa (about 200 miles west of Dar es Salaam), is an upland region of dry brush, the extremely tough roots of which spread for considerable distances in the iron-hard soil. Both soil and roots made cultivation and harvesting extremely difficult. Moreover, stripping the bush encouraged the growth of a variety of weeds that choked the harvesting implements. But the decisive factor was climatic. Peanuts need a total of at least 20 inches of rainfall during the growing season, the six weeks immediately following germination being especially critical. In the Kongwa area, however, rainfall during the growing season rarely reaches this level; more important, the plant-germination period came at a time when the area is actually subject to drought in many years. Eventually, cultivation was abandoned at Kongwa and the cleared areas were devoted to cattle pasture.

The wise farmer is concerned not only with the immediate problem of raising a good crop this year but also with ensuring his future prosperity. Assuming that he does not try (as the peanut scheme did) to raise crops or stock in an unsuitable environment, his main problem lies in maintaining the fertility of his soil. If he succeeds in this—by the use of rotations, by replenishing the soil's organic content with manure or fertilizers, by suitable plowing methods, and so on—his soil may remain productive indefinitely. But if he fails, the soil will gradually become exhausted.

Natural soil and plant life have established a state of equilibrium with climate and with erosional forces over a period of hundreds, perhaps thousands, of years. Grassland soils, for instance, are granular in structure and are more resistant to erosion than forest soils. The latter, however, receive more protection, from both the foliage and the litter on the forest floor. Again, the structure of soils on steep slopes generally offers better resistance to erosion than that of soils on flat land. But the cultivator literally cuts across such distinctions: continuous cultivation tends to reduce topsoils of every type to the fine, powdery form that is especially vulnerable to erosion.

The experiences of the early colonial farmers in North America provide many examples of the danger of using well-established European farming techniques in regions climatically unsuited to them. From the very first, the early plantation settlers on the east coast had to contend with erosion. The farmers cleared the mixed forest in order to plant grain crops and tobacco. Because their economy depended on a massive surplus (earmarked for their paymasters in Europe), they had to farm the land intensively, and even steeply sloping woodland came under the ax and plow. After a few years, the dark-brown forest soils began to lighten in color—a sure indication of deterioration in soil structure owing to loss of fertility. By the end of the 18th century, hundreds of the pioneer farms had been abandoned as their owners were forced to push westward in search of better land.

What went wrong? In the first place, most of the early New World farmers concentrated on crop farming: mixed farms or stock farms were rare, with the

Work on a medieval mixed farm (from a 15th-century Flemish manuscript). Features such as the hand-ax, billhooks, horses' harness, and wattle fence are little changed today. The farmer and his wife beat gongs to make the bees settle.

197

result that intensive manuring of the soil—an essential feature in the European system—was impossible. In the second place, the climate on the American Atlantic coast is less friendly than that of northwestern Europe; in particular, it is subject to violent rainstorms. Every season, when the surface of the ground was left bare after plowing and harvesting, some of the topsoil was removed by heavy rains. This process, called *sheet erosion*, is almost imperceptible at first, but soon accelerates because the soil becomes progressively less porous at depth. So, as each layer is removed, more water is retained at the surface. If the land slopes at all, the water will flow into streams and rivers, carrying off an ever greater amount of soil.

Another area of the eastern United States that has suffered greatly from erosion, especially during the last hundred years, is the Piedmont—the upland region separating the Appalachians from the coastal plain in Virginia, the Carolinas, and Georgia. During the early years of this century huge areas of the Piedmont lost almost all their topsoil through sheet erosion and the development of gullies. Gullying begins on bare soil: rain-water runoff, seeking its natural path downhill, forms rivulets and then, if there is enough water, consolidates by stages into torrents that are capable of carving great chasms in the land. In parts of Georgia, several gullies more than 150 feet deep have developed in less than 40 years, and the whole Piedmont is scarred with tens of thousands of lesser ones.

Gullying is a natural consequence of sheet erosion because it is due to the soil's inability to absorb moisture. Once gullying begins, the usefulness of the soil to the farmer has usually gone beyond the point of no return. It hardly needs saying that water is one of the determining factors in soil productivity. But water is useless to plant life if the soil can no longer absorb and retain it. If this happens, even areas of high rainfall suffer, in effect, from drought.

One of the principal causes of erosion in the Piedmont was *monoculture*—the farming system based on a single crop (in this case, cotton) that was raised year in, year out. Cotton lands are especially liable to erode because the soil between the widely spaced rows of cotton plants must be rigorously weeded throughout the growing season, and the fields must be completely cleared of old growth after harvesting. Monoculture has caused similar disasters in South and East Africa, in Queensland, Australia, and in many other parts of the world. Today, much of the Piedmont is devoted to a variety of other crops, and with the help of rotations and massive applications of fertilizers, its soil is gradually being improved.

Pasture lands all over the world, and especially in semi-arid regions, have persistently deteriorated under the stock-raising methods of European farmers. In America, South Africa, the Soviet Union, and Australasia, damage or outright destruction of range pastures has been due, indirectly, to economic expansion and the steady rise in human populations during the 19th century. More people meant that more food was needed, and this led to increasing pressure on available pasture land and, in many cases, to over-grazing.

Reclamation

The problem of overgrazing and how to restore the productive capacity of depleted grasslands has been studied in many countries. In 1948, for example, the New Zealand Soil Conservation and Rivers Control Council bought some 8000 acres of severely eroded pasture land in the Tara Hills, South Island. Most of the land lies between 1300 and 5000 feet; it has a semi-arid climate, with an annual rainfall of about 20 inches.

When the council took over, most of the land above 4000 feet was bare of vegetation except for isolated tussocks of snow grass. The exposed ground, especially on the sunnier faces, was affected by severe sheet, wind, and ice erosion, and there were numerous gullies. Between 2000 and 4000 feet, the grasses were severely depleted on sunny faces and more continuous but excessively grazed on the shadier sides. Most of the soils below 2000 feet were devoid of any vegetation except scab-weed. The only well-grassed areas were the "fans" of fine, sandy loam that had been washed off the hillside and had formed at the bases of gullies.

The landscape around this Oregon farmhouse consists of dusty sand underlain by infertile hardpan. This dust-bowl has been caused by overgrazing of rangeland, which left the topsoil exposed to seasonal high winds.

Opposite : only a few years ago.
all the area in the foreground
(in Natal, South Africa) was
at the same level as the
"plateau" at center and right,
and was good wheat-growing
land. Overgrazing by sheep led
to denuding and erosion of
steep-sloping hills nearby, and a
few seasons of heavy rains
caused runoff from the hills to
carve these gullies in the valley
below.

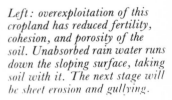

Left : overexploitation of this
cropland has reduced fertility,
cohesion, and porosity of the
soil. Unabsorbed rain water runs
down the sloping surface, taking
soil with it. The next stage will
be sheet erosion and gullying.

Left : carcass of a cow lies in
an expanse of ruined grazing
land in the Kajiado Plains of
Kenya. Destruction of the
pastures is due to overgrazing,
overuse of fire to clear the
ground for new grasses, and a
few seasons of low rainfall.

The key to the problem of these grasslands was that the natural vegetation had evolved in the absence of grazing animals. When the land was settled by farmers early in the 19th century, the mature native grasses proved unpalatable to sheep. The farmers dealt with this problem by regularly burning the vegetation to encourage the growth of young shoots, and by fencing the land into large blocks to separate summer and winter pastures. Because the summer pastures, on the higher ground, could be used for only a comparatively short season, the winter pastures tended to be progressively overgrazed. The situation worsened dramatically during the 1870s, when the pastures were invaded by vast numbers of rabbits. The effect of the rabbits was particularly severe in the overgrazed areas.

It is much easier to destroy pastures and arable land than to restore them: eroded soil will not regain its fertility overnight. The program developed for the Tara Hills by the council's scientists, though experimental in nature, had to take account of practical problems facing the ordinary farmer. In particular, it had to be cheap and it had to permit at least part of the pastures to be used at any one time while the rest was being restored. The first step was to reduce the size of the sheep flock by almost half; cattle were brought in to graze on the rougher grasses, and especially to check the spread of species unpalatable to sheep, and a concerted effort was made to stamp out the rabbits. At the same time, new fences were erected to divide the land into smaller blocks, stocking of each block being carefully controlled to prevent overgrazing. Research was done on various experimental plots at different altitudes to discover which exotic (non-indigenous) grasses and legumes were

Below: map shows location of Tara Hills and Wither Hills, on South Island, New Zealand. Grassland in light gray; forest, dark gray; scrub and moor, mid-gray. Like Tara Hills, Wither Hills have been the scene of intensive work to restore productivity of upland pastures. Photographs show (center) a valley in 1945, when conservation had just begun, and (right) 1966.

NORTH ISLAND

Auckland

SOUTH ISLAND

Wither Hills

Christchurch

Tara Hills

Dunedin

likely to thrive. Water was brought to the flat lands at the base of the hills along an irrigation channel from a nearby river. The flat lands were divided up into small paddocks, which were planted with alfalfa and clover, in order to supplement, and raise the protein content of, the winter feed. Parts of the high pastures up to 4000 feet were sown with grasses and clovers.

The program has been a notable success. With all but the highest land re-vegetated with both native and exotic species, the number of sheep carried has risen from 1053, when the project began, to well over 3000, while the yield of wool from each sheep has almost doubled. In addition, the pastures support about 150 head of cattle, which are not only a useful economic bonus for the farmer but also an important factor in maintaining the quality of the pastures. Erosion and water wastage have been checked by building terraces to guide rain-water runoff into areas where it is most needed.

The lessons to be learned from this quite modest scheme of pasture re-clamation apply to every kind of farming. Agriculture, whether crop-growing or stock-raising, represents an interference with natural processes. Climax vegetation, as we saw in Chapter 10, is in a state of equilibrium: it endures because present output (plant growth) is balanced by input (the materials on which future output depends). Under natural conditions the soil is protected by its permanent cover of vegetation and its fertility is maintained by the continuous conversion of dead plant matter into humus.

In northwestern Europe, farmers have learned in the last thousand years that agriculture is, essentially, a transaction between man and the soil: in exchange for its gifts of foodstuffs, man must provide the soil with food-making ingredients in the form of manure, artificial fertilizers, and various soil-enriching plants. Throughout history, the price of man's ignorance of his side of the bargain has been to turn vast tracts of fertile land into useless wastes. Today, it is a price our hungry world cannot afford to pay.

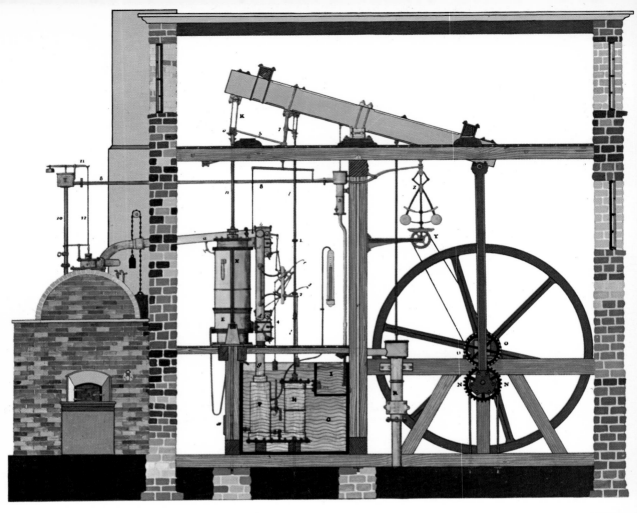

13

The Machine Age

There is a tendency to regard the Industrial Revolution as a rather abrupt watershed—a fixed moment in historical time when man began to use machines to do jobs that, for thousands of years, had been done by hand. It is not, of course, as simple as that. Quite complex machines were in common use long before the revolution began. The water mill, for instance, was used by the Chinese, Greeks, and Romans before the Christian era, and the windmill (a development of the Tibetan prayer wheel) was introduced into Europe during the 11th century. Within Europe, moreover, the revolution began at different times in different countries. Although industrialization began in Britain about 1780, in France, Belgium, the United States, and Germany it did not get under way until the mid-19th century, and in Sweden, Japan, Russia, and Canada until the late 19th century.

The rise of technology, which was to give Europe dominance over the rest of the world, began in the Middle Ages. From the 13th century onward engineers harnessed the power of wind and water for an enormous variety of industrial uses—to drive textile machinery, lathes, trip-hammers, grindstones, and much else. Perfection of the escapement mechanism for clocks during the 15th century made possible the development of many new machines.

Some of the most important factors pointing the way toward the revolution were evident long before this. Coal, which was to power the factories and railways, had been mined on a small scale for hundreds of years. In England, Newcastle-on-Tyne's first city charter, granted in 1239, permitted its citizens to dig coal. The need for large quantities of coal developed in the early 18th century. Until that time timber was used for fuel (as wood or as charcoal) as well as for construction. The rise of England as a mercantile power greatly increased her demand for shipbuilding timber and led to the destruction of many of her largest oak forests. Manufacturers, seeking a substitute for charcoal for smelting iron, turned to coal and its derivative, coke, following the lead given by Abraham Darby at his Shropshire ironworks in 1709. By the mid-18th century cast iron replaced wood in the construction of an ever-growing variety of machines in industry, farming, and transport.

Top left: a James Watt reciprocating steam engine of 1787, adapted to produce rotary motion for driving factory machines. Bottom left: Abraham Darby's Coalbrookdale ironworks, Shropshire, in 1801. By the end of the 17th century the use of charcoal for iron-smelting was threatened in England by shortage of timber. Another fuel had to be found, and in 1709 Darby became the first to smelt successfully with coke. (Coal is unsatisfactory because it contains sulfur, which tends to weaken the metal.)

*Construction of the Manchester-Liverpool railroad.
1831. The inexorable advance of the railroad
lines across the countryside was one of the most
potent symbols of man's control over new sources
of power and of his entry into the machine age.*

The people who were to provide the labor force in the mills and other factories were part of a steady exodus from countryside to town that had its roots in the Black Death. After this pandemic had wiped out whole villages and towns, vast areas of arable and pasture land remained untilled and ungrazed for years on end. One of the effects of this disaster was to encourage the process of land enclosure—the passing into private hands of areas of common land that hitherto had been freely cultivated or grazed by yeomen and commoners. Between 1760 and 1850 alone, some 6 million acres of British common land were enclosed. Land enclosure, though ultimately it was to benefit agriculture, destroyed village industry and created great hardships among the peasantry; from the beginnings of the 17th century, thousands of landless laborers moved into towns to seek work.

All these historical factors made the revolution possible. What made it essential, in economic terms, were the tremendous expansion of overseas trade during the 17th and 18th centuries and the sharp rise in population almost everywhere in Europe from the mid-18th century onward. Britain's dominance of the world's seaways, her colonial riches, and her consequent leadership in world trade enabled her to take the first steps in the industrial breakthrough; commercial wealth, a practical-minded scientific elite, and abundant resources enabled her to sustain this load for almost a century.

Yet, although the cultural, economic, and scientific seeds of the revolution were planted long before the 18th century, there can be no doubt that its consequences *were*, almost immediately, revolutionary. They were revolutionary not least in the extent to which they combined material progress with social upheaval. More significant, however, was the way in which the revolution changed man's ecological status for all time. In replacing the power of muscles with the power of machines, man transformed his capacity to intervene in natural processes of every kind; in harnessing his new power to old and new activities in factory and farm, he changed not only the lives of whole populations, but also their relationships with every other part of the ecosystem.

During the late 19th century, petroleum began to rival coal as a source of power. A fossil fuel, like coal, petroleum is derived from the decomposed bodies of tiny marine animals, and possibly plants, trapped beneath impervious rock on the seabed. Man has known of petroleum for thousands of years. During the sixth century B.C., for instance, King Nebuchadnezzar of Babylon constructed roads made from dried bricks and asphalt (a petroleum derivative). But its large-scale use as a fuel (initially as kerosene for lighting) began in 1859, when the first oil well began commercial operations in Pennsylvania. The use of gasoline to power internal-combustion engines became practicable in 1887 when Gottlieb Daimler designed the carburettor, which measures the correct mixture of fuel and air into the combustion chamber. The rise of the automobile industry, which consumes billions of gallons of gas and diesel oil a year, stimulated research into other petroleum products. Petrochemicals, as they are called, today affect the lives of almost

Man-made fertilizers—especially those based on nitrates, phosphates, and potassium—greatly increased man's agricultural output during the 19th century. Above: phosphate recovery plant in Florida. Left: two lettuces that were planted at the same time; the plant on the right is growing in soil deficient in potassium.

Opposite page, top left: 500-MW. turbo-generators in a British power station. Turbines, used for electricity generation all over the world, are descendants of the wind- and water-mill. The other pictures on this page show a few of the hundreds of derivatives of petroleum—herbicides, aviation fuel, plastics, synthetic fibers, and detergents—all widely used in modern life.

everyone on this planet: they include plastics, synthetic fibers and rubbers, and detergents, and are constituents of fertilizers, paints, and much else.

The term "Industrial Revolution" conjures up a variety of pictures in the mind's eye, from gaslit cotton mills to the modern automobile production line. But the revolution's impact on agriculture was just as profound as it was on manufacturing processes. An important link between the two was the cotton gin, invented by the American Eli Whitney in 1794, which enabled cotton fibers to be separated quickly and efficiently from their seeds. The cotton growers of the American southern states, and the Lancashire cotton millers were mutually dependent on each other: both quickly came to appreciate the value of the gin. In the year of Whitney's invention, cotton-fiber production in the United States was about 5 million pounds a year. By the 1830s, when the gin was widely used, production was 40 times higher.

Machines on the farm

From the early 19th century onward there was a steady advance in farm mechanization. That British agriculture was able to take advantage of these new machines was due at least partly to land enclosure—much of the best arable land in the country was in the hands of comparatively wealthy farmers who could afford to buy the machines. Their use spread more gradually to the poorer tenant farmers, many of whom organized themselves into local groups and hired the machines on a cooperative basis.

One of the first labor-saving devices on the farm was the seed-drill, which not only economized on manpower but, by planting in rows instead of broadcasting, was much less wasteful of seed. By the early years of the 19th century drills had been developed that could plant seed at different depths, according to the nature of crop and soil, and also deposit fertilizer.

Until reaping was mechanized, it required more labor than any other farm job; on the smaller farms, extra hands were usually hired at harvest time. Reaping machines began to appear during the 1780s, but the first successful one was developed by a Scotsman, Patrick Bell, in 1826. This was followed, in 1831, by a more efficient machine manufactured by the American Cyrus McCormick, then only 22 years old. From 1845 onward McCormick effectively cornered the American market by mass-producing his machines. At about this time, too, threshing and cleaning machines began to appear. The first self-propelled combine-harvester, which reaps, threshes, and cleans, appeared during the first decade of the 20th century.

Steam power came to the farm in 1784, when James Watt patented a device by which two static steam-engines drew a plow backward and forward on wire ropes. Steam tractors appeared in the 1860s: heavy and difficult to maneuver, their use for traction was confined mainly to multifurrow plowing on the vast fields of the American corn belt. The earliest efficient tractor to be powered by an internal-combustion engine was built by the American John Froelich in 1892; by 1918 Henry Ford was mass-producing them.

Between 1800 and 1900 the population of Europe doubled, from 188 million to more than 400 million. One reason for this was that even the poorest people benefited to some extent from the increased wealth created by the Industrial Revolution; more important, the revolution stimulated scientific research into an enormous variety of subjects, including medicine, and the conquest of several important diseases led to a reduction in the death rate during the 19th century. The extra people concentrated mainly in the industrial or commercial centers: farm mechanization and the lure of new jobs in the factories accelerated the flow from the countryside to town. Before the Industrial Revolution less than 10 towns in Europe had a population greater than 100,000. By 1900, Britain had 37 such towns and Germany 33.

The land: natural fertilizers

It was not only human labor that dwindled on the farms as a result of mechanization: from the mid-19th century onward, as tractors and other sources of power became available, there was a steady reduction in the number of draft animals, especially horses. As a consequence, less farmyard manure was available. Long before this, admittedly, farmers had supplemented their stocks of animal manure with such things as human and animal bones and saltpeter, both of which were known to contain substances that promote plant growth. Nevertheless, the lack of adequate supplies of manure, added to the need for higher food production to feed the rising populations, would have led to enormous pressures on land use if scientists had not come to the aid of farmers by the development of artificial fertilizers. These chemicals have been both a blessing and a curse, as we shall see in Chapter 20.

The land: chemical fertilizers and pesticides

The development of fertilizers was the logical sequel to research, in the late 18th and early 19th centuries, into the nature of soils, and the discovery of the 16 chemical elements needed by plants for growth. Of these elements, three (oxygen, hydrogen, and carbon) are available to the plant in air or water. The other 13 must be imported from the soil. Of these, nitrogen, phosphorus, and potassium are needed in the greatest quantities.

Fertilizers began to be mass-produced during the first half of the 19th century. These man-made chemicals were important not only because they meant that farmers could henceforward rely on adequate supplies of fertilizer, but also because the chemists were able to produce them in forms that plants could absorb more readily than "natural" fertilizers, such as bones. Most soils contain certain amounts of nitrogen, phosphorus, and potassium, together with the other elements. The point is that food crops such as grains, as we have seen before, concentrate most of these elements in the seed head and so they are lost to the soil when the crops are harvested. The output of almost any soil will be raised by applying fertilizer mixtures. Without fertilizers, crop output in the United States would drop by at least a quarter.

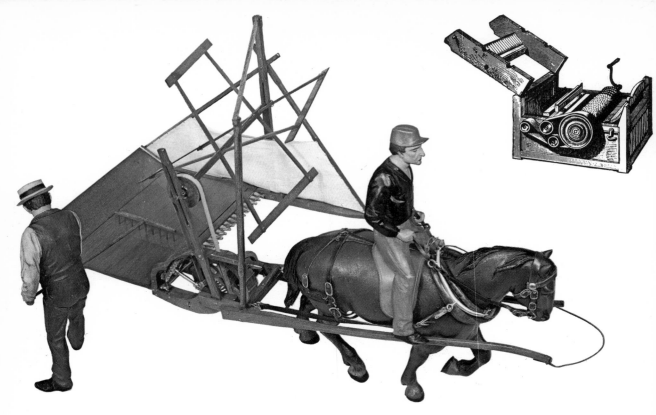

During the present century, chemists have provided the farmer with two more groups of important chemicals—weedkillers and insecticides (the latter are considered in Chapter 16). The definition of a weed varies according to circumstances. In a virgin forest, for instance, all (or none) of the plant species can reasonably be called weeds; in agriculture a weed is, quite simply, any plant that the farmer does not want. On the farm, weeds not only make harvesting more difficult: by competing with cultivated plants for light, water, and soil nutrients, they can significantly lower crop yields. During the early years of this century, chemists discovered that plants contain certain growth-regulating substances that, if present in large enough quantities in the tissues, will kill the plants. During World War II it was found that certain of these substances kill some plants but do not harm others. As a result, a wide range of *selective* weedkillers have been developed during the last 20 years. Some, which are harmless to grasses, are used to keep grain crops free of weeds; others, which kill grasses, are used with legumes, root crops, and so on. Still others are so highly selective that they will kill certain grasses, such as wild oats, while

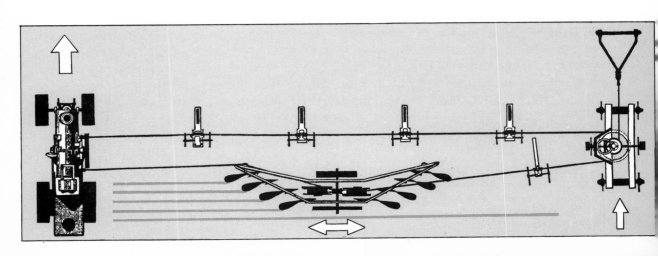

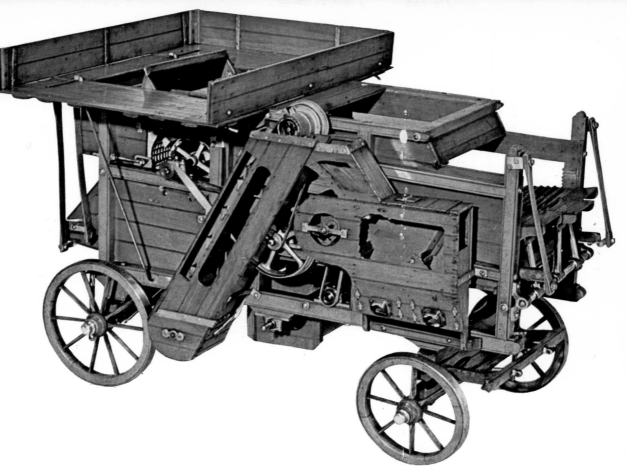

Opposite, top right: Eli Whitney's cotton-gin of 1794 tore cotton seeds from their fibers by means of a hand-rotated drum with pointed teeth. Opposite: model of Cyrus McCormick's reaper of 1831. Full-scale production of improved versions of this machine coincide with a serious loss of farm labor owing to the California gold-rush of 1849.

Above: model of a threshing machine of about 1860. The grain harvest was fed into the top of the machine and dropped onto a beater-drum that detached ears from straw. The grains were then separated from the chaff and weed seeds by rotary screens and a blast of air. The grains were automatically fed into sacks, while the straw was ejected from the rear of the machine.

leaving unharmed closely related species such as wheat or barley. The last are among a variety of *defoliants* that were used to kill off jungle foliage during the Vietnam war in order to deprive the Vietcong of concealment. (What the long-term ecological consequences of this are we simply do not know.)

The invention of the steam engine and its application to industry can be regarded simply as an episode of history. But in ecological terms it is more apt to regard it as an early stage of a process that continues today—a process that is enabling man (both as a species and as an individual) to gain an ever-greater control over his physical environment. It has been said that each individual in the United States today has at his fingertips an annual amount of power equal to that which can be generated by 10 tons of coal or by the work of 100 slaves. Power is morally neutral—it can be used, deliberately or accidentally, for good or ill. The power under our control has increased immeasurably since the onset of the Industrial Revolution; so, too, have the ecological consequences of the use of that power.

Left: steam-plow apparatus made by a British engineer in 1858. Power was supplied by the steam engine at left, which drew the double plow at center in alternate directions across the field by means of an endless steel rope that looped around the pulley wheel on the carriage at right. Slow and laborious, it was nevertheless used quite widely in Europe until the coming of the highly maneuverable petrol- and diesel-engine tractors pioneered by Henry Ford.

Shooting migrating passenger pigeons for sport in northern Louisiana (from a sketch made in 1867). The gigantic migratory flocks of these pigeons provided unmissable targets for anybody with a gun, and by 1914 the species had been exterminated.

14

Man the Predator

So far in this book we have concentrated mainly on man's impact on the soil and on the plants he raises for food. Now we must have a brief look at his record as a predator on animals.

Man hunts animals for four main reasons: for food, for commercial gain, for sport, and to safeguard his "interests." (The last includes not only self-defense but also the need to protect domestic animals against wild animals that either prey on them or compete with them for food.) Sometimes, of course, more than one of these reasons applies. Many of the fiercest carnivores, such as the polar bear and the tiger, have been hunted both for sport and for the value of their fur or skin, as has the harmless Australian wallaby.

Hunting to extinction

Man's record as a predator is nowhere more striking than in the territories opened up by European explorers and settlers from the 16th century onward. The colonization of North America is perhaps the prime example of this because of the size of the subcontinent and the pace of settlement. In many cases, the settlers deliberately exterminated as many "undesirable" species of animals (including human "animals," the Indians) as they could; in other cases, forest clearance and the plowing up of grassland just as effectively wiped out many animals by depriving them of their natural habitats. Among many mammals exterminated by the settlers are two species of moose, three species of grizzly bear, the eastern puma, the plains gray wolf, and the badlands bighorn; extinct birds include the great auk, the Carolina parakeet, the Labrador duck, and the heath hen. Two other species that may well die out are the beautiful whooping crane, which has been hunted for its flesh and plumage for more than two centuries, and the bald eagle, which is, ironically, the emblem of the United States of America.

How was it possible that such species, many of them once very abundant in America, could simply have been wiped out? The American thriller writer Raymond Chandler once wrote that, if his imagination flagged, he would resort to the simple but effective expedient of having a man enter a room carrying a gun. The power of the gun completely alters the relationship between the armed man and his fellows (and other animals). It represents power—destructive power—of a different order of magnitude from the spear, the bow and arrow, and the knife.

Just how different is apparent from the fate of the passenger pigeon. During the early 19th century there were probably 5000 million of these birds in North America. Every fall they migrated southward in immense flocks. In 1806 the ornithologist Alexander Wilson observed a flock that was about 240 miles long and more than a mile wide. He calculated it contained more than 2000 million pigeons. It was this habit of gathering into vast migratory flocks that made the species vulnerable to the gun. Throughout the 19th century the migrants were shot from the sky or from the trees in which they rested briefly on their southward journey. Every fall the toll was so enormous that farmers would bring their pigs from miles around to feed on the corpses that lay thick upon the ground. The slaughter increased greatly during the second half of the 19th century with the growing use of breach-loading shotguns with percussion-fired cartridges. By the early years of the present century the combined effect of the shotgun and the clearance of the bird's forest habitat achieved what had once seemed impossible: the species was wiped out. In 1914 the last passenger pigeon died—ironically, from natural causes—in Cincinnati Zoo.

Slaughtering a species

The massacre of the American buffalo, or bison, illustrates strikingly how a combination of ruthlessness and technological advance can almost obliterate a once-abundant species. Before the arrival of Europeans in America there were probably 60 million bison, mainly in the grasslands but also in the eastern forests. The bison was the central feature of the economy of many Indian peoples—notably the Plains Indians and the western farmers (the latter mainly in what is now the corn belt)—who used its flesh for food and its hide for clothing and shelter. In ancient times the Indians hunted the bison on foot, with bow and arrow. During the late 16th century, however, the Plains Indians learned to ride horses that had escaped, or had been captured, from Spanish explorers and colonists in the south; later, they bought or captured rifles from European settlers. Armed, and on horseback, the Indians took an ever-increasing toll of the vast herds. The final, determining phase of slaughter, however, was carried out by Europeans.

To the European pioneers pushing westward from the early 19th century onward, there were two main threats: the Indians, whom the settlers were unceremoniously depriving of land, and the bison, who fed off the pastures that were needed for domestic cattle or for plowing up for crops. The Europeans—settlers, profiteers, hunters, and politicians—realized that by destroying the bison they would not only free the grassland but would also deny the Indians their means of livelihood. And so the massacre began. During the 1830s and 1840s, great piles of rotting bison corpses were a common sight over much of the Great Plains. No matter that, in this region, bison were probably

Deck of an Arctic seal-catcher. Since the early 17th century, seals have been slaughtered (often with unnecessary cruelty) in their thousands every year. In spite of international regulations the survival of some species is in danger.

217

a better source of protein than the imported cattle: to the Europeans they were a pest and must be eradicated. Many men made a profession of bison hunting—notably "Buffalo Bill" Cody, who, during the 1860s, killed almost 4300 bison in the space of 17 months. (Cody sold the meat to the railroad construction gangs; the irony is that the tracks of several railroads were laid along trails that had been leveled and hardened by generations of bison.

By 1900 the bison was near extinction; even the most experienced hunters could find only a few, usually solitary, individuals roaming the plains. The species undoubtedly would have died out if conservationists had not rescued it. Today, the bison is protected on government reserves such as the National Bison Range in Montana, and the population has now risen to several thousand. Interestingly, it is in these reserves that the native grasses (which elsewhere on the Great Plains have been dangerously overgrazed) are staging a comeback. In turn, this is leading to the development of complex communities of plants and animals very much like the plains ecosystems before the coming of the Europeans.

Fur, ivory, and horn

Early American settlers often competed with Indians for commercially valuable animals. Sometimes the competition led to war—as in beaver-trapping. Before the arrival of Europeans there were probably at least 100 million beavers in Canada and the United States. The opening-up of much of the Canadian hinterland was due in no small measure to the value of beaver pelts, and the prosperity of the Hudson's Bay Company depended on it during the 18th and most of the 19th centuries. By about 1790 the company was exporting at least 50,000 pelts a year to Europe. A hundred years later the beaver was almost extinct in the United States and much of Canada. Ruthless massacre for commercial gain also drastically reduced the elephant and rhinoceros populations in Africa during the 19th and early 20th centuries. The elephant was hunted for the ivory of its tusks, the black and white rhinos for their horn. Today, all are protected in game parks.

The hunting of animals for commercial reasons has often had serious ecological consequences, and with the rise of technology these consequences have escalated. The greater the commercial value of an animal, the more are companies and individuals prepared to invest in developing equipment needed to catch it. And once the investment has been made, market pressures force ecological restraints into the background. Nowhere on this planet has free enterprise slipped so quickly into anarchy as in man's exploitation of the animals of the sea. The North Sea fishing grounds, for instance, have been so seriously depleted during the last 60 years that British trawlers are now obliged to sail much farther afield to make a living. More than half their total catch is now taken in "distant waters" (which may be as far away as Labrador and the Barents Sea), while another large fraction comes from "middle waters," west of Scotland and around the Faeroes.

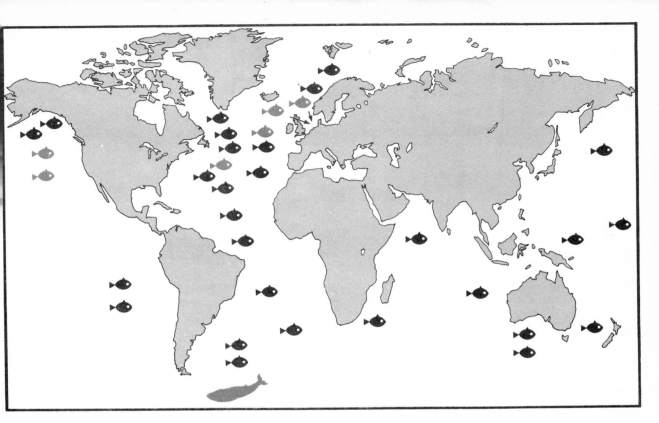

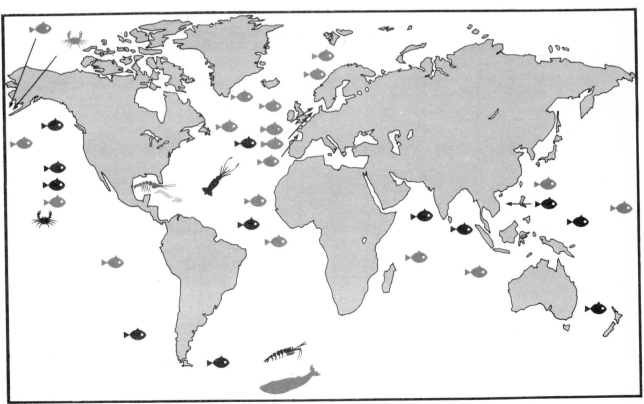

Maps show fish resources (overexploited species in gray, underexploited in black) for the years 1948 and 1968. The great increase in species that have been overfished reflects the fact that world catches almost doubled between 1958 and 1968.

Blubber, oil, and bone

Whaling provides the most notorious example of man's over-exploitation of the sea's resources, and of the ways in which greed allied to technology can lead to the decimation of whole populations of animals. Man has probably been hunting whales for more than 3500 years. In Europe the pioneers were the Norwegians, who by the end of the ninth century were sailing along the coasts of the Arctic Sea in search of catches. A little later the Basques of southwestern France and northern Spain began to hunt whales in the Bay of Biscay. Gradually they ventured farther afield, and by the mid-16th century they were whaling off the Newfoundland coast. These long voyages had become worthwhile because the market for whales had changed. Whalemeat, for which the animals were hunted in the early days, would have decomposed on the long voyage home. But by the 16th century the animals were being

Predation for profit: the price of this red-fox fur coat (made from about six skins) is around $1200. Even today, when many animal species are protected, the law of the market too often takes precedence over the principles of conservation.

hunted mainly for their oil, which was used in lamps, and for whalebone, which was in great demand for use in such things as whips or umbrella ribs, and as stiffening in clothing and stays. The whalebone comes from the large, fringed, horny plates that baleen, or whalebone, whales have instead of teeth. The plates help to sieve out the tiny euphausians (animal plankton) that form the main diet of these animals, which include the right whales and the rorquals (blue, fin, and sei whales). The other main suborder, the toothed whales, includes the sperm whale, dolphins, and porpoises.

Whaling, as an industry, began during the 17th century, after European sailors had discovered that the waters around the islands of Spitsbergen, Novaya Zemlya, and Jan Mayen—all well within the Arctic Circle—had abundant stocks of both right whales and rorquals. Within a few years, the Dutch, Norwegians, and English had large whaling fleets in these waters. By 1680, for instance, the Dutch fleet numbered over 250 ships; and in 1697 almost 2000 whales were taken off the coast of Spitsbergen alone. The main target of these whalers was the slow-moving right whale, which came into the island bays to breed. By the mid-18th century the stocks of these whales were so severely depleted that European ships began to hunt farther westward, around Davis Strait and in Baffin Bay, to the west of Greenland. Here they were competing with American whalers, who by now were also hunting in the Bering Sea off the coast of Alaska. In the Pacific the Japanese (probably the earliest Asian whalers) began to use nets to catch whales near shore. Pelagic (open sea) whaling began in the 18th century, in both the Atlantic and Pacific oceans, and by the early years of the 19th century, sperm whalers were regularly sailing to the coasts of Australia and New Zealand.

Whaling in the Antipodes was sparked off by Captain James Cook. When he returned from his second voyage in 1775 he reported that the great southern ocean was teeming with whales. In the course of the next 50 years the "southern fishery" revealed itself as the richest area for whales. Once again it was the right wales that were decimated, mainly in their summer breeding grounds in the bays and estuaries of Tasmania, Australia, and New Zealand. By 1846 the United States alone had more than 730 whaling ships in the Pacific, most of them in Australasian waters. It will never be known how many whales were killed between 1780 and 1850, but at the end of this period the stocks were so depleted as to make whaling unprofitable. Only the waters around icebound Antarctica had still to be prospected—but these, in time, were to yield the greatest riches of all.

Whaling: big business

The era of modern whaling began in 1864, when the Norwegian Svend Foyn invented the harpoon gun, and steam-powered whaling ships were introduced. Both revolutionized the industry. Steam power made the hunting of rorquals—the blue, fin, and sei whales—a practicable proposition. Rorquals are too fast and maneuverable to be hunted in the open sea by sailing ships; moreover, they do not usually enter coastal waters to breed and calve. But the faster steam-powered whalers were not limited by wind power and could penetrate the pack ice, where the rorquals gathered in their millions. Steam power also helped the whalers in another, equally important way: with the aid of compressed-air pumps powered by ships' engines, they could inflate the carcasses of dead rorquals and so keep them afloat. (Right whales remain buoyant when dead, but rorquals sink to the bottom.) Modern whaling ships, displacing up to 1000 tons and capable of 16 knots or more, are powerful enough to tow several of the largest rorquals to the factory. The harpoon gun is capable of killing a whale with one shot. The harpoon, which weighs up to 170 pounds, has an explosive shell in its head and is fitted with hinged barbs that prevent its withdrawal after it has penetrated the whale.

Until the late 1920s the Antarctic whaling industry was based on land factories, mainly on South Georgia and the South Shetlands. The last, and most decisive, step in whaling occurred in 1928, when the first factory ships appeared in Antarctic waters. The whaling fleets now became completely mobile. Both the catcher ships and the pelagic factories could follow the whales wherever they went, and the entire fleet could remain at sea for a year or more. A modern factory ship, which may displace 25,000 tons, contains a plan, or flensing platform, onto which the whale carcasses are winched via a slipway in the ship's stern. Here the whale-oil (used mainly in the production of margarine) and other valuable products are extracted and the waste dumped overboard. The factory is the mother ship of a fleet that may include a dozen catchers, a transport ship (also used for storing the whale-oil), and a refrigerator ship. The fleet may also be equipped with helicopters or light planes to aid the search for whales.

Whaling: legislation

The introduction of pelagic fleets transformed Antarctic whaling. In the 1927-8 season, just before pelagic factories were first used, the Antarctic catch was less than 14,000 whales; by 1930-1 it had jumped to more than 40,000; and from 1934 until the outbreak of World War II the average catch exceeded 30,000 a year. Since the 1940s there has been a steady reduction in the once-enormous stocks of Antarctic whales, in spite of regulations introduced by the International Whaling Commission to limit the size of catches. Not every country has been prepared to abide by these regulations. The Norwegians, Russians, and Japanese have consistently over-killed, their justification being the enormous capital invested in their fleets.

By 1963, blue whales (the largest animals this planet has ever known) were so scarce that they had ceased to be of commercial significance. They are now completely protected in Antarctic waters; but because they are probably fewer than 2000 left, it may be too late to prevent this species from dying out. Fin whales, which replaced blue whales in economic importance, have also been over-killed. Whereas there were estimated to be more than 110,000 fin whales in the mid-1950s, the stock had dwindled to less than 35,000 by 1963. The stock of humpback whales (another whalebone species), never as great as that of either blue or fin whales, has steadily declined since the 1940s. By 1962 one estimate put the stock at 600, and since 1963 the species has been protected throughout the Southern Hemisphere. It is likely that it will take at least 50 years for the humpback stock to recover sufficiently to make carefully regulated catching an economic proposition.

The Antarctic whaling fleets of the world have effectively cut their own throats. In spite of warnings from scientists over the last 60 years, they have plundered the stock of whales wherever they have managed to catch them. In spite of regulations, and the protection of several species, stocks continue to decrease year by year. Between the seasons 1957-8 and 1965-6, the harvest of whale-oil gathered by pelagic fleets in Antarctica dropped from 313,000 tons to 76,000 tons—a fall of more than 75 percent. During this period such great whaling nations as Britain, the Netherlands, Australia, and New Zealand stopped sending ships to Antarctica. They found it was no longer worth the journey. Technology, aided by greed and stupidity in equal parts, had put them out of business.

Man the hunted

As a footnote to this chapter of ecological disasters, we should remind ourselves that man is the only animal species that is *predatory* against itself. Men have never scrupled to kill other men in order to safeguard their "interests"; and our ingenuity in developing new and ever more potent means of mass extermination is rivaled only by our gift for deceiving ourselves about our motives for doing so.

We are even capable of murdering humans for sport—during the 16th century, the Spanish conquistadors hunted the native Arawaks like game animals in the forests of Jamaica; and the early English colonizers of Australia often treated the Aborigines in much the same way. Today we consider ourselves more civilized. We shrink with horror from the brutal murder of an individual—though we are still able to find convincing reasons and justifications for engaging in genocidal wars.

Throughout human history, wars have been fought for an enormous variety of reasons. Some, certainly, have had strictly ecological justifications—at least for the victors. Even the savage Mongol hordes that descended on the dry-belt farming civilizations (see Chapter 10) were striving to establish themselves at the head of a new food chain. Lacking sufficient pasture for their livestock, they needed to deprive others of their source of primary producers in order to survive. In the past, most major wars have had important secondary ecological consequences—disease, pestilence, and starvation. We shall consider some of these in Chapter 16, but meanwhile we can note that, in terms of misery and loss of life, these secondary consequences have often been more devastating than the wars that caused them.

*The potato and its pest, the Colorado beetle, are
New World natives. The beetle spread from its
home in the American Southwest as a result of
potato cultivation in North America and Europe.
It reached Western Europe during the 1870s.*

15
Animal and Plant Travelers

Since very early times, man has taken other organisms with him on his travels and migrations. Some he has taken deliberately, such as domesticated livestock and the seeds or shoots of cultivated plants; to others he has unwillingly offered a lift—rats, mice, parasites, and disease organisms being among the least welcome of these. One of the earliest examples of animal imports was the dingo dog, which the Aborigines brought with them when they colonized Australia perhaps 15,000 years ago. Other early travelers were dogs and pigs that the Polynesians brought with them to the Society Islands. But the greatest importers of fauna and flora have been Europeans, who since the days of the great navigators in the late 15th century have purposely introduced both plants and animals into new habitats all over the world.

The survival value of an exotic species in a new habitat depends on many factors—chiefly, of course, the availability of food and the presence or absence of predators and parasites. If food is too scarce (or food-getting too competitive), or if predators are too numerous, the exotic species will not survive. If food is abundant and predators are absent, it may rapidly multiply to plague proportions and disrupt the life of every organism (including man) in the habitat. Even if the balance between food and predators is just right, the import may produce important secondary ecological consequences. A good example of this is the introduction of the horse into North America by Spanish explorers in the 15th century. Many of these animals escaped, and they bred and prospered in the wild. Later, they were domesticated by the Plains Indians, and the culture of these bison-hunting peoples was profoundly altered by the greater mobility conferred upon them by the horse.

A plague of rabbits

The European rabbit is probably man's most notorious deliberate importation. This species is a native of the western Mediterranean region, and probably reached England in the 11th or 12th century. It was valued for its flesh and fur, and became a pest in England only in the late 19th century. Oddly, in view of later events, many unsuccessful attempts were made to introduce the rabbit into Australia from England, and it was only in 1859 that the animal established itself. Thereafter, the rabbit population multiplied at a terrifying rate, and within a few decades it had spread over two thirds of the continent. The reasons for this population explosion were the absence of natural predators (apart from the dingo dog) and the rabbit's rate of reproduction (it breeds up to eight times a year, and each litter may contain as many as eight young). The introduction in 1868 of the European fox, which multiplied rapidly, had disastrous effects on the indigenous bird and marsupial populations, but did little to halt the spread of rabbits.

The rabbits threatened not merely the prosperity but the very existence of Australian sheep-farmers. Many of the largest sheep stations then, as now, were on the arid steppes in the inland regions of New South Wales, South Australia, and Western Australia. In all these regions the rainfall is sparse: it rarely exceeds 15 inches a year, and is often as little as 1 or 2 inches. The vegetation typically consists of various shrubs interspersed with patches of native grasses—wiry perennials that are briefly supplemented by annuals after the rains. This vegetation offers the sheep a living—but only just. In areas that have been invaded by rabbits, however, overgrazing has inevitably occurred. The grasses have often disappeared and in drought years both sheep and rabbits have starved. In many of the greatest sheep-farming regions, the size of flocks has declined rapidly in the last 80 years. In western New South Wales, for instance, the sheep population fell from 15 million in 1891 to 7 million in 1950.

The rabbit disease myxomatosis was deliberately introduced in 1950 and within a few years had reduced the rabbit population by 80 or 90 percent. There was hope that the disease would permanently keep the population in check. This hope is now fading, however: the rabbits are becoming more resistant to the virus, and it is evident that direct control measures will still have to be used on a large scale.

Much the same thing happened in New Zealand, where rabbits became established in 1864. The second half of the 19th century was in fact a period when a vast number of animal and plant species were introduced into new habitats for sport or simply for pleasure—to satisfy the desire of immigrants to be surrounded by things that reminded them of home. It is a rather grim irony that the first rabbits introduced into New Zealand were nurtured by acclimatization societies, whose members welcomed the animals as pets, for sport, and for the value of their skins.

Another import into New Zealand during the 19th century was the Scottish red deer. Climate and the absence of predators enabled them to multiply rapidly, and within a decade they had become a pest. They were driven off the most valuable valley farmlands and established themselves in the more inaccessible hills of North Island and in the Alps of South Island. Here they overgrazed the native grasses on the steep slopes and were directly responsible for erosion on much land that was needed for sheep pasture. Eventually, the government appointed cullers to keep the deer population in check.

Since World War II, however, this purely destructive answer to the deer problem has been modified. It has been discovered that there is a ready market for venison in the wealthier countries. In 1965 a New Zealand government committee recommended that deer-farming should be organized on much the same lines as conventional cattle-farming, with the proviso that deer ranches should be confined to areas where the animals were abundant and where the land was not vulnerable to erosion. Several deer ranches are now established and the scheme looks like being a success. There are plans to extend the idea to include other pests such as hares, wild pigs, and wild goats.

Remedies: trials and errors

Almost all vertebrates imported into new habitats have been introduced deliberately. The most obvious and important examples, of course, are domesticated livestock. Others, as we have seen, are introduced for sport; still others are imported for purposes of biological control—that is, to reduce or stabilize the populations of indigenous or other imported species that man regards as pests. In 1872, for instance, the mongoose was introduced into Jamaica from India in order to control the hordes of rats (another import) that had become a serious pest on the sugar plantations. The mongooses multiplied rapidly and within a decade they had so reduced the rat population

Plains Indians hunting bison. The Indians got their first horses from wild herds that had escaped from Spanish settlements in the Southwest. The Spanish had introduced the horse into the New World early in the 16th century.

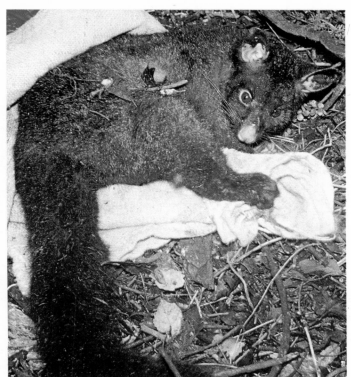

Left: reindeer, introduced into Alaska from Lapland in the 1890s, multiplied to more than half a million in a few years. Since then the number has declined catastrophically owing to overgrazing of lichen, their main winter food. Top right: brush-tailed opossum, imported into New Zealand from Tasmania for its fur, spread rapidly and now causes great damage to vegetation. Above: European carp, a 19th-century importation into America and Australia, ruined the river habitats of important native species.

that they turned, with equally devastating effect, upon several harmless native species of mammals and birds. In the Virgin Islands, to which the mongoose was introduced from Jamaica in 1884, much the same thing happened. But recently observers have reported that the rats have staged a comeback and are now almost as numerous as ever. It seems that the Virgin Islands rats have adapted to the threat to their existence by building their nests high off the ground, in trees and buildings, where the mongooses cannot get their young. Meanwhile, throughout the Caribbean and in the Hawaiian Islands, the mongoose is itself now regarded as a serious pest.

Another vertebrate introduced into several tropical areas for biological control is the fish *Gambusia*, an insectivore with a taste for mosquitoes. The species has been put into many freshwater lakes, ponds, and backwaters with the aim of controlling the malarial mosquito. The idea is perfectly sound, though there is as yet little evidence to suggest that *Gambusia* has noticeably reduced the incidence of malaria outbreaks in any area. Many other species of freshwater fish have been introduced into rivers and lakes for sporting purposes. The most notorious example of accidentally introduced animals is the lamprey, which has seriously depleted the population of game fish in the American Great Lakes. The cause of introduction was almost certainly the completion in 1932 of the enlarged Welland Canal, which links lakes Ontario and Erie and bypasses the barrier of Niagara Falls.

Pests and parasites

With the exception of certain insect species used for biological control, and "domesticated" insects such as the honeybee and silkworm, almost all *invertebrate* importations have been accidental. The chief imports are those insects and other arthropods that commonly live in some kind of association with man—house spiders, ticks, lice, houseflies, mosquitoes, and many others—and those that are pests or parasites of his cultivated plants and domesticated livestock.

Accidental importation is always likely, because most arthropods are small and easily overlooked. Today, the opportunities for importation are greater than ever, owing to the speed of modern transport. An aircraft may carry an undetected insect halfway around the world in a few hours. Not so long ago, however, the chances of the same insect surviving a long sea voyage would have been much less good. Speed, moreover, is not the only factor. The sheer density of traffic plying scheduled land, sea, and air routes offers greater opportunities for "illegal immigrants" than ever before. During the period 1937-47, almost 29,000 of some 80,700 aircraft inspected by the United States Public Health Service contained various species of arthropods. In spite of rigorous quarantine inspections, there is still a great danger of potentially harmful invertebrates migrating from one country or continent to another.

One of the most famous examples of this occurred in 1930 when an entomologist in Natal, on the northeastern coast of Brazil, discovered a malarial mosquito, *Anopheles gambiae*, which is a native of Africa and had hitherto been unknown in Brazil. Within a year, *A. gambiae* had spread 115 miles along the coast, and by 1938 it was so abundant that it set off a malaria epidemic that killed 14,000 Brazilians in six months. The Brazilian government launched a campaign to eradicate the mosquito and, to the surprise of many experts, finally succeeded in 1940. *A. gambiae* was undoubtedly an importation. Exactly how it was introduced has never been known for certain.

Biological Control

Insects have scored some notable successes in biological control. One of the earliest and most famous examples of the method was the campaign against the cottony cushion scale, a pest of fruit trees. This small insect was accidentally imported into California from Australia in the 1860s and within a few years threatened the output of citrus orchards over vast areas of the state. All the available insecticides were useless against it because the insect's waxlike covering is highly resistant to spraying. The problem puzzled entomologists because in its native land the scale was not a serious pest. Evidently, in Australia its numbers were kept in check by a local predator unknown in California. The United States Department of Agriculture sent an entomologist to Australia to identify the predator, which turned out to be a small ladybug, *Novius cardinalis*. About 140 of these beetles were despatched to California, where they were released onto an infected orange tree that had been enveloped

Opposite: sea lampreys attacking trout.
Lampreys established spawning runs in lakes
Huron and Michigan during the 1930s. Since
then they have multiplied enormously throughout
the Great Lakes and have decimated the
populations of lake trout.

in a gauze tent. The beetles rapidly ate all the scale insects. Later, they were released from the tent, and they spread throughout the citrus areas of California. Within two years, the scale insect had been brought under control, and by 1890 had ceased to have any economic significance as a pest. Since then, the ladybug has traveled far in the service of man, and has dealt with the scale problem in many countries. Its success has prompted the use of other insects against other pests, with variable results. The U.S. Department of Agriculture has to date imported about 500 parasitic species for biological control. About one fifth of these have become established—that is, they have succeeded in bringing their target pests at least partly under control.

Most farmers are plagued, now and then, by insects that eat or contaminate their crops; most insects, after all, are herbivorous. Early in the days of biological control, entomologists began to consider the possibility of suppressing useless or harmful plants (in other words, weeds) by the planned importation of insects known to feed on them. The dangers of trying this were obvious: unless the chosen insect species were highly selective in their diets, they might destroy not only the weeds but also any cultivated plants they were intended to protect. In spite of this danger, the method has had some success.

One of the most persistent and widespread weeds overcome by this method was the prickly pear cactus, *Opuntia inermis*, which was introduced into Australia from its native South America in the late 18th or early 19th century. During the next 100 years the cactus spread with ever-increasing momentum over

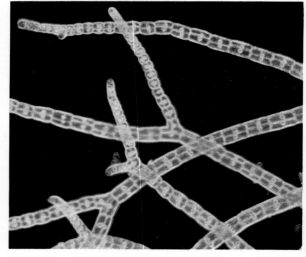

Left: Asparagopsis armata and (above) its asexual phase, when it attaches itself to other marine plants. A native of the south coast of Australia, this seaweed was accidentally introduced (probably on a ship's hull) to French waters in 1923, and has spread along many coasts of Western Europe and the western Mediterranean.

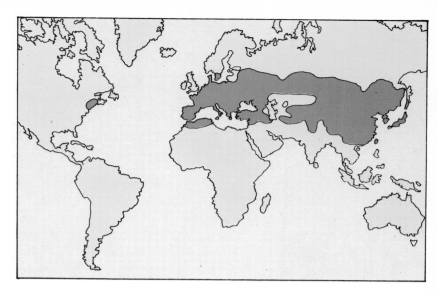

large tracts of good farming land in Queensland and New South Wales. By 1925, at the peak of the infestation, it covered some 60 million acres of land, and was spreading at the rate of 1 million acres a year. A special Prickly Pear Board was set up by the Australian government, and entomologists were sent to various countries to discover any species of insects that used these, or related, cacti for food.

Some 145 species of insects were collected, including 56 moths and 50 beetles. All these species were tested on the pear. Many that ate the pear had to be rejected because they also ate important fruit crops such as tomatoes, apples, and peaches. Of 18 species finally selected for extensive trials, one quickly showed itself a formidable opponent of the cactus. This was *Cactoblastis cactorum*, a moth native to Argentina. The female lays its eggs in long "chains" that it attaches to the spines of the cactus. When the caterpillars hatch, they burrow into the cactus and devour its soft internal tissues. The damage caused by the caterpillars is often intensified by bacteria and fungi that develop in the exposed tissues.

In 1926, $2\frac{1}{4}$ million eggs were attached to cactus plants throughout the two states; the following year, a further $6\frac{1}{2}$ million were attached. By 1929, thousands of acres of land were strewn with dead cacti; millions of moths perished in the absence of their food supply. But the hardy cacti staged a recovery, new shoots springing up from the roots of partly eaten plants. By 1933, the situation was almost as bad as in 1925. However, the moths also made a comeback: as the cacti spread, so too did the rapidly multiplying population of moths. Finally, in 1935, almost all the cacti in Queensland and New South Wales were totally destroyed, roots and all. Today, with few cacti available, the moth population has shrunk to only a tiny fraction of the 1935 figure. But it has shown that it is capable of increasing with extraordinary rapidity if the cacti begin to spread, so that plant pest and insect exercise control over each other, to the benefit of man.

Man's reasons for importing plants, other than food crops, into new habitats are many and various. In 1787, for instance, a Brazilian cactus related to the prickly pear that was later to become such a pest was introduced into Australia by the governor of New South Wales. The cactus was intended as food for cochineal insects, imported at the same time, which were to provide the red dye needed to color the tunics of the colony's soldiers.

Hitching a ride

It is quite impossible to make a realistic guess at the number of plant species that have been unwittingly introduced by man; such accidents are even harder to prevent than the inadvertent importation of insects. Sir Edward Salisbury, a former director of the Royal Botanic Gardens at Kew, once raised no fewer than 300 individual plants from the dust he found in his trouser cuffs. It is not too fanciful to suppose that such European weeds as couch grass, convolvulus, groundsel, and thistle all reached America in the clothes and baggage

Slavery, in which humans are reduced to the status of a commodity, provides another example of introduced species. Tens of millions of Africans were exported to the Americas between the 16th and mid-19th centuries. Officially outlawed throughout the world, slavery persists in the 20th century. These two Ethiopian slaves were photographed in the late 1920s. Even today slaves are still imported into Arabia from Africa and parts of Asia.

of the passengers aboard the Mayflower. Moreover, many accidental importations that we do not normally regard as weeds provide extremely difficult problems of detection to botanists and ecologists. Some of the world's grasslands were quite accidentally created by man after he had cleared away the forests. It is almost certain, for instance, that the grasslands of California are due to early Spanish settlers.

Finally, we should note that one of the most significant introduced species is man himself, in the guise of slave. Many past civilizations have practiced slavery, but never on such a scale or with such total disregard for humanity as did the European slavers—mainly Portuguese, English, Dutch, and French—from the 16th to the 19th century. In this period, at least 15 million slaves arrived in the Americas from the Guinea coast of Africa alone; probably another 4 million failed to survive the appalling conditions on the voyage across the Atlantic. The ecological effects of slavery were profound. In the Americas, including the West Indies, it made possible the transformation of vast areas of land into plantations, mainly of cotton, tobacco, and sugar (the last having been imported from India). In Africa, the systematic draining away of men, women, and children led to the collapse of tribes and even nations, and left deep cultural scars that remain to this day.

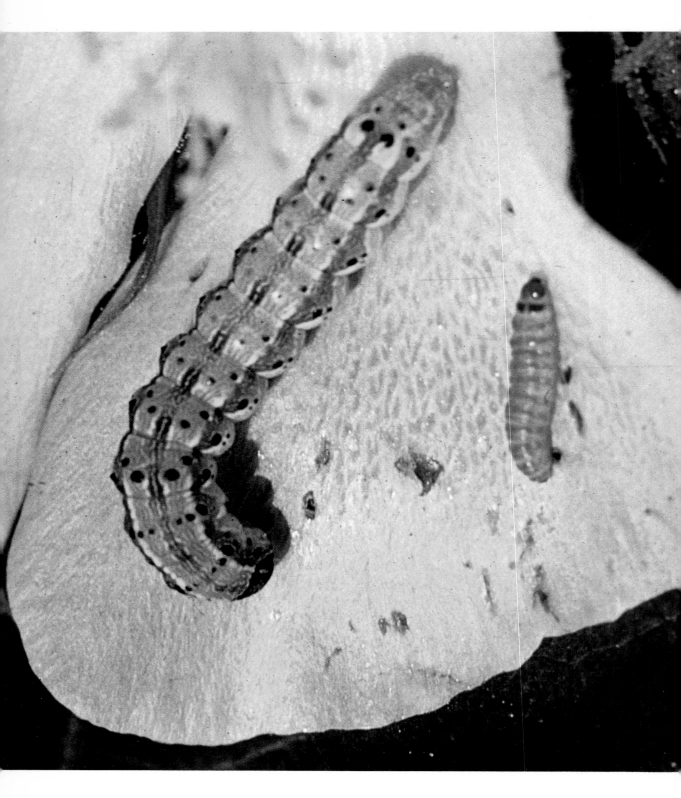

Larvae of (left) cotton boll weevil and (right) pink bollworm moth on a cotton flower. Both types of pest are capable of destroying the cotton harvests of whole regions by eating the seeds in the bolls and preventing growth of the fibers.

16
Pests and Pestilence

Insects are the most numerous animals on our planet. They are vital to nature's economy, for many of them help to fertilize flowers by transporting pollen from one plant to another. They are also valuable as scavengers, and many species eat dead animals, including mammals and birds. Some insects, such as the honeybee and silkworm, are directly beneficial to man; others have been exploited for biological control. Yet we commonly regard insects in general as being enemies of man—and, in the case of many species, with good reason. For insects also steal or spoil our food, and they are instrumental in spreading many of our deadliest diseases.

The interesting thing is that insects have become serious pests only since the dawn of agriculture. Man has, in fact, unwittingly brought his insect problems upon himself. As destroyers of human food, insects made their mark only when man began to set aside particular areas of soil for particular crops: a wheat pest, for instance, is likely to multiply to plague proportions only if it is confronted by a large acreage of soil devoted entirely to wheat. As spreaders of disease, the insects' opportunities really began when man started to concentrate in urban communities. Overcrowding, and the appallingly unhygienic conditions of urban life in its early days, were prerequisites for the spread of many of our worst epidemic diseases.

Apart from locusts, which have indiscriminately ravaged cultivated and wild-growing plants for thousands of years, crop pests have become a serious world economic problem only within the last two centuries. Before then, most agricultural areas of the world were made up of relatively small farms that raised a variety of crops. Because most crop pests attack only one type (sometimes only one particular species) of plants, insect depredations were usually local affairs—serious for individual communities but not, usually, for whole regions. But since the early years of the 19th century both the techniques and the economics of farming have changed. One of the most important of these changes is that, in many regions, vast areas of land have become devoted to the cultivation of a single crop. A familiar example of this is the great wheatlands of the United States of America, Canada, and the Soviet Union. Another is the cottonlands of the American southern states.

The latter is a classic case of the destruction and economic ruin that an insect pest can bring to whole regions. By the late 19th century, cotton plantations occupied more than a quarter of the available cropland in the Carolinas, Georgia, and Alabama. Cotton was also the principal crop in the flat, alluvial soils of the Mississippi basin in the states of Mississippi, Louisiana, Tennessee, Arkansas, Missouri, and Kentucky, and there were other important cotton-growing areas in Oklahoma and northern Texas.

The villain of this story is the boll weevil, which was first reported, in Central America, during the late 19th century. It entered the United States in 1892 and within two years had spread over large areas of southern Texas. The weevil continued to spread northward and eastward until, soon after World War I, it infested the whole of the cotton belt. The insect feeds and lays eggs both on the cotton flower and inside the boll; or seed case. Within the boll, the larvae eat the seeds and so prevent development of the slender threads (cotton fibers) that are borne by the ripening seeds.

In attempts to eradicate the weevil, farmers found that the available insecticides were almost useless against larvae within the bolls. Moreover, the period from the larval stage to the fully adult stage, when the weevils breed, is often less than three weeks. Thus, in any one season, the bolls could be infested with several generations of weevil larvae. Throughout the 1920s the weevil took an ever-greater toll of the cotton harvests. And since the economy of the southern states was founded almost entirely on this one crop, the pest affected the lives of almost everyone in the region, from bankers to farm laborers. Hundreds of farmers went bankrupt, and many thousands of Negro plantation workers were forced to migrate northward in search of work.

The pest was finally defeated not by science but by its own destructive effects. Most of the smaller farmers ceased to raise cotton and turned to other crops. Within a few years the economy of the south was transformed. Although cotton was still grown by many farmers, the plantations were separated from each other by large areas of arable land on which other crops, not vulnerable to the weevil, were raised. This helped to limit the spread of the pest. In addition, only the best soils were used for cotton growing, and new, early-maturing strains with a high yield were developed. Finally, thorough cleaning away of old plants after harvesting was enforced, so that the weevil was denied one of its most important places for winter hibernation.

Insecticides

The development of insecticides, like that of weedkillers and fertilizers, is one of the most important products of the Industrial Revolution. There is no question but that insecticides have enormously increased the effective yields of crops all over the world. Insect pests can reduce our total agricultural production by at least 10 percent; for certain crops, or in certain years, the percentage has been much higher and has led to famine. Damage is not confined to crops growing in the field. In 1949, for instance, about half the

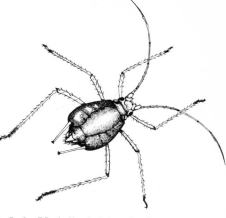

Left: Vedalia ladybug beetle eats fluted-scale pest of orchards—an example of biological control. Above: Myzus aphid, which transmits over 30 viruses to plants, including potato and tobacco mosaics. Below: sterilizing screw-worms (a cattle pest) with gamma rays. On release, the insects will mate, but produce only sterile eggs.

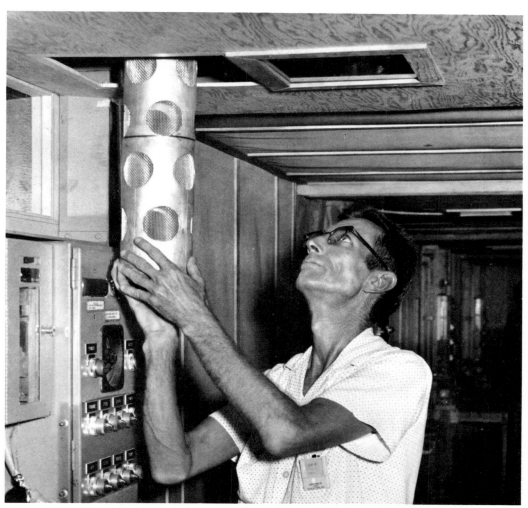

corn crop of Honduras was destroyed in storage by various pests; in India at least a million tons of stored grain are lost every year.

DDT, the first really effective man-made insecticide, was first synthesized in the 1870s, but its insecticidal properties were discovered only in 1939. Since then many others have been developed—some as a consequence of research into chemical warfare. Before DDT became commercially available, farmers had used insecticides from naturally occurring substances, including arsenic, strychnine, copper, zinc, lead, pyrethrum (from plants of the chrysanthemum family), nicotinic acid (from plants related to tobacco), and rotenone (from legumes such as derris and barbasco). But the new man-made insecticides have much greater biological potency than the natural poisons, and their effectiveness is constantly being increased.

One of the principal problems with insecticides, and the main source of public unease about their use, is that they are with few exceptions unselective: they are poisonous not only to the pests they are designed to destroy but to other species as well. For instance, a spray used to destroy corn pests such as the corn borer may kill many other harmless insects that normally inhabit corn fields. Often, an insecticide will kill not only the pests but other insects that prey on them. In tests carried out on plots of alfalfa in Washington state between 1956 and 1959, it was discovered that DDT was actually *more* toxic to the predator of an alfalfa pest than to the pest itself.

The use of insecticides has, however, profounder dangers—dangers that Rachel Carson dramatically brought to public notice in her book *Silent Spring*, which was first published in 1962. The principal danger is that many insecticides retain their toxic properties over a very long period, in some cases for several years. This is especially true of the chlorinated hydrocarbon poisons such as DDT, DDD, aldrin, heptachlor, endrin, and others. Thus, although TEPP (an organic phosphorus insecticide) is probably the most poisonous of all agricultural chemicals—one ounce is sufficient to kill 500 people—it is probably less dangerous to non-pests than the chlorinated hydrocarbons because it breaks down rapidly and loses its toxicity within a day or two.

The sustained toxicity of chlorinated hydrocarbons is hazardous not only to pests but to all animals, including man, that are part of the same food chain. The hazard has been exemplified by comparing the effects of DDT with that of the drug aspirin. Both have about the same toxicity, and it is well known that an overdose of aspirin tablets can kill a person. There is a striking difference, however, in the effects of long-term exposure to each. Most people can take doses of up to 20 grains of aspirin per day for many days without harm, because the drug breaks down and is eliminated by the body. A certain proportion of DDT, however, remains in the body, where it concentrates in the fat tissues, so that even small intakes, over a period of time, have a cumulative effect.

Owing to this cumulative effect, the concentration of a chlorinated hydrocarbon poison increases at every stage in a food chain. In 1949 an attempt was made to eliminate various species of gnat, mosquito, and midge that had

Many insecticides are capable of harming, and
even killing, a wide variety of animal species, and
their persistent toxic residues are capable of being
passed along a food chain, building up the
concentration of poisons at each stage. In this
photograph, taken at the Nature Conservancy's
Experimental Station, near Huntingdon, a
postmortem is carried out on a barn owl.
Chemical analysis shows that the bird's liver
contains significant quantities of residues of DDT
and dieldrin, both of which are powerful organic
chlorine pesticides. The probable source of both
poisons is grain, treated with pesticides, and later
eaten by mice that, in turn, fell victim to the owl.

become serious pests at Clear Lake, California. DDD was applied to the water of the lake at a concentration of about 0·015 parts per million (DDD was preferred to DDT because the latter is known to be more toxic to fish). The midges were almost entirely wiped out, and it was only in 1953 that the DDD treatment was repeated in response to another build-up in the pest population. After the 1949 campaign, investigations showed no apparent harm to the fish or other wildlife dependent on the lake. In 1954, a year after the second application, the situation changed dramatically: the lake and shore area became littered with the bodies of hundreds of western grebes—fish-eating birds that are abundant in this area.

A full investigation showed that the grebes were at the highly toxic end of a local food chain. Plankton in the lake were found to contain about 5 parts per million of DDD. Small fish that eat the plankton had about twice as much, while larger fish contained considerably more. The most extraordinary findings, however, concerned the grebes: many of them contained concentrations of as much as 1600 parts per million of DDD. Moreover, the actual death-count of birds, high as it was, did not reveal the true extent of the loss to the grebe population, for it is known that a significant fraction of the survivors of 1954 failed to breed.

It is clear that farmers and nurserymen need to know much more about the possible side-effects of using certain insecticides, if only to safeguard their own interests. In Britain, for instance, persistent spraying of fruit trees with DDT during the last two decades has had curious ecological consequences in some areas: it has encouraged the development of a new pest, in the shape of the red spider mite. These small mites have always been present on fruit trees, but, until about 20 years ago, in insufficient numbers to cause much damage. Evidently DDT has wiped out many of the insects that prey on the mites. Moreover, some entomologists believe that low concentrations of DDT actually stimulate the hatching of the mites' eggs.

The public concern roused by the ecological side-effects has led to the prohibition of some insecticides and to much greater care in the selection and use of others. Chemical manufacturers are continually developing new insecticides that are more selective and have less-persistent residues. Much ecological damage can also be prevented by careful application of these poisons—notably by treating plants with only the minimum quantity of insecticide necessary, and by avoiding contamination of the soil and other parts of the immediate environment. Care of this kind (taken, no doubt, for economic reasons as much as for ecological ones) at least partly explains why the insecticide damage to wildlife has been proportionally much less in Europe than in the United States. When all is said and done, insecticides are here to stay; they are vital aids in the battle to produce enough food for hungry peoples. We cannot be surprised if the undernourished Asian peasant is indifferent to the fact that an insecticide poisons a few wildlife species if, at the same time, it helps to save him and his family from starvation.

Man and disease

Man's battle against infectious diseases has been a long and frightening one. Until quite recently he had no prospect of preventing or curing them, for the good reason that he had no idea what caused them. For one thing, the disease organisms seemed to be invisible. We know now that many of these organisms cannot be seen with the naked eye, while some are too small to be seen even under a normal microscope. Most of the deadly infectious diseases of man are due to microorganisms such as bacteria, fungi, viruses, protozoa, and rickettsias. Others are due to various species of worms. Most of the disease organisms are *parasites*, that is, they live on or in another organism (a *host*), depending on the latter for food and a suitable environment for existence.

Most animals contain microorganisms from one or more of the above groups. Bacteria, for instance, live in the gut of cattle and other ruminants and help to break down the cellulose in the host's grass diet into sugars that are easily digestible. Humans harbor abundant bacteria and viruses; the functions of many are still unknown, but most of them are harmless if the host is in normal health. Even the streptococci that inhabit our throat and nasal passages do us no harm until we are attacked by the common-cold virus.

Although a parasite may harm its host, it is clearly in the parasite's interest that the host survives their association. Indeed, in the long run, evolutionary and ecological pressures tend to force both host and parasite species to adapt to each other—though many individuals on both sides may succumb.

In evolutionary terms, man is, perhaps, still in the middle of this journey toward accommodation with disease parasites. In ecological terms, however, he has made the journey much harder for himself. By the development of agriculture he has brought himself into association with "new" parasites against which he has no natural defenses; by crowding into cities he has created conditions in which these parasites can spread with terrifying speed.

In the host, new parasites cause harm in various ways. They may release toxins, which poison the host's tissues; or they may multiply so rapidly that they overwhelm the host's body processes by weight of numbers. The host has natural defenses against many disease organisms. It is able, for instance, to produce *antibodies*—substances that attack the parasite or its toxins. Often these antibodies last only as long as the host is exposed to the parasite. Sometimes, however, they continue to be present in the body for years, in which case the host becomes immune to the effects of the particular disease organism. Populations that are regularly exposed to certain disease organisms tend, over several generations, to acquire immunity (though the diseases may affect some individuals more than others). This is why measles, for instance, is no longer regarded as a serious illness in the northern temperate zone. Yet many Polynesians and Eskimos, lacking previous exposure to the disease, died when Europeans accidentally introduced measles to these peoples.

Natural immunity to diseases may be undermined by malnutrition or general ill health. In parts of West Africa diseases and malnutrition combine to produce a vicious circle in which local populations have been helplessly imprisoned for hundreds of years. In The Gambia, for instance, malaria (due to a protozoan parasite spread by mosquitoes) and hookworm (which enters the body through the skin) are endemic to many rural communities. The malaria mosquitoes breed soon after the spring rains—the time when food is scarcest and most laborious work has to be done in the crop fields. Many of the farmers, made listless by new malarial infection, abandon distant fields and restrict the crop areas to small patches of soil around their dwellings. But here sanitary conditions are bad: the ground is contaminated by human and animal excrement, and this greatly encourages the spread of hookworms, which also weaken the people and make them more vulnerable to malaria and other diseases. Under these circumstances, the choice for the people is stark: if they fail to grow enough food, they starve; if their meager crops enable them to survive, they may hang on in a wretched debilitated state for several generations. This is, indeed, one of the reasons why swidden farming (Chapter 11) has persisted in parts of the wet tropics: the usual population pressures, which elsewhere have disrupted this mode of farming, have been kept in check by malnutrition and disease.

Malaria is one of the oldest tropical diseases known to man, but its importance dates from the beginning of agriculture. We know this because in Africa the spread of the disease is closely connected with the spread of farming peoples. The insect that transmits the disease in Africa is *Anopheles gambiae*—the most dangerous of all the malarial mosquitoes, owing to its marked preference for human blood. *A. gambiae* requires stagnant water and the absence of shade in order to breed. The early farmers, migrating westward across Africa, provided ideal conditions for the mosquito. By clearing the forests they removed the shade afforded by the trees and deprived the soil of much of its capacity to absorb water. Malaria is the great killer in the tropics—in India alone about one million people die from it every year. But, historically, it may be more important as a weakener of peoples than as a killer.

Diseases of the temperate zone

A good case can be made from the argument that malaria and other tropical diseases have been the main force preventing tropical peoples from making a social and technological breakthrough equivalent to that of the civilizations of the temperate zone.

This is not to say that the temperate zone, especially Europe, has been free from devastating diseases: far from it. But, unlike malaria, sleeping sickness, and bilharziasis (a disease of the dry tropics), the diseases of the temperate zone have usually struck swiftly, caused great havoc for periods of a few weeks or months, and then died away. They have not persisted as a permanent feature of the civilizations that they have attacked.

Pandemics (that is, epidemics on a continental scale) came to the temperate zone as a consequence of the development of cities, international trade, and

Antimalarial spraying with DDT in northwest Argentina. A campaign to eradicate the disease-carrying Anopheles *mosquitoes begun by the World Health Organization in 1955, includes preventive measures such as this even in regions where malaria is no longer a serious threat.*

war. The comings and goings of merchants and soldiers helped to spread them; the appallingly insanitary conditions in which both city dwellers and soldiers lived increased their virulence.

The Black Death

Plague, a disease caused by a bacterium, shows how some diseases thrive on urban overcrowding and the breakdown of social life. Originally, plague was probably a disease of the gerbil family of rodents; today, at least 50 species of rodents in various parts of the world are known to carry the disease. The bacterium is transmitted by a flea common to rodents. During the Christian era, the great epidemics and pandemics of plague have been associated with the black rat, which migrated westward from central Asia and spread the disease among all the human communities through which it passed.

Anywhere that man concentrates in large numbers, be it a city or an army encampment, provides opportunities for rats to breed in large numbers unless ruthless measures are taken to stamp them out. The point is that, where large populations of rats exist, the plague bacterium is able to pass quickly throughout the rodent population. As a result, large numbers of rats die of plague, and the myriad disease-carrying fleas seek another host. They find it in the rats' unwilling meal-ticket—man. Once humans are infected, plague can spread in two ways: by the rat fleas and by droplet infection—that is, by healthy people inhaling the breath of infected ones. In crowded communities, the

disease spreads like a bush fire. About half the population of the Roman Empire died in a pandemic that raged between A.D. 542 and 594. During the Black Death, which spread into Europe from eastern Asia in the 14th century, about 43 million people died.

The eight centuries between these two pandemics were more or less free from major outbreaks of plague. The sixth-century pandemic had not only killed millions of people, but caused towns and cities to be abandoned, and had brought interregional trade to an end. In short, the plague bacterium, by disrupting the ecosystem in which it had thrived, sabotaged its future, at least temporarily. (An exception was Venice, whose rise to supremacy in international trade during this period was bought at the cost of more than 60 plague epidemics. Later, the Venetians, having learned their lesson, became the first corporate group to introduce quarantine laws). By the middle of the 14th century, however, conditions in Europe were again ripe for plague. The rapid rise in urban populations, coupled with food shortages, wars, and international trade, set the scene for the Black Death.

Typhus

Epidemic typhus, another deadly scourge of man, is interesting for the complex system of routes by which it has come to infect its human victims. Originally, the disease organism, one of the rickettsias, was a parasite of ticks, which later became immune to it. Then it passed, via the ticks, to wild rodents.

Right: clearing dense riverside vegetation for crop-growing in parts of tropical Africa disrupts the habitat of the tsetse fly, carrier of the sleeping-sickness parasite. The disease affects more than 10 percent of the population in some African states.

Opposite: dipping bags containing copper sulfate into an irrigation canal near the White Nile in Sudan. The poison kills the snails that spread bilharziasis, a weakening disease affecting millions of people in the dry belt of North Africa and the Near East. The development of water resources for irrigation and hydroelectric generation has greatly increased the spread of this disease.

Occasionally humans also became hosts at this stage, contracting the rare disease known as spotted fever. But the most important next link in the chain was the transmission of the disease from wild to domestic rodents, such as rats, by means of lice. Murine typhus, which lice transmit between rodents, is also transmitted to man by rat fleas. Finally, epidemic typhus spreads when human lice pass from an infected person to another person. This form of typhus is one of the more recent human diseases. The fact that the disease is fatal to human lice suggests that these have not yet had sufficient time to acquire immunity.

The typhus organism thrives on filth and is likely to spread wherever hygiene is disrupted for any length of time. At one time it was rife in prisons, whence the English custom for assize judges to carry a posy of flowers into court to guard themselves against infection. But typhus has had its greatest success as a wrecker of armies. One of the earliest typhus epidemics in Europe struck the Spanish army in 1490, when it was besieging the Moorish stronghold of Granada. The army lost 20,000 men—3000 killed by the Moors, and 17,000 by typhus. The Thirty Years' War, during the 17th century, established

Left: rats found in the kitchen of an Indian house being sprayed with insecticide to kill possible plague carriers. The successful eradication of plague depends as much on promoting sanitary living conditions as on use of vaccines. The point is emphasized in the clinically clean laboratory (opposite), where a technician is handling solutions containing kidney cells, on which new plague vaccines will be tested.

centers of typhus infection all over Europe; Napoleon lost enormous numbers of troops to the disease during the retreat from Moscow in 1812. During World War I, hundreds of thousands of soldiers died from typhus on the eastern front, mainly in Russia. During World War II the disease was responsible for a great many deaths in the German concentration camps.

Influenza

The poor health of populations suffering from the food shortages and social disruption of World War I undoubtedly helped toward the devastation caused by the influenza pandemic that lasted from 1917 to 1920 and spread over much of Asia and Europe. This, the last of the great pandemics, killed about 20 million people—more than the combined losses of all the armies during the war. Influenza is one of the most striking examples of a virus that is capable of developing new strains. Hardly a year passes without the disease causing local epidemics in many countries of the temperate zone. Often these outbreaks are due to new strains of the virus and people who have acquired

natural immunity are usually defenseless against the new strain. Immunity to this virus, and to most other disease organisms, is highly specific; moreover, immunity to influenza is usually short-lived. For this reason, there seems little prospect at present of bringing the disease under complete control.

Cholera

We have seen that the havoc caused by our most deadly diseases is due, in large measure, to man himself—to his desire to crowd together in large communities, and to his neglect in matters of cleanliness and hygiene. Plague and typhus are but two examples of this. Cholera, a bacterial disease spread when the excrement of infected people comes into contact with supplies of drinking water, is another. This disease escaped from its endemic focus in China and India in 1816, and in little more than half a century it spread through Asia and Africa to Europe and North America.

Poliomyelitis

These diseases have been conquered as much by improved standards of public hygiene (backed by laws) as by the development of drugs and vaccines to treat or prevent them. It is our hard-won appreciation of this that makes the case of poliomyelitis one of nature's more cutting ironies. This disease is

most common in countries where standards of hygiene are *high*; moreover, its high incidence is due directly to these high standards. The reason why virtue is so harshly rewarded is as follows. Although the disease virus attacks children of all ages, the first three years of life are a "safe period" for almost every child. If, during this period, a child is exposed to the virus, it will acquire immunity. In societies where hygiene is neglected, many infants—and often the entire population under three years of age—will become immune. In contrast, the infants of countries where hygiene and child welfare are high social priorities are much less likely to come into contact with the virus until after the safe period. They will usually be exposed to the disease for the first time at school, at an age when they are most likely to become seriously paralyzed. Fortunately, there are now vaccines against poliomyelitis.

The pests of our crops and the disease organisms that prey on human populations both pose similar ecological problems. The overcrowding of urban man, and his consequent vulnerability to diseases, are analogous to the farmer's "crowding together" of individual plant species in crop-fields, which enables pest populations to explode and destroy his food supplies. Today, we have reached a point where most infectious diseases have been (or can be) conquered. Yet this very success brings fresh problems. The most important of these—of overpopulation—we shall consider in Chapter 19.

Right: Indians on their annual pilgrimage are stopped at a checkpoint to show evidence of inoculation against cholera or to be inoculated before being allowed to proceed. Cholera is especially liable to spread if large numbers of people gather for many days in places lacking proper sanitary facilities.

Opposite: this Afghanistani boy was one of thousands dusted with DDT in a successful antityphus campaign sponsored by the World Health Organization during the 1950s.

Above: the original wild man of Borneo, an orang-utan whose name means "man of the woods" in Malaysian. It is the rarest of the great apes, threatened with extinction because of its popularity and charm. Despite stringent controls, many orang-utans are smuggled out and sold abroad every year.

17

Wildlife Gives Way

As our own civilization spreads across the earth, wildlife is often the innocent victim. Large numbers of animals have become extinct since records first began and the extinction process is known to be speeding up.

Many books on conservation begin with statements such as, "During the past 2000 years, at least 100 different forms of mammals have become extinct in various parts of the world," or, "Over the last 300 years we have exterminated more than 200 forms of birds and mammals." Claims such as these are rather meaningless in themselves and are usually intended to shock the reader. Throughout this book we have seen that man is largely to blame for the plight of other species, and that we have an obligation to prevent this plight from getting worse, if we can. But unless we understand how and why animals are threatened, we cannot hope to preserve wildlife.

Extinction

When a species consistently fails to produce enough young in each generation to keep pace with the death-rate, it eventually becomes extinct. The rate of extinction can be fast or slow, depending on the causes. There are, as we know, no hard and fast rules about either extinction rates or the probable life span of a species. But the following tentative calculation has been made: in 1680 a bird species could expect an existence of 40,000 years, but in 1964 its life expectancy would be only about 16,000 years.

This is yet another generalized statement. It is designed to make the reader aware that the probable life expectancy of a species (not of an individual) is greatly reduced today as compared with 300 years ago. The implication is that we are responsible. As we cannot argue against this, the best thing we can do is to set out the facts as they are known.

There are two kinds of extinction. One is by direct assault, the other is by indirect interference with an animal's habitat. First, we shall deal with extinction by direct assault.

For thousands of years we have killed animals for food and clothing, and indeed for sport. But in fact we can trace an ascending curve of slaughter that is directly related to technical improvements in sailing ships that enabled men to make extended voyages to all parts of the world, thus opening up new opportunities for killing, although still with crude weapons. But with the Industrial Revolution the curve of slaughter became even steeper, as mass-produced fire-arms and ammunition became available. Also, in the 19th century, one of the by-products of the Industrial Revolution was a sharp increase in the population of Europe, which in turn led to massive colonization of new areas, especially in North America. The invention of the locomotive and the steamship speeded up the slaughter in more ways than one, and today helicopters are used for spotting whales.

This is the technical background against which the carnage of recent years must be viewed; it provided means and opportunity, but what of the motive? In the following account of extinctions and near-extinctions, we make no attempt to be comprehensive, but only to show some of the ways in which these disasters have come about; for ease of remembering, however, we make use of a simple mnemonic—all the motives for killing begin with the letter F. They are Food, Finery (Fur and Feathers), Fun, Financial gain, and Fear. Sometimes these categories will overlap, but in general they are a good guide.

Killing for food

The target has normally been animals that congregated in such large numbers as to make commercial exploitation worthwhile; often their numbers were so great that no one believed that wholesale killing could reduce those numbers appreciably. The extinction of America's passenger pigeon and the near-extinction of the bison and the big whales (Chapter 14) are cases in point. More recently we have almost wiped out the various species of giant turtle.

Giant turtles are very primitive reptiles and they are found today on the Seychelles, Mauritius, and Aldabra islands in the Indian Ocean and the Galápagos Islands in the Pacific. They were much more widely distributed in the past, but they have long since disappeared from other parts of the tropics and subtropics. On these remote islands the turtles survived because they had no predators before man came. So numerous were they on the Galápagos Islands during the 16th century, that the discoverer of the islands, the Spanish priest and explorer Fray Thomas de Berlanga, named them after the turtles (*Galápago* is the Spanish for turtle). These islands were also called by the name "Encantadas"—the enchanted islands.

The sailors were most pleased with their discovery—the turtles were delicious to eat, and the English buccaneer and navigator William Dampier said of them in 1697: "The land turtles were so numerous that five or six hundred men might subsist on them alone for several months without any other sort of provision." The turtles' great misfortune was that, if kept damp and cool, they could survive without food or water down in the holds of ships for up to 21 months. Sailors used to collect 600–900 of the smallest and most succulent and take them aboard. The American Captain Porter, describing

The bas-relief (right) from
Nineveh, Iraq, shows that
lions were hunted for sport in
the time of Ashurbanipal,
king of Assyria. This
probably contributed to their
disappearance from the Near
East, where they were once
common. Below: polar bears
are menaced by modern
hunting methods, such as using
helicopters for spotting them on
ice floes.

Land, freshwater, and marine turtles are very primitive reptiles and of great interest to biologists. They are vulnerable as long as man continues to indulge his taste for them. The Galápagos giant turtle (left) was killed only for its liver; the rest of the beast was left to rot. Below right: the green turtle; its eggs are such a delicacy that they must be protected by being transferred to special hatcheries.

victualing operations at Galápagos during the early 1800s, said: "In four days we had as many on board as would weigh about fourteen tons. They were piled up on the quarter deck for a few days with an awning spread over to shield them from the sun, which renders them very restless, in order that they might have time to discharge the contents of their stomachs; afterwards they were stowed away below as you would stow any other provisions." The cook would simply kill them as required.

Charles Darwin found the Galápagos turtle still quite numerous when he visited the islands in HMS *Beagle* in 1835. He noted that they got their fresh water by eating spring cactus leaves, because the only fresh water on most of the islands was in the form of temporary pools. He was struck by the sight of "the huge reptiles, surrounded by black lava, leafless shrubs and large cacti, that seemed to my fancy like some antediluvian animals."

At about the same time as the Galápagos Islands were discovered, explorers landed on the Seychelles. There they found giant turtles living in similarly large numbers. The French explorer Lequat, writing in 1691, said of the Island of Rodriguez in the Mauritius group: "There are such plenty of land turtles in this isle that sometimes you see two or three thousand of them in a flock, so that you may go above 100 paces on their back." One hundred and fifty years later the giant turtle had been completely wiped out from Rodriguez and other islands of the group. Since man has settled on almost all other islands where the giant turtles live, they have become even rarer. They are still prized for their meat and of course for their eggs, and sometimes an animal is killed by piercing its shell simply in order to extract the liver, which is a special delicacy. The younger animals are pestered and killed by introduced dogs and rats, and they are in great need of protection. They have always been infrequent breeders, because they did not need to produce many young in an environment free from predators. But in this new situation, where there are predators on all sides, it is doubtful whether they can survive. Seychelles islanders keep a few as pets, and zoos are attempting to breed them in captivity. We shall be talking about this in the next chapter.

The eggs of many reptiles and birds are good sources of protein, and are much sought after. Not only land turtles suffer: freshwater and marine turtles, too, are hunted, both for their flesh and for their eggs, and turtle soup is a delicacy that is becoming more and more scarce.

Many birds are easy victims of the nest robber, because the birds themselves are almost always easy to shoot or to scare away from the eggs they are incubating. Take the case of the Laysan albatross, which breeds on the Hawaiian island of the same name. This is one of the many kinds of sea birds that breed on remote Pacific islands. Each female lays only one egg a year, which the islanders collect, carting millions of eggs away in trucks and barrows. This annual harvest brings the Laysan islanders a living; but, with no long-term planning for allowing at least a proportion of the eggs to hatch, how long will it continue to be profitable?

Because of their fearlessness, island-dwelling animals are often much easier to trap than mainland ones. The dodo (Portuguese *doudo*, meaning stupid) was a victim of its own fearlessness; sailors wrote incredulously of the way in which the dodos on Mauritius just sat and allowed themselves to be captured

and killed. The birds (and indeed the reptiles and other animals too) of the Galápagos and Aldabra islands are equally fearless, so that these islands are favorite places for organized egg-hunting. Another island where organized bird-nesting is carried out is Nightingale Island, of the Tristan da Cunha group in the South Atlantic. Every September the Tristan islanders cross 25 miles to Nightingale Island to collect rock-hopper-penguin eggs and guano. They return each April to collect the birds, which are a source of oil. They have to make this dangerous journey because their own native rock-hopper, once so numerous on Tristan, is now reduced to two rookeries.

Killing for finery: fur

Some seals are hunted for their fat, but more are, or have been, hunted for their fur. The most highly prized fur seal, the Pribilof or Alaska fur seal, is a native of the Bering Sea. It was far more numerous at the turn of the century than it is now; almost 4 million were slaughtered between 1908 and 1910 by Japanese, Russian, and American sealers, and as a result the population was reduced to a mere 200,000 animals. Because of its commercial importance, the U.S. government assumed control in 1911, and after a resting period to enable the stock to recover, the killing is now regulated at 65,000 seals a year, so that the animal is no longer in danger of extinction.

The Guadalupe fur seal is still in great danger, however, though it is no longer exploited commercially. The chief period of exploitation of this seal was in the 18th and 19th centuries, when enormous numbers of them lived off the Pacific coast of America. In 1805, about 80,000 animals were slaughtered for fur off the California coast. This kind of ravaging continued without control until, by the end of the 19th century, harvesting was no longer profitable. By then, the Guadalupe fur seal was very rare. After 1880 it was sighted only occasionally until in 1954 a breeding colony was discovered in a cave on its native Guadalupe Island; and the population now appears to be in the region of 200–500. But this seal is by no means out of danger; in fact it is doubtful whether the species can survive. Its close relative, the Juan Fernandez fur seal, is even closer to extinction. In 1792 it was estimated that there were about 3 million of them on Juan Fernandez Island, off the coast of Chile, but extensive hunting almost exterminated them and the latest estimate is put at about 50 individuals.

Harp or Greenland seals are found in the open Arctic seas; they migrate north in the summer and come south in the spring for the breeding season. They are hunted in enormous numbers for meat and oil as well as for their fur and leather. But it is the coat of the baby seals that is most prized—either the first "whitecoat" of the newborn pup, or the gray and softer "beater" that replaces the whitecoat. The largest kill on record was in 1831, when 687,000 seals were killed. The present annual kill is about 80,000 adults and about 180,000 pups. The young are known to the sealers as "bedlamers"; this has nothing to do with Bedlam, the infamous British lunatic asylum, but comes from the French "Bête de la mer," the name given to harp seals by 15th- and 16th-century Breton settlers in Canada.

One of the principal objections to sealing is the way in which it is often done. The public imagination has recently been caught by what it feels to be the brutal way in which many types of seal are hunted. One of the methods used is to club them on the head, which often merely stuns them, and then they are skinned alive. More often, however, they are shot; detailed instructions are given in the pamphlet *Sealing in U.K. and Canadian Waters* (1968), of the exact caliber of rifle to use and of the precise position on the head to aim at. A very close range is recommended—one inch, if the animal is approachable. Seals are very awkward on land and it is quite easy to cut off their access to the sea and then to shoot them.

No animal, whether a land- or water-living species, can expect to be left alone if it sports a fur coat of any beauty. Many of the great cats are endangered for this reason; they are the quarry of both fur hunters and trophy seekers. Since the middle of 1971 the International Fur Trade Federation has been operating a voluntary ban agreed with the International Union for the Conservation of Nature. Under it, Federation members agree not to use furs of any of the following species: tiger, snow leopard, clouded leopard, La Plata otter, and giant otter. They also agreed to a temporary ban on leopard and cheetah skins, subject to a survey conducted by the IUCN and the World Wildlife Fund, financed by the Federation. A survey is also being made of the jaguar and ocelot to determine whether or not they should also be protected.

The World Wildlife Fund's former executive director, Herbert H. Mills, emphasized the problem when he said: "Women of the world hold the future of such wild creatures in their hands. . . . Status symbols being what they are, and with the buying power of today, fashion trends can spell complete destruction to any wild thing that becomes the whim of fad and fashion."

The great cats breed fairly frequently; one pair may produce up to 20 young in a lifetime. Many of them have no natural enemy except disease, so that—in theory at least—there is no real reason why the excess should not be exploited by man for fur coats, rugs, or even bath-mats. But the trouble is that field conservationists cannot control hunting to within such fine limits. It is probably wiser therefore to put complete rather than partial protection on the killing of cats for fur. Even so, complete protection has very little effect on any animal when poachers are determined and cunning enough. Such are the poachers who roam the Florida Everglades in an increasingly difficult search for alligators. There are an estimated 1000 poachers still at work in spite of a law forbidding the killing of the American alligator. Unfortunately, alligator skins are highly prized in the leather and souvenir trade. One report says that over 50,000 were poached in the USA in 1966. Unless this particular status symbol is superseded by another less destructive one the alligator is unlikely to survive. An ex-poacher is on record as saying, not long ago, "I wouldn't give the 'gator more than 3 or 4 years. . . . They'll kill until they get the last of them."

Above: The Laysan albatross, one of the victims of the Hawaiian Islands egg industry. Above center: a similar industry flourishes—for the moment—on several of the Seychelles Islands; eggs collected from the nests of the sooty tern, boxed ready for marketing. Like the Laysan and even rarer Steller's albatross, the sooty tern lays only one egg per year.

Far right: man's technical skill has triumphed over the blue whale, largest of all mammals. The dead whale has been towed alongside the boat and air is being pumped into it to keep it afloat. Below: diagram of a blue whale shows blubber (gray), muscle (red), and skeleton, all exploited commercially. A man (corner) gives an idea of size.

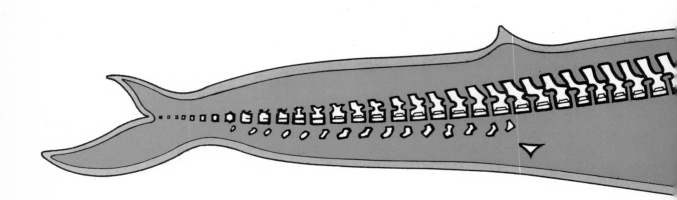

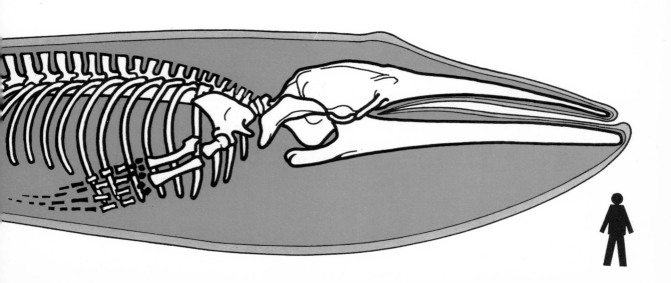

Killing for finery: feathers

Another animal whose skin can be turned into rather smart handbags is the African ostrich. At one time, however, the ostrich was not sought after for its leather, but for its tail feathers, which became a tremendous fashion fad that lasted over the turn of the century right up to World War I. In 1912, about 160 tons of feathers were sold in France alone. They were made into hat trimmings, floating feather boas, and fans for fan dancers and great ladies. Ostriches in the wild are vicious beasts, and can give a very nasty kick with their long legs. So it was naturally easier to shoot the birds than to risk severe injury by capturing them and plucking out their tail feathers. But a few enterprising businessmen realized at about this time that farming the ostrich would be profitable—after all, tail feathers can be plucked and they will grow again. Ostrich farms grew up in a few places; there is still a famous one at Uitshoorn in the Cape Province of South Africa. Unfortunately for the farmers, the fashion suddenly changed, as is its whim, and the bottom dropped out of the market in ostrich feathers; they were too expensive for most people during World War I. But the ostrich, which had been brought to the brink of extinction, was saved.

Another bird persecuted for its tail feathers is the bird of paradise, a native of the Molucca Islands. Legend had it that this bird never alighted, but spent its whole life on the wing, feeding on air and the rays of the sun. For this reason, it was named *Paradisa apoda* (the specific name means "without feet"). The bird of paradise has feet like any other bird, but it also has very beautiful tail plumage that, if plucked, will grow again in about a year. Naturally, as with the ostrich, it is much easier to shoot it. This is done with special blunt-tipped arrows so as not to damage the plumage.

While only the Moluccans coveted the feathers for head-dresses, this bird was not in danger, and although there was a trade in feathers up to World War I, the bird of paradise was rigidly protected. But since 1957 the Moluccas have become a strategic keypoint for the movements of Indonesian troops, and the bird of paradise has suddenly and unwittingly become a status symbol. When an official tour is made of the Moluccas by any high-ranking army official, the first thing he will do is ask for a "Bouroung Tjendrawasih"—a bird of paradise. The subordinates, anxious for promotion, will naturally do all they can to obtain one, even though they are well aware that the bird is protected by law. Recent figures show that a single skin will fetch up to 670 U.S. dollars; a stuffed bird mounted on a perch can be sold for as much as 2200 dollars at the port of Djakarta.

If the bird of paradise does manage to survive this slaughter it will be for a rather comic reason. The local hunters appear to know nothing about the sexual dimorphism in this species. (This means that the male and female look quite different.) In fact it is doubtful whether they even know that the modest brown female exists. They sell the mounted male birds as females if they have bent necks, and as males if they have straight necks. There is also another hope for the survival of the species; the full adult male plumage develops only at the end of the second year. This allows the male birds to breed before they are coveted and shot.

We cannot hope to talk about all the wildlife on the danger list today. We cannot even list all the reasons why certain animals are persecuted until they cease to exist. Some of these reasons lie deep in tribal lore and legend. We cannot know why a particular beast or some part of it should be said to be endowed with magical or healing powers. In some cases there is a definite factual basis for these powers: for instance, the head fat of the dugong or sea-cow does seem to cure the Malagasy people of headaches, and the efficacy of certain kinds of snake venom as a blood-clotting agent is beyond doubt—but whether there is always sound medical evidence for healing powers is an open question. While people continue to believe in them, their unfortunate sources of supply will continue to be in danger. Of all the beasts to be endowed with mysterious powers, the rhinoceros is of course one of the unluckiest. The belief that the horn, and almost any other part of the rhinoceros, is a powerful aphrodisiac, has meant that the three Asiatic species have been hunted almost to extinction. Suffice it to say that an animal whose horns are worth up to 2500 dollars in many parts of the East is in dire need of care and protection.

Killing for fun

In some affluent societies, hunting and fishing have become status symbols, particularly in countries where wildlife is restricted to small areas; but it should be said in defense of the status seekers that, apart from undersized fish that are thrown back, most of the animals they kill make a useful protein supplement to their diet. But the souvenir industry has become a major racket. Alligators are stuffed and sold as souvenirs; elephants are killed for their tusks, and their feet are hollowed out and turned into novelty elephant-foot umbrella stands or wastebaskets.

In the heyday of colonialism, administrators were responsible for the near-extinction of rhinos, the tiger and several other great cats, and a host of other beasts, so that the newly independent countries are left with a somewhat depleted heritage. It is to be hoped that they will all realize the importance of their native faunas and try to control further irresponsible killing.

Killing for financial gain

The live animal trade accounts for very large numbers of animals every year. Poachers are alert to the markets, and the rarer the animal species, the greater its market price to collectors. Perhaps the most lamentable case of wild animal smuggling is that of the orang-utan. Although there is now a complete ban on the export of orang-utans, and although they are very carefully guarded in the wild, the smuggling still goes on.

The orang-utan is not the only primate to be threatened by the wild animal trade. Various species of monkeys are required for medical research, and several types are being depleted by this new trade. Since 1952 the veterinary

Above: a prospective buyer in a warehouse in Nairobi, selecting from the beautifully mounted skins of zebra, leopard, and antelope. The trophy industry is a good source of revenue for Kenya. Below: the whitecoat pup of the harp or Greenland seal, the main target of commercial exploiters.

Above: the snow leopard,
which lives in the high
mountains of China, India, and
the USSR, is one of the most
beautiful of the great cats. This
does not go unnoticed by fur
traders; their pelts fetch high
prices where they are not
forbidden imports. There are
probably no more than 500 left
in the Himalayan part of their
range. Left: the alligator gets
its name from El lagarto, "the
lizard" in Spanish. But
alligators and crocodiles are not
lizards, but archosaurs—
related to dinosaurs and to the
remote ancestors of birds. They
are the last of the "ruling
reptiles."

department of London Airport has recorded some 750,000 monkeys passing through their care.

Killing through fear

Many carnivores attack man as well as other animals and are—if there are weapons to hand—shot on sight. Crocodiles, tigers, wolves, bears, and great cats have been reduced enormously in this way. Even herbivores such as rhinos, elephants, and African buffalo can become extremely fierce if approached, especially when they are nursing their young. No one ventures into wild game territory without a rifle, if he can help it, and a man may be forgiven for shooting a killer before it attacks him. This is why it is so difficult, especially in the wilder regions of the world, to put dangerous animals on the list of protected species. Not only is there a risk that a carnivore will attack a man; there is also a constant fear that livestock will be assaulted. A farmer's stock is his livelihood and this must be protected before the predator.

One of the world's rarest animals is the thylacine or Tasmanian wolf, also called the Tasmanian tiger. This animal—which is in fact neither a wolf nor a tiger, but a marsupial—is the largest of the Australian carnivores, and was very common over the whole of the Australian mainland as well as Tasmania, but was probably decimated by the smaller dingo back in the Pleistocene. It was

Above left: fans fashionable at the time of Queen Victoria's Golden Jubilee. They were trimmed with birds and butterflies, tortoiseshell, ostrich feathers, and elephant ivory. Later, in 1930, the ostrich feather fashion reached the most dizzy heights. Above: a huge headdress being "supported" by glamour-girl Jessie Matthews.

later recognized as a sheep killer and from the early part of the 19th century it was the victim of highly organized persecution by farmers. As far back as 1832, an Australian newspaper reported it as rarely seen. The government, quite unaware of the uniqueness of the thylacine, offered bounties of £1 for each adult animal brought in, and 10 shillings for each subadult. Between 1888 and 1914, the government paid out 12,268 bounties. There is thought to have been an epidemic in 1910, perhaps of distemper, and this reduced the thylacine population still further. When the Australian government eventually gave it protection in 1938, it must have been nearly extinct. It has now quite disappeared from the Australian mainland, although occasional authentic reports of it still come in from mountainous localities in Tasmania. Even in Tasmania, not one has been trapped since 1922. During the 1950s there were hardly any sightings of the animal itself, but there has been evidence that it still exists, from its tracks and characteristic kills. (The victim's chest is ripped open and the heart and lungs eaten; the thylacine also has a preference for

blood sucking.) Recently sightings have become more frequent, though we cannot yet estimate its numbers. Over 1½ million acres east of Macquarie Harbor have been set aside as a sanctuary embracing a large part of the region in which the greatest numbers are thought to have survived. Dogs, cats, and guns are forbidden there—a ban that also bodes well for another threatened animal, the rare Tasmanian ground parrot. Other areas where thylacines are known to exist include the Cradle Mountain National Park, in the center of Tasmania, and at Sandy Cape on the west coast (where one was accidentally killed in 1961).

Since the Boers settled in South Africa, the native ungulates (hoofed animals) have suffered the consequences. No fewer than five forms of antelope and zebra have become extinct or almost extinct. Unfortunately, three antelopes and two zebras—one of which was the partially striped quagga—were competing with the Boers' own herds for the pasturelands of Cape Province, and were systematically destroyed. Actually two of the three

These hunters (left) kill the elephant for food, but manage to make a reasonable sideline out of selling off its inedible parts to the souvenir industry. In workshops such as the one on the right, elephants' feet are polished and turned into "useful objects." This year's fashion seems to be coffee tables and stools, with zebra-foot ashtrays a close second.

Above: to be hunted to near-extinction for meat and hide (and because its tail makes an excellent fly whisk) was the fate of the South African white-tailed gnu. The last 1000 or so survivors now live on a few reserves and farms.

antelopes—the bontebok and the white-tailed gnu—do survive today in small numbers, but only in captivity.

Indirect extermination

A large proportion of the wildlife today in danger of extinction is threatened almost by default. Animals are the victims of our own species' ever-increasing need to expand, and this expansion often takes place without anyone giving a thought to its by-products. Clearly a lot of this is unavoidable—the spread of human industry, urbanization, and agriculture is inevitable; but in its wake many of the world's rarest, often most beautiful and curious, species suffer and become extinct.

One of the major causes of danger to wildlife is the felling of primeval forest. In former times, forest spread densely over many areas of the world—not only in the tropics, but in temperate zones too. This climax vegetation has remained stable for a very long time and has allowed a network of stable ecosystems to develop; each one has its own complex of plant and animal life. We know very little about the habits of many of the inhabitants of the more remote jungles, because they tend to be shy and sometimes nocturnal, and are therefore very difficult to catch for study in captivity.

At the present time the extension of agriculture threatens a very large number of species with loss of habitat; already many of the rarer forms have been squeezed into a tighter and tighter habitat by the gradual cutting back of their natural forests, and several more face total extermination as the last of their forest ecosystems are being destroyed.

This is certainly the bleak picture in Malagasy (formerly Madagascar), which has a unique biological history. On this island live some extremely interesting animals found nowhere else in the world. These are the lemurs, which are relatives of the monkeys but much more primitive. They are adapted to life in the dense tropical forest that, until recently, covered most of the island. At one time there were lemurs on the African mainland, but they could not compete with the true monkeys, which are more agile and intelligent. The lemurs of Malagasy have been free from predators for many millions of years, ever since the island broke away from the mainland.

The rarest lemur of all is the tiny aye-aye—so rare that there may well be no more than 50 individuals left. (Incidentally, the aye-aye suffers not only from the destruction of its habitat, but also from being hunted as a witch and herald of death by the local people). When numbers fall as low as this, it is very doubtful whether a species can be revived at all. However, successful attempts have been made with several other species and perhaps the present efforts to save the aye-aye will also have good results.

In South and Central America much of the tropical forest is being cleared; in Brazil 85 percent of forest has given way to agricultural land. Great cats cannot survive in a constricted space; the jaguar, for instance, needs 60,000 acres in which to roam on its nocturnal hunts. Other dense-forest dwellers,

Right: the coyote is hunted as a pest because it kills man's livestock; it is pictured here with a calf it has killed. Above: one of the finer pictures of the thylacine. Its amazing similarity to a true dog is an example of parallel evolution. Right top: slash-and-burn of tropical rain forest in Peru. Right below: conservation's dilemma—the soil erosion that follows deforestation for agriculture.

Below: the Lacandon Indians of Mexico are of great interest to geneticists studying the effects of inbreeding. Below left: a young man being tested for his reactions to the chemical PTC (phenylthiocarbamide). High proportions of non-tasters of PTC and albinos (below right) are genetical indications of inbreeding.

such as the tapirs of Malaysia and Mexico, are threatened, as are also the pygmy chimpanzee of Zaire, the Central American spider monkey, and countless birds. The list of species threatened by loss of habitat is very long.

The draining of swamps and lakes also causes loss of habitat. Water-birds have some chance of survival if they can reestablish themselves elsewhere; but amphibians such as frogs, toads, and newts are far less mobile, and unlikely to find new ponds in which to breed. For example, much of Upper Galilee in Israel was a fetid malarial swamp 30 years ago. When it was drained to make way for agriculture, the most likely victims would have been the amphibians and turtles; but fortunately these were spared by setting aside part of the region as a reserve.

In other parts of the world, water-dwelling species are threatened by man-made disturbance. For instance, in Guatemala the Lake Atitlán grebe is on the danger list mainly because its life is being made unbearable by speedboats.

Pollution

Another major cause of indirect extermination is pollution. There are various types of pollutant, and farmers and conservationists are increasingly concerned about the effects of chemicals used in agriculture. One well-known pesticide is DDT, which has recently been found to be indestructible; a

The bonobo or pygmy chimpanzee lives on the south bank of the Zaire (Congo) River, in humid forest. As it was not discovered until 1929 we are uncertain whether it is rare because its forest habitat has been partly destroyed by man, or because it has always had a very limited range.

penguin caught as far away as Antarctica was found to contain DDT in its fat. Other aspects of pesticide pollution will be mentioned in Chapter 20. Some of the by-products of industry should, however, be mentioned here. For instance, the industrial effluent from the town of Irkutsk is pouring into Lake Baikal in the USSR. This lake is the largest and deepest body of fresh water in the world; it contains one fifth of all freshwater reserves and in the lake lives the world's only freshwater seal as well as over 700 other species that are *endemic*—that is, they live nowhere else. Not only is pollution destroying some of this fauna, but it also threatens to cause climatic changes that could result in the advance of the Gobi Desert, as well as making deep incursions into our water reserves.

Although no actual extinctions have as yet been caused by radioactivity, the dangers of radiation must be stressed, and further contamination of the Earth by irresponsible people should be prevented.

When a non-native species is introduced into an area, there will often be disastrous results that were not predicted at all. Many instances of this were described in Chapter 15.

Natural extinction

Sometimes a species dies out for no obvious reason; the usual causes of extinction—such as over-exploitation, disease, or failure to compete with other animals, or to adapt to a changing environment—do not seem to fit. But there is one other reason, and this is a condition known as *racial senescence*, in which the ability of a species to survive seems to have become played out. This senescence is connected with the organization of the enormous numbers of genes that determine heredity. Every species needs a constant reshuffling of its genes if it is to produce a big enough variety of offspring to ensure its adaptability to a changing environment. This supply of genes is commonly called the *gene pool*. If the pool is too small, the variations that enable a species to survive are much reduced, especially in a small natural population. Then the weak characteristics in its genetic make-up, such as a tendency to disease or to poor hearing, are no longer masked by the stronger features, and the inevitable result is that the species will die out.

A good example, among humans this time, is the Lacandon Indians of Central America, an isolated tribe numbering no more than 300. None of them has married anyone outside the tribe since the break-up of the Mayan Empire, 1000 years ago. Today the Lacandones are extremely inbred so that their gene pool is much reduced. They suffer from serious disabilities, their resistance to disease is very low, and unless they enlarge their gene pool by marriage with outsiders they are doomed to eventual extinction.

This is an example of racial senescence in humans, but the laws of genetics apply not only to humans, but to the rest of the animal and plant kingdoms. It is evident therefore that a knowledge of genetics is a necessary part of the equipment of a wildlife conservationist, and, as we shall see in the next chapter, this knowledge is being used in interesting and unexpected ways.

Above: this Chartley bull belongs to a herd that has been kept and bred in captivity since the year 1248. They are probably the descendants of the original white cattle introduced into Britain by the Romans, possibly as sacrificial animals.

18
Man to the Rescue

Clearly, much of the world's wildlife needs immediate care and protection; the question is how. This is one of the main concerns of the International Union for the Conservation of Nature and Natural Resources (IUCN). Its headquarters are in Switzerland, and it operates a department, called the Survival Service Commission, that maintains contact with conservationists all over the world. The commission collects and collates information from field officers working on fauna preservation, from amateur naturalists, and from anyone who has the time and ability to make detailed and accurate surveys of the state of any one or more species.

Whatever information can be gathered is sent at intervals to the headquarters of the IUCN, and the Survival Service Commission compiles a loose-leaf record called the Red Data Book, which is sent to subscribers. Fresh information sheets are sent out when there is anything new to report, and every effort is made to keep the book up to date. But if their information (and hence ours) is sometimes out of date, it is because surveys cannot always be made often enough, and the fortunes of many species may change considerably over a short period of time.

At present there are two Red Data Books, one on mammals and one on birds; further volumes on amphibians, reptiles, and plants are in preparation. Invertebrates are too numerous to be cataloged in this way, and no Red Book is planned for them, although there is a long and varied list of rare invertebrates of interest to zoologists as evolutionary mysteries.

There are three main ways in which attempts may be made to preserve wildlife. The first is to remove a breeding stock from the natural habitat and breed it in captivity under controlled conditions: the captive-bred offspring may be returned to the wild if its numbers have recovered sufficiently to give a reasonable chance of survival under natural conditions. The second is to surround the habitat with a fence in order to protect the habitat plus the wildlife inside it. The third is by legal protection of the species.

Captive breeding
When a species is in danger of extinction it may be necessary to transfer a

breeding stock to a zoo. This is not ecologically ideal, because zoos, however good, provide an artificial environment; but it is preferable to losing a species altogether.

Because many of the larger zoos rely on the public, they are bound to exhibit the popular, large furry creatures that draw in the crowds. In the background, however, zoo staff are perhaps more interested in the shy, nocturnal beasts and the rare species, especially in inducing them to breed. It is difficult to keep some of these in captivity for any length of time even without attempting to breed them, and breeding successes are still rather patchy. (Basel Zoo has one of the best records, with 47 of its 79 species breeding regularly).

What are the problems of breeding in captivity? We know that if there are few individuals to breed from, the gene pool is so small that the risk of inherited weaknesses is very high. Zoos realize this and try to introduce their breeding stock to strangers of the same species in other zoos.

Diet is important if a species is to be healthy enough to breed successfully. It is not always essential to provide an animal with its natural diet, but it is important to make sure that each type of animal gets its correct balance of protein, carbohydrates, fats, minerals, vitamins, and trace elements. Some animals, however, do need specific food; the koala, for example, feeds only on

Left: London Zoo in 1866—tiny cages, crowds of visitors—far-from-ideal conditions for all concerned. Right: Longleat Lion Park, England. Spacious, and full of attempts to imitate African game reserves—but primarily for profit, not for conservation.

eucalyptus leaves, which have to be of a particular species and of a certain age, because older leaves contain prussic acid, which is toxic.

Large carnivores, such as lions and tigers, eat almost their entire "kill" in the wild. Scavengers and carrion feeders feed on the remains of the carcass, but the carnivores consume flesh, entrails, liver, glands, brains, and small bones. The bones are important for maintaining the animal's calcium content; without a regular intake of bone, the carnivores eventually suffers from spinal disorders. Zoo dieticians therefore make sure that the great cats receive enough of the type of bones that they can crush and swallow.

Wild animals usually have a number of internal parasites that do them little harm in their natural environment but may, in captivity, respond to changes in the environment and diet of their host and become lethal to it. The okapi, which lives in the Congo jungle, has to undergo a long course of deworming before being exported, and on arrival at a zoo a further period of quarantine is necessary before it can safely be introduced to other okapis.

In the days when small square cages were the rule in zoos, keepers often wondered why many animals could not be induced to breed. One reason is that animals need exercise to keep healthy enough to breed. Another is that some birds, especially birds of prey, need to indulge in elaborate pre-mating display flights. Without space for these instinctive behavior patterns, breeding will not take place. The Japanese white stork is being bred in captivity not because it is rare but because conservationists are alarmed at the detrimental effect on these beautiful birds of toxic chemicals used in agriculture. The storks are bred and reared in large "flight cages" where they can perform their mating ritual, and it is hoped that their offspring will be returned to the wild eventually, when the use of toxic chemicals has been discontinued.

There is an increasing move among the larger zoos to specialize—that is, not to try to show specimens of every animal, but to limit numbers and to set up group exhibits. Although this is an exhibition venture, it may well have a fruitful by-product. As a group exhibit takes up more space and several species are grouped together, the natural habitat is more nearly reproduced than is possible with each species in isolation. One of these exhibits may be seen in the Bronx Zoo, New York. A variety of grazers and other herbivores is grouped in an enclosure, and in the middle, surrounded by a moat, is an island on which lions live—separated from their natural prey. Thus no damage is done but both the public and the animals can see and feel something of the atmosphere of an African savanna. Of course, there are always problems, even with captive animals that breed regularly. In the wild state, where natural selection operates, weaker individuals would not survive to maturity. It is thus important that in a captive breeding stock the weaker progeny be weeded out.

Effects of isolation

It is advisable to allow animals that have been in captivity for a long time to meet an individual from the wild. Without wild companions, caged

279

individuals (of some species) not only do not breed but may even forget their natural habits. Not all behavior is instinctive or innate; some is learned from contact with members of the same or of different species. The song of certain birds is a good illustration. The wren seems to have an inborn knowledge of its own song, whereas the starling and the mocking-bird incorporate phrases from the songs of other species.

Even the outward appearance of a bird, altered by selective breeding, will often revert back to that of the wild population after a few generations back in the wild. An interesting example is the budgerigar—an Australian parakeet. For a long time it has been popular as a pet throughout Europe and the United States. Budgerigars often escape from their cages but are usually eaten by a predator or die of cold. In Florida, however, the climate is similar to that of their Australian habitat and those that escape do well in the wild; there are an estimated 10,000 feral budgerigars in Florida. Selective breeding had produced a wide variety of colors including blue, mauve, gray, and yellow, but the feral birds are quickly returning to their natural coloration —green with yellow heads.

"Doctor of the Arabs"

Let us look at some examples of breeding in zoos. In some cases the animals concerned would not have survived in the wild, and in others they were— surprisingly—already extinct in their original habitats.

The Arabian oryx was once very common in the Middle East. After 1800 it probably became restricted to the Arabian and Sinai peninsulas. Although its range was gradually reduced over the 19th and early 20th centuries, there was a disastrous change in the 1930s when the discovery of oil made the sheikhs extremely wealthy. With automobiles, the sheikhs began to hunt the Arabian oryx intensively; the rough, gravelly desert habitat of the oryx presented no problem for motor cars.

The oryx had a reputation among the Arabs for bravery, strength, and endurance, so the Arabs believed that by eating its meat they would absorb its qualities. A local name for the oryx is "Doctor of the Arabs."

Eventually stocks of the Arabian oryx fell so low that a drive was mounted in 1962 by the British Fauna Preservation Society, assisted by the World Wildlife Fund, to save it from extinction. An expedition was sent to capture a small stock from an area east of Aden. The expedition did not succeed in getting any, but Sheikh Jaber Abdulla al Sabah of Kuwait gave three beasts from his private herd. One of these survived the journey and joined an existing collection—two males and one female—at Phoenix in Arizona, where the climate was thought to be sufficiently like that of the Arabian peninsula. King Saud of Saudi Arabia presented four more.

By January 1972 there were more than 30. Oryxes take well to captivity; they are docile, handle easily, and breed normally. The herd at Phoenix, another at Los Angeles, and at least two small captive herds in the Middle East

Above : a nine-year-old African elephant posing with his rations of food and water for one week (and keeper). The sacks on the left contain root vegetables, and the bin on the right was full of colorful biscuits before the photography session. His food costs the London Zoo $3000 per year.

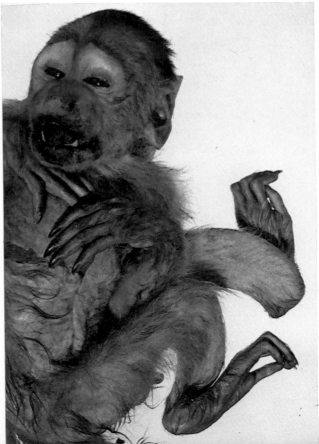

Right : zoos are always receiving for treatment domestic pets that have been badly cared for by their owners. This weeper capuchin monkey was brought in suffering from rickets, a deficiency disease caused by a shortage of vitamin D.

augur well for the future of this animal. In the wild probably no more than a few hundred survive. The outlook and customs of the nomads who share their habitat make their existence precarious.

Breeding back

Another success story starts in Mongolia. The plains near the Mongolian border with China are the home of one of the last true wild horses. This horse, named after its discoverer Przewalski, is smaller than the domestic horse but its head is larger in proportion. Neither Mongolian nor Chinese nomadic herdsmen go there much nowadays—because of improved methods of animal husbandry there is no need for nomads to wander so far in search of pasture. The Przewalski horse is, nonetheless, rare in the wild, but there have been numerous sightings until quite recently. Since the end of the last century, expeditions have been sent out to the Mongolian plains. From 1899 to 1907 twelve specimens were brought back, but only three survived.

Between 1925 and 1935 four more were transported from Russia to Germany (with a permit) and then two more (without a permit). There the breeding experiments began. It was long thought that the Mongolian form was closely related to a European wild horse, the tarpan. The tarpan figures in many prehistoric cave paintings but became extinct recently: the last one died in captivity in 1919. The theory was that a large part of the genetical makeup of the tarpan survived in the Asiatic Przewalski race; it was believed that some tarpan qualities survived in the domestic horse but had become diluted through selective breeding. On this basis the Heck brothers of the Munich and Berlin zoos started experiments to breed back the tarpan. By 1967 there were 10 specimens (eight females and two males) of the so-called bred-back tarpan in the Hellabrun Zoo at Munich.

Meanwhile, efforts continued, mainly at Prague Zoo, to increase the numbers of the pure-bred Przewalski horse. By January 1972, there were 182 in captivity altogether and a system of keeping studbooks had proved successful.

Ancestral cattle

The gene pool was used by the Heck brothers to breed back another animal, extinct since 1627. In theory, every creature has in its genetical makeup some of the characteristics of its ancestors; each one of us has half those of our mother and half of those of our father; a quarter of each of our grandparents, one eighth of our great-grandparents, and so on. Of course, it is not as simple as that, because genetically controlled characteristics operate by passing on one of two alternative characteristics from each parent. This is why you may have red hair when neither of your parents has it. One of them may have a gene for red hair masked by one for brown; it is a matter of chance which gene you receive.

The Heck brothers decided that within the genetical makeup of the world's domestic cattle must be the complete formula for the extinct aurochs—from which all today's stock were bred. They set about proving it. The appearance of the aurochs was known only from a few pictures. It was known to have longer legs and bigger horns and to be as large as or larger than the biggest cattle of today. It was black with a yellowish-white stripe along its back, and grew a shaggy winter coat, shedding it for a smooth coat in summer. The Heck brothers also studied ancient Egyptian frescoes of the Mediterranean breed, to get background information.

Around 1921, Heinz Heck took one breed with the right horns and bred it with another of the right color. He saved time by using ready-made hybrids, and success came quickly. By 1932 he produced the first new aurochs. The calf was brown, as in the original description, and developed the stripe as it grew older. (The American buffalo and the Indian wild ox, called the gaur, also have brown calves that change color later.) The original aurochs was known to have certain mental characteristics and some of these returned with the breeding; they were ferocious and short-tempered in captivity but, when returned to the wild, Heck's aurochsen were shy and hid deep in the woodland. The aurochs is quite immune to the usual diseases of cattle, so there may be a practical outcome to these fascinating experiments.

The question we have to ask about breeding back is: how can we know whether the whole formula of an extinct animal has been reassembled? We can judge only by external characteristics and a few records of temperament, about which we can know little. This does not detract from the interest of the experiments, but we must be wary of accepting that an extinct animal has been "brought back to life."

Infertile turtles

It is preferable to try to breed endangered species before they become extinct. San Diego and other zoos undertook the task of breeding the Galápagos giant turtle. These creatures have always been infrequent breeders, even when they were being plundered from the islands. Zoo experts had no false hopes that in captivity the turtle would step up its breeding rate. Even so, early experiments were not very successful. Collection for establishing a breeding stock started in the late 1920s and colonies were set up at several zoos.

The first hatchings were in Bermuda in 1939, where five eggs hatched out of eight; meanwhile the San Diego colony produced nothing. There were about 20 specimens there in the early 1930s, and from time to time more were brought in from the Galápagos Islands. The larger colony should have had a higher fertility rate than before, but the early eggs proved to be infertile.

Eventually it occurred to the curator that what was lacking was a suitable area for copulation. The floor of the pen was made of a rather unyielding earth; the curator thought possibly the male could not get into the proper position on a hard floor (though it seems more likely that the female was unable to dig her nest). So in 1957 the San Diego Zoo modified the enclosure and put down an entirely new floor of soft sand. Soon, one of the females laid a batch of eggs. Among turtles nesting is quite complicated. It begins with a close examination of the nesting area by the female; she wanders about and stops frequently to smell the sand. When she has selected a site, she starts to excavate, squirting urine to moisten the sand. The nest, which she digs with her hind legs, is jug-shaped—wider at the bottom than at the top. The eggs are white and spherical and an average clutch is between 10 and 20. One female at San Diego produced 153 eggs over 6 years and another laid about 110; so it is not the lack of eggs that caused poor hatching results. One possible reason is that the eggs are laid from a height and some are cracked in the process. The female also rolls the eggs about within the nest, and this inevitably cracks a few more.

When she has finished laying she covers the eggs with sand and stamps it down vigorously; more may be cracked at this stage. Despite this rather wasteful process, about which the zoo can do nothing, there has been a great increase in the numbers of young born alive. Although the San Diego colony, as well as other smaller ones in Bermuda and Honolulu, now breed successfully, fertility remains low (6.98 percent of 257 eggs producing live young at San Diego). The conclusion that the zoo authorities draw from this is that, even in the wild, Galápagos giant turtles have always produced many eggs that never hatched.

Breeding of rare animals, especially those that need space, is not exclusive to city zoos. There are private collections such as the one at Woburn Abbey in England, owned by the Duke of Bedford. This collection is very varied and includes rheas, bison, and Manchurian sika deer. But the pride of the collection is the famous herd of Père David's deer, named for Père Armand David, who discovered them: the mi-lu—as this deer was called in its native China—has been extinct in the wild for thousands of years. One herd was kept in captivity, however, in the Peking hunting park, for the Imperial Court.

When Armand David visited the park in the late 1800s, he saw the deer, a species unknown to Western science. He arranged for some to be shipped to the Paris and Berlin zoos, where the first breeding herds were established. It is fortunate that he did so, because very few of the original herd survived. Opinion varies as to what happened: one authority states that floods breached the wall of the park. All but a few of the deer escaped and were drowned or eaten by starving peasants. The other authority maintains that the wall was breached during a political insurrection, with the same result. The Woburn herd now numbers 400 and there are another 100 dispersed in other zoos.

Our last example of captive breeding is the golden hamster, a very popular pet. Every golden hamster alive today is descended from a single pregnant female found in Syria in 1920. Despite numerous collecting trips, no other golden hamster has ever been found. It cannot now be considered rare, although apparently extinct in the wild; on the contrary it is very prolific and, though the mother tends to eat her young if disturbed, hamsters breed well in captivity. Incidentally, this example shows that a small gene pool is not neces-

Right: elephant tusks recovered from a poachers' cache. The game warden's helicopter has, in recent years, put an end to the hunter/poacher's dream of making a good living out of elephants.

Before animals become so rare that recovery is
impossible, conservationists try to keep an eye on
numbers, by marking individual animals or by
removing them to an area of greater safety, as is
being done with this polar bear.

sarily harmful, always provided that the parental stock is genetically sound.

Many animals discussed here can be seen in the zoos of larger cities. This is the only chance most of us get to see some of these animals. But conservationists realize that zoos are not ideal for preserving rare species. It is important not only to preserve a species in isolation in a zoo, but, if possible, to preserve the habitat as well. This is perhaps the biggest part of nature conservation.

Reserves—a reprieve for nature

Reserves can be designated for many purposes: for preserving a particular type of habitat; for the protection of vegetation or of animals in their natural environment; for scientific research; and—implicit in the ideals of the National Park—for public education and recreation. Many reserves can become an economic asset, but it is difficult to persuade governments that it is worthwhile setting aside land for conservation. The public, too, are oblivious of the complexities of putting conservation ideals into practice but they are certainly willing to use National Parks once they are provided.

A reserve can be as small as a pond, or land millions of acres in area. There are wildlife reserves such as the Gir Forest in India, areas of magnificent scenery such as the Engadine National Park in Switzerland, crocodile-inhabited swamps such as the Florida Everglades, lakes and rivers for fishing and boating, steep mountains for hiking and skiing, extinct volcanoes such as the Ngorongoro Crater in Tanzania, and live volcanoes in many parts of South America. There are also smaller reserves, such as the Mosi-oa-Tunya (formerly the Victoria Falls) in Zambia. There are famous buildings, religious shrines, archaeological and anthropological remains—for example, the mysterious sculptured heads of Easter Island—and aboriginal cave paintings in Australia. There are geological phenomena such as the Grand Canyon, and deserts and valleys carved out by glaciers. The Cerro de Comanche National Park in Bolivia is devoted to the protection of a flower that blooms regularly every 125 years, and a minute reserve in the middle of the Negev Desert in Israel protects a cluster of breathtakingly beautiful irises. In short, an astonishing variety of different types of scenery, habitat, and natural phenomena have been designated as reserves and National Parks. Their names can be confusing, however: a reserve can be called a sanctuary, a refuge, a game park, a game reserve, a national park, or a national monument. Some are private and devoted to research, others are open to the public; a few are in urban areas, others are all but inaccessible.

All have one thing in common: land has been set aside by planners and developers for some special purpose. It is impossible to be comprehensive, so we shall look at some widely different types of conservation areas—beginning with some of the problems involved.

When a reserve is set aside, an embargo is put on human settlement within the area if possible, and where settlements already exist, further growth may be prohibited; every effort is made to keep the area stable in accordance with ecological laws. But is such an aim really practicable, or merely an ideal?

When an area is made a reserve, the ecological balance is disturbed as surely as if the whole had been surrounded by a wall. Nature is not static; an ecosystem is in a perpetual state of flux: plants, animals, soil, and atmosphere are in constant interaction. To put an artificial boundary on these interacting processes upsets the equilibrium around the edges. Gradually, depending on the size of the reserve, subtle changes in ecological processes will filter inward and ultimately affect the whole reserve. The larger a reserve, the longer this encroachment will take and the greater the chance of the whole being self-supporting. In a small reserve not only is there greater interaction with the outside, but the habitat is also more vulnerable to disease and to human interference.

In the case of wetland and marshland reserves, a small area is useless unless there is some guarantee that the water table, which determines the environment, will remain stable. This seldom happens, because drainage or some other development even quite a distance away can easily upset the water level.

Islands can be turned into sanctuaries with smaller danger of interference, but even with an island there are risks. Uncontrolled fishing or pollution of surrounding waters can affect the littoral zone (seashore) and this rather specialized habitat is in contact with the food webs of the interior; damage can thus occur even if people do not go ashore with rats, dogs, or diseases.

We mentioned the encroachment of outside influences *into* reserves; the opposite situation also exists, where it is difficult to prevent the ecosystem inside

Right: the Pasque flower is fast disappearing from its limestone habitat in many of its British sites. Conservation of wild flowers in Britain is often the concern of vigilant local communities who fight for the protection of rare species. Far right: all wild flowers are protected in the Swiss National Park. Gentians (the blue flowers shown here) are, however, picked for making liqueurs.

Left: a gannet and guillemot sanctuary at Sula Sgeir Island, off the Outer Hebrides. This species of gannet is confined to the north Atlantic, and the rock, a suboceanic island, is the only British gannetry on which numbers seem to be decreasing.

reserves from spreading outward. Even in the largest inland reserves, to contain the ecosystem indefinitely presents great problems. Migrating birds cannot be contained all the year around in a reserve; even lions wander away from the Nairobi Park into the city, and walk around sedately before returning to the park where—apart from occasional boredom—they seem content.

Many grazing animals cover long distances in search of food. If they do not have a very specific diet, the problem is less acute. But if animals graze only one type of plant, they follow their food into new areas; the plant life of the reserve is on the move too. It may colonize new areas in a mosaic (filling in empty spaces) or in an advancing "front." If the front advances outside the reserve, it becomes impossible to contain primary consumers that feed on it. The conservationist is then faced with two alternatives: to allow the reserve to run itself, and remain ecologically intact while running the risk of dissipation at the borders and some loss of game; or to try to prevent undue losses, by imposing man-made judgments and thus destroying the natural balance. This dilemma causes a basic conflict in conservation theory. The two schools of thought can affect policy in the management of reserves.

Drought

So far we have mentioned problems that arise in normal conditions. But what happens in extreme conditions, when the balance of nature is seriously disturbed? During a drought many animals die; others go miles in search of

water. Water birds and amphibians seek new lakes or ponds, and a great deal of pressure is put on the water sources that are left.

Several game reserves in Africa have collected money to provide boreholes, windmills, and drinking tanks. The whole installation can cost well over $10,000, and must be well protected (by fences made from old railroad track) against elephants. If the water in the tank is used up, elephants become enraged and destroy windmill, tower, tank, and all. Lack of money limits the number of boreholes and windmills but some game parks—such as the Albert Park, in Zaire—do not permit interference with the ecological balance, and to provide extra water certainly amounts to interference. The Albert Park policy is based on the fact that drought results in competition for water, which in turn results in natural selection of animals that can survive on less.

Food shortage

Even during stable conditions, overgrazing in a game park is a problem. Overgrazed land takes time to recover and drought invariably aggravates the situation. The question is whether to provide extra food and salt licks. Land surrounding the game reserve is often tempting to wildlife—especially when food is short. There are many tales of deer wandering into suburbs to nibble at hedges and shrubberies, and of small animals raiding trash cans. This generates conflict between those responsible for the reserve and those who live outside it.

Disease

Wild game animals, especially in the tropics, are storehouses of disease. Many tropical diseases are caused by parasites carried by mammals and spread by insects. There have been many attempts to stamp out this type of disease, and vast numbers of game animals have been slaughtered in order to exterminate parasites. To leave disease carriers in nature reserves must surely negate much of the effort. The question is whether any type of disease control, whether slaughter or immunization, is ecologically sound in a reserve: endemic disease could be viewed as a mechanism of natural selection.

Research

Reserves that provide facilities for research are valuable to the scientist because an area kept as natural as possible forms an ideal outdoor laboratory.

Administrators and game wardens undertake systematic surveys both by means of aerial photographs and by observation of the area plotted, as well as detailed topographical maps and large-scale geological survey maps. Meteorological stations are being set up on the larger reserves. With this background information it is possible to plot changing factors such as the quantity of food consumed in different areas and the fluctuations of ecologically dominant animals. Fertility and breeding rates and the rate of turnover from one generation to the next are plotted by a process called "marking." A number of animals are captured and marked with coded colors so that their subsequent

Right: as the waters of the Kariba Gorge rise, a rhino, immobilized by drug-darting, is strapped to a barge to be translocated to safety.

movements can be plotted. Game marking is done on the same principle as the banding of migratory birds; banded birds may be picked up thousands of miles away but game marking helps to study movements and feeding pathways within the reserve. Once birds are banded, the public is asked to cooperate in sending the details required (shown on the tag) to the research station. With game surveys, too, the public can help once the animals are marked. "Operation Necktie" was carried out in the Kruger National Park to find out more about the grazing movements of zebras. The organizer immobilized a proportion of the zebra population by drugs, and tied around their necks one of 12 different patterns of tie. An antidote to the immobilizing drug was then administered; the zebras soon trotted away suffering no ill effects. The antidote is important because, without it, the animals could become victims of a predator while still dazed. Visitors to the park were asked to complete a questionnaire stating which patterns of tie they saw in different parts of the park. A great deal of useful work was done, and at very little cost.

Game counting is also done by *telemetry* (measuring from a distance). At first only large animals such as giraffes and elephants were counted in this way but the technique is being extended to smaller beasts such as rabbits. Tele-

Left page: a Kob antelope, immobilized and marked (above). After it has been set free, its movements will then be traced by game wardens. Below left: one of the marked animals has been eaten by a lion. Its coding enables wardens to trace life span and movement in the reserve. Below right: a royal tern is marked by banding so that its subsequent movements can be recorded.

Right page: the rare Cape mountain zebra (below left), immobilized by drug-darting, being measured before it is set free again. Bottom left: lions in the Amboseli National Park, watched by research workers from a camouflaged vehicle. Above right: cropping hippos to keep numbers down. The operation is carried out strictly according to the game warden's calculations. Center right: every part of a cropped animal is used for food or research. A fetus has been dissected out of an elephant and will be preserved and kept in the reserve's collection for study. Below right: loading a crated Hunter's hartebeest into a helicopter in one of Kenya's big reserves. Helicopters are invaluable in game management.

metry consists in fixing a battery-driven radio transmitter onto the animal. This transmitter emits signals that can be picked up on a directional receiver at the research station; the movement of marked game can then be plotted until the battery runs down, which may take several weeks.

Researchers take advantage of regular game-cropping programs. When cropping is about to begin, anyone interested in a portion of the animal comes along. The beast is killed, and carved up; some researchers take specimens from the liver, pancreas, stomach lining, and so on; others take blood samples; almost every part is utilized, including the muscle (meat), which is distributed among the helpers and sold to nearby villages.

Back at the laboratory, researchers fix and mount tissue samples on microscope slides for detailed study, not only for their own use, but to send to scientists studying similar problems elsewhere. Only if a biologist demands the killing of rare animals for research purposes is there likely to be friction with the game wardens. Usually they work with researchers very amicably.

Zoning

One aim of conservation is to make the best use of land, and this can be applied to reserves by a system of zoning. Under this system an inner zone, untouched by researchers or visitors, acts as a reservoir—a "gene bank"—providing new genetic material for the rest of the reserve. Other areas are zoned according to function; in one, research is permitted; in another, visitors move freely on foot, in another, there are roads for motor traffic; and so on. Thus, a single reserve functions successfully for a number of purposes.

Wildlife management

One of the biggest tasks is regulation of the predator-prey-pasture balance and to do this it may be necessary to transfer animals from where pressure on food is high to areas where it is lower, either as cowboys drive cattle or by *translocation*.

In its broadest sense translocation means moving animals from one place to another. In the sense used here, years of experiment have gone into perfecting its techniques, and it has revolutionized largescale game management. First, for translocation to succeed there must be a detailed understanding of the ecological and physical requirements of the animals. Secondly, moving animals can involve them in a good deal of trauma, and this problem has been solved by the technique of drug-darting, first used on a big scale from 1959 to 1963 during construction of the Kariba Dam on the Zambezi River. As the waters of the lake rose, animals were left stranded on shrinking islands; the rescue operation, "Operation Noah," was put into effect to remove large numbers of animals to the mainland, there to set them free or to translocate them to a reserve. Drug-darting techniques are also used in game marking; it saves chasing animals to exhaustion, which could reduce their chance of survival.

Translocation experiments since "Operation Noah" have been very successful and techniques have improved so much that few animals die in the process. In Zululand, in the Umfolozi reserve, white rhinos have been bred and redistributed to other parts of Africa where stocks had become depleted. As a result of efficient and timely conservation, the Umfolozi white rhino, once listed in the *Red Data Book* as being in danger of extinction, is being saved for posterity.

The Asian rhinos, on the other hand, are in great danger. The largest Asian form is the Great Indian rhinoceros; this beast, which has deep body folds, may reach a length of 14 feet and its horn may be up to two feet long. This rhino is confined to six reserves in India and the Rapti Valley of Nepal. The most severely endangered forms, however, are the Javan and Sumatran rhinos. The Javan rhino looks like the Indian form: its deeply etched body folds give it a somewhat armor-plated look. There may be no more than between 25 and 40 Javan rhinos, and they are not breeding fast enough to ensure the survival of another generation. These are confined to the Udjong Kulon reserve in western Java, but a small area on the Thailand frontier also seems to contain a few, to judge from the occasional sale of horns.

The Sumatran rhino, the smallest species, is covered in early life with short hair. This rhino is protected over only part of its range in the Loser reserve in Sumatra. Unfortunately, although the animal is protected by law, in Burma it is still legal to sell rhino blood for medical purposes.

Russian National Parks do a good deal of work in reintroducing animals into areas where they have become reduced or extinct. The sable, the musk-rat, various ungulates, and birds are often translocated to new areas tens on hundreds of kilometers from the reserves in which they were reared. Beavers are trapped and translocated from the Voronezh, the Berezin, and many other reserves, with the result that there are now an estimated 10,000 beavers in the USSR, where quite recently they were practically extinct. Most of the translocation in the USSR, however, is for restocking hunting grounds. In New South Wales, Australia, over 7000 young koalas were translocated to restock areas such as Mount Kosciusko State Park where their numbers had become drastically reduced.

Ethics of poaching

For many tasks in a large reserve, the airplane is invaluable. One of its principal uses is the enforcement of the park's regulations, and this means detecting poachers and their caches of ivory, skins, rhino horns, biltong, and so on. The Tsavo Royal National Park was recently equipped with an airplane, and game wardens found enough poached elephant ivory to pay the running costs of the plane for the first year. Conservationists denounce poaching so universally that it is perhaps worth taking a closer look at the problem, to see both viewpoints.

Hunting is a way of life for many peoples all over the world. To these peoples, conservation laws are irrelevant when pitted against their own needs for

The Oka Terrace National Park is one of Russia's smaller reserves. It is set in forests and marshes, and gives protection to a variety of animals and plants. Right: a food rack for deer. Below left: the rare European bison, now being restocked in the wild. Below right: a beaver dam.

survival. Unfortunately for those who happen to live in and near reserves, "hunting" changes its name to "poaching." They are told that hunting is forbidden in the game reserves to those without a permit; and to shoot game without a permit is punishable by fine or imprisonment. The question is, can everyone get a permit? In East Africa, where in principle anyone can get a permit, the price is in practice so high that the local tribesman cannot afford it. Permits are issued on demand to hunters and hunting tourists, however, because the permit system is in fact a method of earning foreign currency. This practice is defended by many game-park administrators, who say that hunters operate under strict supervision (after paying a considerable fee, which is put to good use) and are permitted to shoot only a specific quota of wildlife. This is certainly regulation according to strict rules of conservation, but is it ethical? How does one convince an African that it is fair for him to go to prison for a year for "poaching" while someone else is, not only allowed, but actually encouraged, to "hunt."

Reserves and tourism

Land is not set aside only for wildlife and ecologists. A National Park is a place where people can enjoy themselves and learn to appreciate the natural environment. Conservation is concerned with men as well as wildlife, and places emphasis on man's need and right to return periodically to his pretechnological habitat. Some National Parks, particularly those of America and Africa, specialize in wide open spaces and scenic grandeur. The United States has National Parks and State Parks as well as other types of reserves; altogether these were visited in 1969 by 163,990,000 people. Currently, over 200 million man-days are spent in the National Parks each year. The official American attitude to National Parks is laid down in a statute preventing "occupancy or sale under the laws of the United States . . . and dedicated and set apart as a public park or pleasuring ground for the benefit and enjoyment of the people."

But the advantage of a national pleasuring ground is enhanced if it can be profitable as well; most areas devoted to conservation cannot exist unless they bring in revenue. The income to the State of California was $850 million in one year from spending in and around the park (part of this money finds its way into the exchequer through taxation). Most countries cannot operate on such a scale: Japan has a special type of National Park to suit the country's small size and high population density. Unlike Africa and America it cannot afford to set much uninhabited land aside; instead, the Japanese have a system whereby people live and work in these areas almost as they do in non-conservation areas, but the scenery and wildlife are protected by the public's good behavior. Also, where there is enough wild land they have adopted a system of zoning. With a combined area of 4,360,000 acres, Japanese National Parks occupy nearly five percent of the area of Japan, and more than 60 million people visit them annually. National Parks can be designated irrespective of who owns the land, and limits on building, deforestation, advertising, and other activities can be laid down by the government.

The Australian National Parks are visited by several million people a year. They are under government control, and are important but not vital to the economy. But the nature reserves and other conservation areas are under control of individual states, and the few aware of the need to preserve the fauna, flora, and habitats are far outnumbered by the "if it moves, shoot it—if it doesn't, well, shoot it anyway" school. Nature reserves, as opposed to National Parks, can at any time be handed over to private interests for commercial development without recourse to higher authority.

In the Soviet Union, reserves are under national control, and are mostly not publicized as pleasure grounds; the aim is to keep conservation areas quiet so that natural conditions are not destroyed.

It is in Africa that both some of the least enlightened attitudes and some of the most practical attitudes to conservation are found. In an underdeveloped and largely tribal country the conflicts and pressures against environment protection are tremendous, and many countries cannot cope with the economic demands that a conservation program would make. Thus it has been, and still is, important to convince heads of state that National Parks can eventually (after initial outlay) be immensely profitable as a renewable natural resource.

Above left: Mesa Verde National Park, Colorado. "Cliff Palace," largest of the cliff dwellings built by pre-Columbian Indians, dates from the 13th century. It is said to have had more than 200 living-rooms plus ceremonial and storage rooms. Visitors are allowed to wander over the ruins.
Opposite: a diorama model of the cliff dwellings is reconstructed to give a more rounded idea of the way of life of the pre-Columbian Indians. Above right: visitors can tour the over-hanging, water-smoothed cliffs of Glen Canyon recreational area, Colorado, in boats hired out from the Park's Marina.

Man and beast

To open a game reserve to the public does mean interference with nature in that roads must be built, and rest centers and camping sites must be enclosed for protection from nocturnal prowlers. Visitors are usually instructed not to feed the animals and they do not make the wardens' job easier by disobeying. It would not seem to be doing much harm to feed a banana to a baboon—especially when it seems so tame. But it is doing harm because a baboon that relies on bananas during the tourist season will not be efficient at finding food during the summer, when heat or tsetse fly make the reserve unfit for tourists. Even visitors who understand the ecological repercussions their actions might have will often be won over by appealing eyes. In the United States and Canada there are National Parks where both bears and people roam freely. Rangers labor endlessly to convince tourists that bears are dangerous. But the public cannot resist feeding, coaxing, and encouraging the bears. There have been some severe casualties and a large number of cases of scratching and mauling, usually after provocation—but the public remains undaunted.

Camera safaris

Some parks provide motor or boat services for game spotting, known as "camera safaris"; photography is the only type of "shooting" permitted.

Transport in Indian National Parks may be by elephant. One guide book to Indian sanctuaries gives hints on how to relax when riding on elephant-back so as to go on for hours without getting tired.

Game spotters are accommodated at camps or guesthouses. Some larger game parks have tree hides where visitors can watch game at night, by searchlight or by moonlight. These hides are usually built in a strategic place—next to a water hole, for instance, where animals come to drink, play, and kill each other. In Thailand, experienced game spotters can be hired (with searchlights) to take visitors on a nocturnal safari around the Khao Yai National Park.

But when most people think of National Parks, they do not think of game and game spotting, research, or staring at archaeological ruins; they think of wide open spaces, getting away from it all, hiking, camping, and so on. Enjoyment of the outdoors is what brings most visitors to the United States National Parks every year. Even in underdeveloped countries there are huge areas set aside for public enjoyment. Crowds throng to the parks for holidays, weekends, or day trips, to get away from it all—or are there just too many people? In America they say: "The Americans love their National Parks so much, they're ruining them!"

Take the great National Park at Yosemite, with its thundering waterfalls and wooded mountains. On an average summer weekend, Yosemite is visited by 40,000 people—almost all of whom come by car, blocking roads for miles around. In Yosemite valley the problem is to find a space in the concreted acreage of car park. There are hotels, cabins, a hospital, several souvenir stalls, two paved roads, shops, and an outdoor amphitheater. All these have been added to Yosemite's natural splendors because members of the affluent society are not prepared to rough it, and demand home comforts in the wilderness. In the evening the valley lies beneath a pall of exhaust fumes and campfire smoke. Tents are pitched end to end; pots and pans clatter to the sound of transistor radios and even television sets. Only after campers have gone to bed can the roar of the waterfall and the sound of the wind be heard. In contrast, Isle Royale National Park, an island of half a million acres in Lake Superior—has no roads. It is visited by only 13,000 people per year.

The American National Park was conceived in an age when places such as Yosemite would be penetrated only by two days on horseback. Now the ideal is collapsing in the face of the car and sheer numbers of people. Although as we have said, 200 million man-days are spent in American parks each year, most visitors try to take in as many National Parks as possible in a single trip, spending only a day in each, viewing the scenery through the windscreen, stopping only to take photographs, and keeping to roads and camping sites. These people never wander into real wilderness. Is this the end-product of the great heritage of which Americans are so proud? Are the National Parks in other countries doomed to become an extension of suburbia? If the increase of human populations threatens the existence even of National Parks—the most carefully controlled environments on earth—what about its effects elsewhere?

Right: the Ojcow National Park, Poland, is famous for its rock formations, such as this one, known as "The Club." Ojcow is one of Poland's lesser-known parks.

Below: rock formations in the National Park at Zion Canyon, Utah. This forms part of a very much larger system that extends into Colorado, including the Grand Canyon. They are formed by the action of rivers that began to cut into the land during the Miocene period, at least 12 million years ago, and eventually cut exceptionally deep valleys because the land itself was rising slowly and continuously.

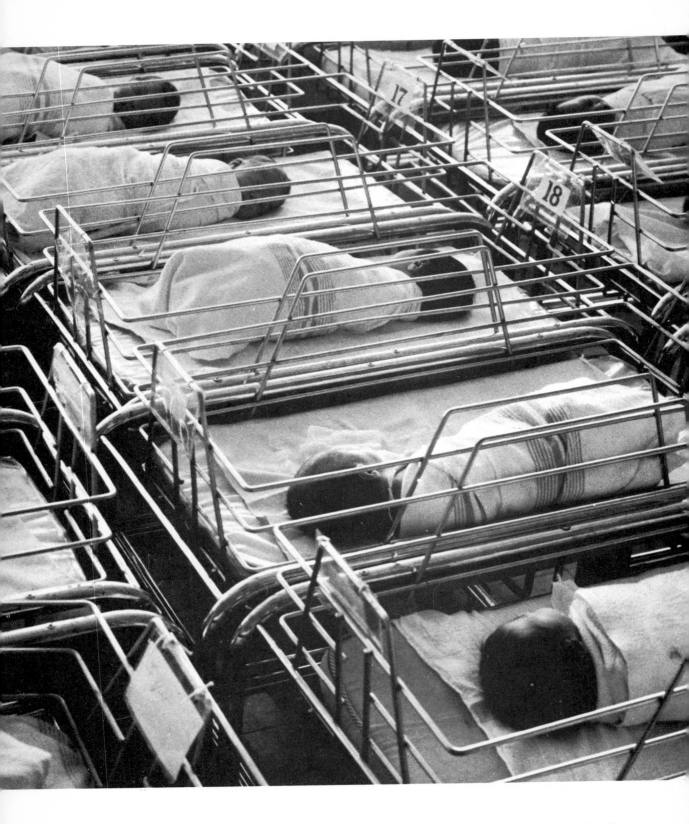

A crèche of newborn babies in a hospital in Mexico City. Today the human family is increasing at the rate of about 74 million babies every year—more than two every second—while the death-rate is almost everywhere in decline.

19

Man Overpopulates His World

Today we are constantly bombarded with advertising, stop-press news, and sensation—on posters, in the press, on radio, in the cinema, but especially on television. Every few minutes, it seems, someone demands that we "consume"; triumphs and catastrophes large and small are paraded before our eyes. The result is a surfeit, a maze we can no longer penetrate. It is also one of the strange effects of surfeit that we can watch scenes of unspeakable horror and misery that we know to be happening here and now—and we hardly feel a reaction any more; we remain detached and uninvolved. So when we hear the words *population explosion* or see pictures of starving children, these images are already so familiar that they take their place alongside the countless trivialities that occupy us for a few moments before our attention is directed to something else.

Until a very few years ago we were lulled into believing that man, aided by his technology, could expand and progress indefinitely; the Earth, we were told, could contain and feed limitless numbers of people. There are several reasons why experts have changed their minds; but the principal one is that, until quite recently, nobody predicted the rate of population growth in any realistic terms. Meanwhile we have quite outstripped even the most daring prophecies. Also food supplies and other resources are not keeping up with demand; but although the provision of adequate food for the present and future must be looked on as a top priority, it cannot be regarded as the only target; more food means more people to breed and still more to feed, and this can lead only to ecological disaster. The need for population control is imperative. But human society is based on a variety of traditions and values, and the changes that are needed strike at the most deeply ingrained customs of marriage and family structure. These changes must come if the Earth is to remain habitable, but the problem is how to achieve them without destroying the values that hold individuals and societies together, and thereby making life meaningless.

Let us first set the scene by considering where people live, and why. The study of populations is called *demography*, and on the next two pages a demographic map of the world shows population distribution and densities.

Population controls

All species are capable of breeding at a rate that could overpopulate the Earth in a few generations. This is because an unchecked population increases by *geometrical progression*. The simplest example of this progression is to be found in a bacterial culture because bacteria breed by dividing in half. But in natural communities there are many ways in which, by natural selection and other mechanisms, plant and animal populations are regulated. In certain types of community, where the balance between food and population is delicately poised, there is an automatic response to increased numbers and food shortage, and this response operates over and above a reduction in population through starvation. What happens is that certain animals such as lemmings, voles, and rabbits are able to adjust their own rate of breeding in order to meet an emergency. This process is called *internal fertility control*. For a time, when there is plenty of food, the birth-rate is fairly high, but eventually pressure of food and lack of space causes fewer young to be born. How this is achieved is a mystery; overcrowding in rabbits was studied under laboratory conditions and it was discovered that embryos that had already begun to develop were reabsorbed and were never born. Another interesting fact is that, in years when the fertility of these animals is high, the fertility of their predators is also high. But when natural fertility controls suppress the population growth, the predator temporarily checks its own breeding rate in response: for

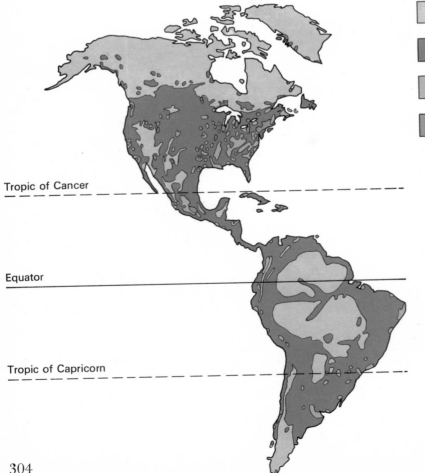

under 2

2 to 64

64 to 512

over 512

Tropic of Cancer

Equator

Tropic of Capricorn

It might seem, from this map of world population density, that the Earth is underpopulated. Red and blue areas are thinly scattered and vast regions support fewer than 62 people to the square mile. Most of these areas, however, are either too high, too dry, or too cold; the soil is permanently frozen or too poor to support many people. The population density an area can support depends not only on the productivity of the soil, but also on the extent of industrial development. An industrialized city may not be overpopulated although its population density is very high, whereas an agricultural area may be considered overpopulated at a much lower density because its potential is lower.

304

instance, the snowy owl of Canada fails to lay eggs if there is a shortage of lemmings. This pattern of fluctuating fertility is also known as *cyclicism*. Man, lacking the benefit of internal fertility control, must work out his own solution. His dilemma is not made easier by the fact that he is not prepared to harden his heart and do to his own kind what he does to other animals that overpopulate their habitat: he cannot, or will not as yet, crop his own species. He has to face his crisis without recourse to mass euthanasia, selective nuclear bombing, or any of the other ghoulish "solutions" that have sometimes been suggested.

The present crisis is actually quite recent in origin; in the past, human populations were, to a large extent, kept under control by natural checks such as famine and disease. Famines occur from time to time because of crop failure, and their efforts are all the more drastic if the human population in the famine area is already excessive. Disease takes two forms, and the more conspicuous is the epidemic: the Black Death, for instance, killed one third of the population of Europe in the 14th century, and the cholera epidemic of a hundred years ago killed 45 out of every 1000 Londoners. Wars in themselves have seldom acted as effective checks, but the aftermath of war has often taken the form of massive famines and epidemics, in which many more people have died than were killed in battle. But though the effect of epidemics is more spectacular, the biggest single control of human population has been the *endemic* disease, which has taken a steady toll of human lives from infancy on.

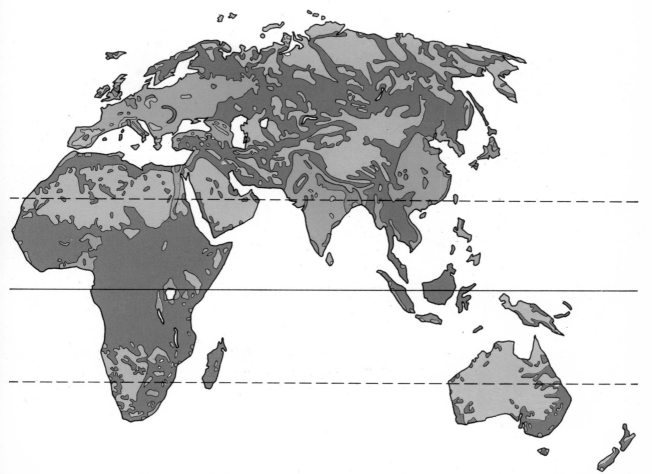

It is difficult to realize how effectively these natural checks regulated human populations and how recently the change has come about. It is only in this century that changes in medical science, communications, and technology have increased man's ability to control famine and disease, and the result has been a sharp increase in population, and hence a food shortage in many parts of the world.

Europe's ups and downs

To understand the background to the present problems we must go back to 17th-century Europe, where a complex series of slow changes, now known as *demographic transition*, was beginning to take place. Following on the beginning of the Industrial Revolution, the pace of industry began to increase right across Europe and beyond. New continents were opened up, and from them came additional food, and raw materials; commerce expanded and with it came improvements in transportation that allowed food and capital goods to be carried longer distances than ever before. Over the next two centuries technology moved into agriculture and industry, and food began to be mass-produced for the first time. Improved housing and sanitation in the mid-19th century gradually led to control of some diseases, and small improvements came also in the medical sciences.

These developments were welcomed wholeheartedly, but they upset the balance that had kept population down during the previous centuries. So, following the drop in the death-rate, life expectation went up, infant mortality went down, and the population of Europe rose. However, after about 1875 there was a noticeable reversal of the former upward trend; the rate of increase *declined* dramatically and this decline lasted until World War II.

What brought this about? On the superficial level one could say that it was due to the increased production of contraceptives and their increased use by the poorer classes. But this was only the means to an end, so to speak, and the

Left: the dead being carted off in Palermo, Sicily, during the cholera epidemic of 1835. In the past, disease was one of the natural checks on population growth. In this century, we have seen such progress in medicine that many of the fatal diseases have been eradicated over much of the world. With the decline in the death-rate, however, populations have greatly expanded.

trend had its roots in the deeper social changes that were taking place due to the Industrial Revolution. Why is this relevant to our present population crisis? Because there are some similarities between the social structure of 18th- and 19th-century Europe and that of today's underdeveloped regions. Europe's history does not, however, give us all the answers, because there are also many very basic differences, as we shall shortly see.

Before the Industrial Revolution, Europe consisted of agricultural communities, with here and there a town that had grown up around natural resources such as coal or iron deposits. Most families were tied to the land and they had to have many children because sons were needed to work on the farm and daughters were needed for the dairy and other indoor work. Also, disease took such a heavy toll that large families were necessary for survival; only two children might survive out of, say, ten born; there was no social security and parents relied on their children to keep them in their old age. Thus large family units consisting of several generations all lived and worked together.

But, as industrialization spread through Europe, more and more people left the land for work in the towns. The traditional rural family structure broke up; urban communities developed and these became based on the new and smaller units of parents and children. As the urban children grew up they trained for trades or professions, but did not usually need to stay to help their parents as they did in the farming society. All this meant that there was a general reduction of emphasis on early marriage and continuous childbearing and there was also an increased burden on the father to support his wife and family (every child had to be fed, clothed, and educated at the father's expense). This was a tremendous factor in reducing the number of children per family. In fact it has been said that materialism is the greatest incentive to planned families. When a family is very poor, and has low expectations from life, the tendency is to have many children. When the income is slightly higher and is coupled with the thirst for capital goods and possessions, the number of children decreases. Only in very wealthy families with the means to gratify greed for material possessions does the family size go up again.

There were, during the whole period, new areas of land to colonize; America became the largest overspill country in the world, receiving a very large number of Europeans. The colonies of Africa and South America also received a fair-sized quota of immigrants from overcrowded Europe and, one way and another, this all helped to prevent a demographic disaster. Even so, Europe remains one of the most densely populated regions on Earth.

Although emigration may have solved Europe's problems in the 19th century, it does not solve anything today, because any major movement of people is largely blocked by political boundaries and immigration policies. There is still a fair amount of physical space available but the Earth is no longer a single reservoir, open to all.

Because of these barriers, the population problem has become acute in some areas before it has affected other areas at all. If there were no barriers to population redistribution there might be no problem. As it is, there is no single solution that can apply equally to all nations and peoples.

How many people?

Today there are only a few places on Earth where the birth-rate is stable or decreasing; everywhere else it is increasing to some extent. By the middle of 1971 the overall population of the world was 3706 million. Two babies are born every second; the world population goes up by 6 million a month (over 1 million of these are in India alone). By the year 2000 there may be around 7000 million people on Earth. Several demographers have tried to dramatize the recent increase in world population by estimating the maximum numbers the Earth could hold provided existing resources were fully utilized. These estimates range very widely; one of the lowest, 2800 million (made in 1945), was overtaken about 10 years later. Several others, ranging from 5000 million

	1940		1950		1960		1970	
Mexico	44·3	23·2	45·5	16·2	45·0	11·4	42.0	9.0
Costa Rica	44·6	17·3	45·9	12·2	42·9	8·6	45.0	8.0
Chile	33·4	21·6	34·0	15·0	35·4	11·9	34.0	11.0
Venezuela	36·0	16·6	42·6	10·9	49·6	8·0	41.0	8.0
Sri Lanka (Ceylon)	35·8	20·6	39·7	12·4	37·0	9·1	32.0	8.0
Malaysia	40·7	20·1	42·3	15·9	37·7	9·5	37.0	8.0
Singapore	45·0	20·9	45·4	12·0	38·7	6·3	25.0	5.0
Japan	29·4	16·8	28·2	10·9	17·2	7·6	18.0	7.0

The above table shows changes in birth-rates (left-hand column) and death-rates (right-hand column) over two decades in selected countries. Except for Japan, the birth-rate has tended to rise but only a little, while the death-rates have fallen staggeringly, as a result of medical progress. It is death control, not any slight rise in birth-rates, that has caused the population avalanche. Right: the process shown diagrammatically (birth-rate purple, death-rate red).

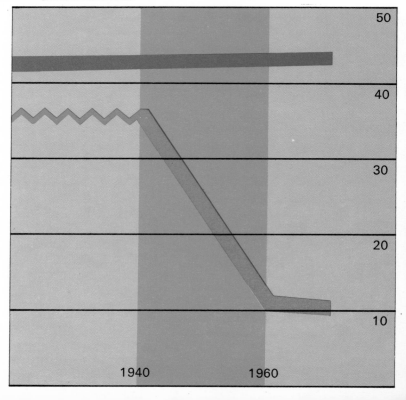

to 7000 million, will almost certainly be exceeded by the end of this century. The largest estimate—50,000 million—once thought to be in the realms of fantasy, will be reached at the present rate of increase in only 150 years. All the estimates except the last were made of the maximum number that could be supported without substantially lowering the standard of living. But it is almost inevitable that before very long the numbers of people in the world will be supported only by a general lowering of standards all round, including those of the affluent nations.

How are these estimates made? Let us look at this in more detail. Demographers usually calculate the fertility index of a population by estimating the *crude birth-rate*—the number of babies born for every thousand of the population in a given period of time, usually one year. This figure gives a rough idea of the increase of population due to births, but it can be very misleading. For instance it takes no account of the sex or age composition of the population and this can make all the difference. A birth-rate of, say, 30 per 1000 only tells

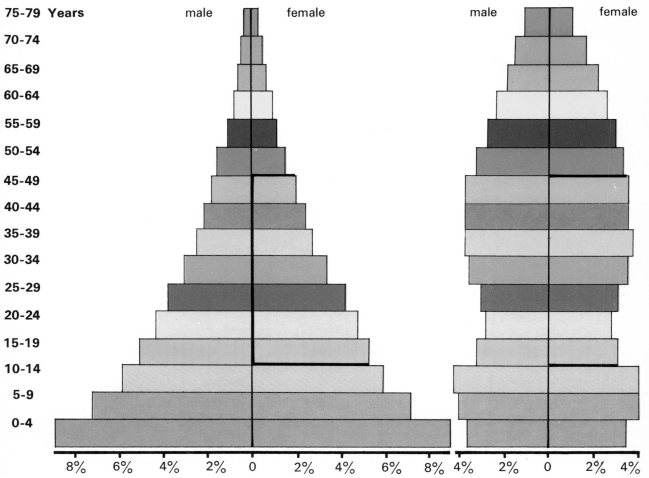

Pyramid diagrams show features of population at a glance. Left pyramid: Costa Rica in 1955 shows the high birth-rate (broad base), the high death-rate (tapering apex), and general increase in population (widening toward base). Right pyramid: Sweden in 1956; lower birth-rate (narrow base), postwar "bulge" (broadening at age groups 10–14 and 5–9).

you that for every 1000 of the population there is an average increase of 30, making 1030 after one year.

But in rapidly growing populations there is a proportionately large number of children. In any sample of 1000 people, one half will be males, and we can disregard these in calculating the birth-rate. If 30 percent of the female population are less than 15 years old, and 10 percent are too old to breed, it means that only 300 women are of breeding age, so the birth-rate actually reflects an increase of 30 children per 300. Another country may have a far smaller proportion of the population above and below breeding age—and the number of fertile females will be proportionately higher. Thus the same crude birth-rate figure will in fact, reflect a smaller percentage of children being born.

A better way of indicating fertility would be to relate the number of children born to the number of women of child-bearing age (15–48) and to ignore the rest. Unfortunately, the first method is used because it is easier, so it is as well to understand the method and to recognize its pitfalls when comparing one country with another.

World patterns of fertility

Birth-rates vary from 10 to 60 per thousand. Unfortunately one cannot rely on data from many of the countries where the birth-rate is probably highest. This is because people do not always register all births and also because accurate censuses are not easy to make in areas where villages are tiny and remote. Even so, some countries in Africa have recorded very high birth-rates—Guinea 62 per thousand, Niger 61, and Ghana 56.

In South America, where the population increased from 91 million in 1920 to 275 million in 1970, there is an average overall birth-rate of 45: in Asia the average is 41. These birth-rates are clearly very high when compared with the average for Europe, which is 16. Although Europe has a much lower average than some of the more underdeveloped parts of the world, even there the picture is not simple. For example in Eastern Europe both very high and very low birth-rates are found. That of Hungary is relatively low (15) and this seems strange, because Hungary is classified as a Catholic country. The main reason for this low rate is that abortion was legalized in 1956. The legislation states that every married woman can make a conscious determination on the size of the family she wants, and it permits her to interrupt an undesired pregnancy by means of induced abortion. Similar laws have since been made in the USSR, Bulgaria, Czechoslovakia, and Yugoslavia—all between 1955 and 1960. In the United States, it has been estimated, the population would stabilize if all unwanted pregnancies were terminated.

The highest birth-rates are in Albania (35.6), Romania (23.3), Ireland (21.5), Iceland (20.7), and Spain (20.2). The lowest, apart from Hungary, are in Sweden (13.5) and West Berlin (10.0). But simply knowing the birth-rate tells you little about the growth or decline of a population unless you know the death-rate as well.

The decline in mortality

The phenomenon we know as death-control is the direct result of recent medical progress, and, because it is universally more acceptable than birth-control, it has far more effect on population changes than declines or fluctuations in the birth-rate. The decline in mortality affects people of all ages; more babies survive and people live longer and breed for a longer period. It is thus the drop in the death-rate that is almost entirely responsible for the present "explosion"; the birth-rate itself has remained almost unchanged. Although the *crude death-rate* is—as with the birth-rate—calculated per thousand of the population, it is, if anything, more difficult to measure, partly because it is difficult to distinguish between still-births (and late fetal deaths) and deaths of persons born alive. Also there is a tendency for people (especially in remote areas) not to register all deaths. The United Nations estimates that only about 33 percent of the world's deaths and 42 percent of births are ever registered.

There is an increasing tendency for the drop in mortality to level out, in the more developed countries at any rate. Declines that began first in Europe (excepting Russia) have certainly leveled out there in the present century. In other parts of the world, however, death-control has begun only in the last two decades and, because of the youthful age structure, the declines have been spectacular.

Mortality has not declined everywhere equally, and unfortunately in those countries where it has done so most dramatically there has all too often been no corresponding rise in the standard of living. It also tends to be higher in rural areas than in towns, because most doctors and hospitals are based on centers of population.

There are two main causes of death; one is environmental, the other degenerative. The environmental deaths occur mostly in underdeveloped countries where infectious diseases are still rampant, living conditions are poor, and food shortages occur. Degenerative deaths—such as heart disease, cancer, and sheer old age—on the other hand, occur mostly in countries in which death-control is well advanced. In fact, people who die of these ailments usually do so because they have not died of anything else first.

Death-control has been heralded as an unequaled triumph of medical science over nature, but it is no longer looked upon with such naive optimism as it once was. It is now seen to be a two-edged sword, as many people are beginning to realize, and it cannot be welcomed wholeheartedly *unless it is matched by birth-control.*

As the population map shows, the areas colored red are those where the population is most dense and where the problem is generally most urgent. Much of Europe, however, despite its high population density, is no longer in a state of emergency. What can Europe's ups and downs teach the rest of the world? In fact, Europe is the foundation on which all of today's programs are based. But in certain ways its value as an example is limited. First, its area is comparatively small. Secondly, the population increased from 120 million in

1650 to only 424 million by 1960; Europe was never dealing in the sort of numbers that Asia is dealing with today. The population of Asia today is 2104 million—more than half the world's population—and by A.D. 2000 it has been estimated that it will probably be between 3000 and 4800 million as compared with Europe's projected 570 million. Thirdly, we must remember that changes came slowly in Europe, over more than 200 years of slow industrialization, slow improvement of communications, steady emigration, and increasing awareness of the advantages of reducing population size. But the problem that faces the underdeveloped countries today cannot possibly be solved in the same leisurely way. Fourthly, Asia may have an enormous area in contrast to compact Europe, but Asia is already overcrowded and has nowhere to spill to. Finally, the resources of many of the most overpopulated areas are played out, partly plundered by Europeans during the colonial period, partly mismanaged through ignorance, and partly exhausted through constant use.

Top: posters like this one from Turkey are a vital part of any Family Planning Campaign. Above: a mobile clinic in Kuala Lumpur, where women can get advice on family planning and be fitted with contraceptives.

The cult of fertility

It is difficult for most of us, who take for granted the influence of Western culture and tradition, to visualize the ways of life of communities that are quite different from ours. We are very possibly in danger of underestimating the extent to which traditions and religions provide a framework that holds non-industrialized societies together. In fact they play a much stronger part in the way of life of agricultural and rural people than they do in urban societies. Although almost everywhere Western influence is encroaching and bringing change on many fronts, really basic changes are not so easily made, for the traditional roots go deep.

As we already know, the need for many children in rural societies has always been a matter of survival. Not only is a plentiful supply of children needed for a labor force, but there is also a demand for sons as warriors in societies where manpower means military strength. The survival element becomes inextricably bound with tradition and religion. Children become a

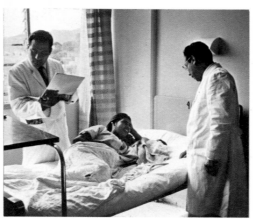

Left: a group of women being instructed on contraceptive techniques by a midwife at a Chilean clinic. Top: a social worker visits a family living in one of Chile's most overcrowded urban areas. Above: a ward in a new Mexican hospital that specializes in planning and maternal welfare.

313

Top left: the fertility cult Venus of Lausell, Dordogne. She holds a horn in her hand—a very common phallic or male fertility symbol. Top right: the cult of fertility continues; Chief Khotsa of South Africa, pictured with his wives.

Bottom picture: in European agricultural societies today, large families are still needed to work on the land. These Sicilian farmers all live together in a small farmhouse: there are about 20 of them altogether, including the children.

314

status symbol: mothers of large families are feted, and the father is honored for his virility, and his status in the community increases.

In societies where there are established structures of marriage and paternity, great emphasis is often put on procreation; indeed many religions, among them Catholicism, look on procreation as being the basic reason for marriage. Some allow one man to have only one wife; others, such as Islam and Hinduism, permit several—a type of marriage known as *polygyny*, which is practiced in many of the most densely populated parts of the world, and which encourages a high birth-rate.

Just how embedded the idea of fertility is in the social and religious life of India can be judged by a traditional greeting to a Brahmin girl—"be the mother of eight sons and may your husband live long." The other side of the coin, the shame and suffering of the childless, is dramatically shown by the South African writer Ronald Segal in his book *The Crisis of India* (1965): "The plight of the childless woman is only less pitiful than the spinster or, even worse, the widow, and there can be few sights in India as moving as that of a tree bearing the offerings of the barren like a crop of prayers. The childless endure any humiliation or trial in pursuit of divine intervention. In one south Indian ceremony held every year, the barren women, bathed by their families in the village tank, lie face downwards on the main village road, their arms extended above their heads, their hands together and holding plantains, coco-nut, and betel leaves. At last the priest, accompanied by shouting and drumming, and bearing a tall phallic symbol and covered with marigolds on his head, walks along the road stepping on the backs of the prostrate women, who are then lifted up by their husbands and sometimes carried further down the road for a second 'step.' On the other hand, the pregnant woman is regarded as auspicious and receives respect in public, while the woman who has produced a child, especially a son, acquires a new value both within the family and in Hindu society at large."

Although, in India and elsewhere, death-control has caused a phenomenal fall in infant mortality, somehow this fall has not been comprehended fully. It takes longer than one would think for a population to realize that far fewer of its children are dying. Furthermore even if this fact is grasped it still takes a very big effort to throw off the traditions that go with the cult of the large family. Fertility rites, phallic symbols, and the rest have always been part of the way of life, and art and architecture are not easily eradicated—certainly not without the loss of a good deal besides unwanted children. In such a setting what chance has birth-control? Can its importance be made clear enough to people so that it becomes more meaningful to them than their traditions and religion?

Birth-control

In spite of all we have said about the importance of fertility, we must add that birth-control is by no means unknown in preliterate societies (societies that have no written language). Unfortunately methods have never been really effective, but they have sometimes been inventive; they range from a variety of herbs that induce abortion or reduce fertility, to a favorite Chinese method as follows: "Fresh tadpoles hatching in the spring should be washed with pure well water and then swallowed whole three or four days later after menstruation. If a woman swallows fourteen live tadpoles on the first day and ten more on the next, she will not conceive for five years. If the dosage is repeated the following month she will be forever sterile. This method, the manual says, "is safe, effective and cheap; its only defect is that it can be used only in the spring."

The Chinese began using this method when their other chief means of population control, infanticide—by exposing girl babies at birth—was stopped. This practice—which, incidentally, is quite painless—is probably still going on in Tibet and possibly in other parts of the world as well. In the West, too, infanticide was quite common—more common, perhaps, than is realized. Babies with physical deformities were almost invariably exposed (remember Oedipus); this was an extremely efficient alternative to natural selection, which does not operate in humans nearly so efficiently as in other animals and in plants. But, resourceful though people have been, they have not conquered the problem that now faces us. A world program is needed; there are still far too many countries not reached by any family planning activities whatever.

As far as possible, it is better to plan a birth-control program with the support of the government and religious authorities. Of the most influential religions in the world today—Islam, Hinduism, Confucianism, Judaism, Buddhism, and Christianity—only a few sects have any doctrinal objection to artificial methods of birth-control. Of these sects, only the Roman Catholic Church has much real influence, and the 1968 papal encylical *Humanae Vitae* has remained obdurately opposed to any change in the law. In Catholic countries, family planning activities must and do continue without the church's sanction. Elsewhere, traditions still present a tremendous barrier, but family planning programs can at least be launched and carried out under government auspices.

The general attitudes of society to women as chattels and child-bearers must be changed first, and this means aiming propaganda at men as well as women. But the fight against rising population is not solely a fight to overcome religious and traditional objections to birth-control. It is a fight to improve general education levels and living standards of the majority of the world's people. For at the very roots of the population problem is the feeling of helplessness and despondency that prevails among the poor, hungry, overcrowded, and wretched of the earth. Any program of population control must take this into account, and indeed it must begin by encouraging people to find a reason for making a change. It is no easy matter to instill a sense of purpose into people who have lived with their misery for so long. Fatalism is built into many

Above: when a birth-control campaign initiated from outside is conducted in a country long acquainted with colonialism, there may be reactions like this one at Kingston, Jamaica, in 1956.

Right: use of contraceptives and family size are broadly related to educational level. On this chart the dark green bars represent the number of children women said they would have liked; light green bars show the number they had. Red bars show the percentage of each group practicing contraception. The figures are taken from the Taiwan survey carried out in 1962 and 1963.

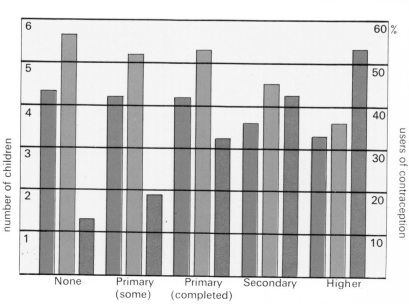

number of children

6 — 60 %
5 — 50
4 — 40
3 — 30
2 — 20
1 — 10

users of contraception

None   Primary (some)   Primary (completed)   Secondary   Higher

religions, especially those of the East, and the here-and-now must be made to matter, even if it undermines reliance on the world to come.

Japan's success story

In Asia one country has succeeded in substantially lowering its birth-rate—Japan. Small and overcrowded, Japan made a bid in 1941 to extend its frontiers and failed. The problem was still there, and indeed the returning armies made it worse than ever. By 1948 the population jumped by 8 million in three years, while the birth-rate was 35.5, far higher than that of any other industrial country. Japan seemed to be heading again for disaster, but instead a most remarkable thing happened. In 16 years the birth-rate was reduced to 17.0, a figure approaching that of Europe and North America.

This very considerable achievement was brought about, at first without government support, by the Japanese Family Planning Association, which mounted an efficient propaganda campaign to reduce fertility. Articles appeared in newspapers and women's magazines; talks were given on the radio. There were, of course, occasional outbreaks of counter-propaganda, in which birth-control was objected to for religious or ideological reasons; but the proponents of birth-control were more than a match for the objectors.

To begin with, the reduction in births was largely brought about by illegal abortion, though this was largely replaced by chemical contraceptives, which became available in 1949. In 1951, however, the government became really concerned with the continuing high rate of abortion and its effects on the mothers' health, and began to sponsor the family planning program officially, but with more emphasis on contraception. The budget for family planning in Japan increased from $59,000 in 1952 to $184,000 in 1965. This big program called for medical and what is known as paramedical staff (such as nurses and midwives). The latter had to be trained, and the midwives especially had to be helped to overcome the feeling that their primary function was that of assisting at births.

The most conspicious results have been since the war, but the preliminary work was done as early as the 1920s: Japan's high standard of literacy has been a major factor for success, and so has the high level of urban development and industrialization. Japan's example is of only limited use when applied to vast regions such as India, but it was the basis for another campaign that we describe in some detail because it gives some insight into the techniques used in fertility surveys, and also into the small returns for great expenditure of effort.

Trial campaign

In another small over-populated Asian island, Taiwan, a survey was carried out in 1962–3 to discover people's attitudes to regulating their own families, as a preliminary to a family planning program. Taiwan has certain similarities with Japan; it is relatively urbanized and industrialized, and literacy and general education levels are quite high. Also Taiwan, like Japan, has good communications and transport and a solid network of medical facilities.

The survey showed that most women had borne more children than they wanted, and that they realized that infant mortality had declined in their communities. People wanted to reduce their family size; all they needed to know was how to do it. A pilot experiment was carried out in the city of Taichung to determine how feasible and costly a nationwide program would be. First the whole city was exposed to a general distribution of posters and leaflets on the advantages of family planning. Next, research teams (consisting of local health officers and a cooperating team from the United States) divided the city up into over 2000 segments called *lins*. Some were called *nothing lins*—and in these, nothing further was done after the original distribution of posters and leaflets. There were also *mail lins*, in which a direct mail campaign was directed to newly-weds and also to parents with two or more children. In the last category, the *everything lins*, a major effort was made to promote family planning.

Every married woman of 20–39 years old was visited by an especially trained nurse and fieldworker who made an appointment for her to visit a health station provided with a wide variety of contraceptives. The health station also answered any questions and gave advice about family planning. In one half of these everything lins visits were made to wives only. In the other half, both husbands and wives were seen separately or together. The research teams did not encounter very much religious opposition, and, though it was then too early to judge long-term effects, the number of pregnant women went down from 14.2 percent at the end of 1962 to 11.4 percent at the end of 1963—a decline of about one fifth. By this time home visits had ceased, but the team kept up a follow-up program of meetings, and contraceptives remained available at the health centers.

The question is how well did the campaign succeed? Indeed how would one measure success? At the outset, good intentions could be measured only by the number of people who took advice, bought contraceptives, and expressed their intention of practicing contraception. Out of nearly 12,000 homes visited, a total of 5297 women accepted some form of planning. Over 4000 of them were from within the city of Taichung itself, the rest came from outside; this showed that news of the program had traveled effectively by word of mouth. But why did so few women from the homes visited take up voluntary birth control as a result of the program? About 16 percent were sterile or had been sterilized; 9 percent were pregnant, and 3 percent were breast-feeding. Some women actively wanted another child—young wives, or those who had not yet had a son. So in one way or another the final figure for new users of contraception was only about one tenth of the women visited.

The team maintained that if one quarter of the sample can be induced to practice family planning after such a program, then it has been a great success. They pointed out that the impact of such a program is not felt immediately; but after an initial period, when word-of-mouth campaigns are well established, the acceptance rate goes up. But even so it stays at a peak for only a short period (in Taichung it was for about four weeks). By then those who are really interested have taken action and the rate declines again (possibly due in part to a surfeit of propaganda).

The Taiwan program shows cause for both optimism and pessimism. On the credit side, people did not in principle oppose the program, and 10 percent of the people visited did respond; should one say "only 10 percent" or "as much as 10 percent"? On the debit side, Taiwan, like Japan, is small and compact; the city of Taichung is hardly typical of the vast and squalid rural areas of South-West Asia where communications and medical facilities are practically non-existent.

Perhaps the lessons most worth drawing from the experience of Japan and Taiwan as well as that of Europe, are that a campaign to reduce the birth-rate must coincide with a general betterment of living standards and education, and that motivation must come from within each individual country and not from outside.

This is not to say that help from outside organizations such as the International Family Planning Federation is not essential, but the main task of putting across the ideas and involving the whole populace can be done only by people who understand the local ways and customs. These people should, if at all possible, be nationals; outsiders may be regarded with suspicion if they try to introduce a family planning program. Look, for example, at India, during the time of the British Raj; attempts at regulating population were regarded as a plot to increase the ratio of whites to Indians. To take another instance, if Israel were to implement a birth-control plan throughout her present territory she would be accused of trying to increase the proportion of Jews to Arabs for strategic reasons.

Family planning in India

Even a program that is organized from within takes time to show results, and no one must expect changes to come overnight. India, in her 15-year plan (1961–76) to reduce her birth-rate, has put forward a three-point plan. The general aim is to reduce the birth-rate to 25 as soon as possible and the general approaches to this goal are as follows: (1) popularizing the various existing methods of contraception and any new ones that might be acceptable in India for mass application; (2) stimulating social changes that have a direct bearing on fertility, such as raising the marriage age of women and the general level of education, increasing security in old age, and eliminating child labor; (3) accelerating basic economic changes in order to increase the average income.

This growing awareness of the imperative need for general social and economic development as a companion to any birth-control program is more than offset in most of the crisis areas by a bottleneck—the lack of doctors and trained personnel. As long as education into methods of birth-control, and the

Many of the developing countries have an official family planning program, others have at least some government support. Among the countries where vigorous efforts to limit the size of families are being made is India. The program includes setting up of permanent-based and mobile clinics, educational courses and propaganda campaigns (top). Above: families are happiest when each child arrives because it is wanted, is the theme of the family planning movement. The emphasis is on the health, educational, and economic advantages of a planned number of children.

actual fitting and doling out of contraceptives, requires a doctor to see each patient, then not enough people can be reached in time.

The teething troubles of Taiwan have shown how important it is to train even non-medical people in family planning and health reorganization. In India every effort is being made to train one male and one female "family planning leader" for each village or group of 1000 of the population. The aim of the plan is to teach the people themselves, and to this end additional

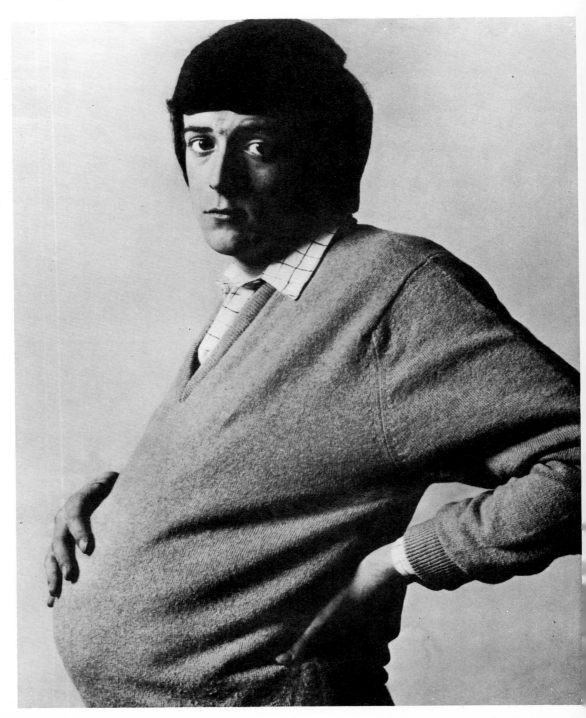

"education leaders" have been co-opted to help mobilize public opinion.

Pamphlets, posters, film strips, and exhibitions are used, and these, by all accounts, leave no one in any doubt as to what is needed to prevent conception and how it can be done. At first, due to almost complete lack of organization, the focus was on birth-control by the safe period method. Subsequently mechanical and chemical contraceptives were brought into use, followed finally by voluntary sterilization, as soon as there were enough doctors to be able to cope.

Even so, progress has so far been very small; India is still only scratching at the surface of the problem. There is some significant improvement in certain towns, notably in Bombay; but Calcutta is a nightmare city where three quarters of the population live in huts or on the streets without tap water or sanitation. Outside the cities, 80 percent of the population live in remote rural areas and many of these have not yet been reached. On the credit side, the program is organized by the government, which also gives its full support to the independent Family Planning Association. The will to conquer the problem is there; there is a certain amount of optimism, but whether it is justified or not, time alone will show.

New World problems

Latin America is another danger spot, with many problems quite different from those of India, and many that are similar. The general education level in much of Latin America is very low and there is as yet not much motivation to reduce the size of families. Unlike India, however, no government is directly sponsoring family planning programs, and the responsibility for this is left, in general, to the doctors, who are extremely scarce. In several countries there are still legal restrictions on the sale and distribution of contraceptive devices; the Catholic Church favors only "natural methods" of family planning, such as the rhythm method.

Only in Chile (with a birth-rate of 26.6 and death-rate of 9.0) and Colombia (38.0 to 10.0) are well-coordinated family planning services attempting to cover the whole country. In Chile there is an extremely high illegal abortion rate—reported to be half the birth-rate; it has quadrupled in 30 years. But there is a great deal of research being carried out on the problem of abortion, and it is possible that Chile will make strides in controlling and coming to terms with it—something that Japan did not wholly succeed in doing.

Colombia has mobilized a large-scale training program for over 1200 doctors and 800 paramedical personnel. As in Japan, public opinion has been stimulated through the mass media and is very receptive to new ideas on contraception. In several other countries, family planning associations exist, and so do private contraceptive services, but so far they are reaching only the urban communities. The problems that are specific to South America are concerned partly with the power of the Catholic Church, and partly with the physical nature of the continent. Almost everywhere in South America, except

Most modern conventional methods of birth control are designed to be employed by women, who also have to bear the burden of an unwanted pregnancy, should they fail. This British poster, captioned: "Would you be more careful if it was you that got pregnant?" sums up a widespread feeling that, because men are spared the physical burden of childbearing, they assume a less than equal responsibility for preventing unwanted pregnancies.

the relatively literate republics of Argentina and Uruguay, population is accelerating alarmingly, but, unlike India and even Japan, the actual density of population is as yet very low, and the untapped resources of the continent are enormous. Distribution of population is uneven, and growth is much more rapid in the urban areas. The "bands of misery" around many Latin American cities grow thicker, as people leave the rural areas as a result of economic pressures. Improvements in housing, schools, and so on, are absorbed by the rising population as fast as they are completed and in all spheres economic growth is impaired But this does not mean that Latin America is over-populated—at least, not in the sense that population density is too high, or that there is a lack of resources. A controlled and slower rate of population growth would result in a universal improvement in education and living conditions; many South American countries are potentially extremely productive, and could become major exporters. But today they are still dismally far from this goal.

Emergent Africa

The population of Africa today is about 340 million. It has risen very steeply

since the 19th century, when missionaries infiltrated the continent, bringing medicine and some death-control, but preaching vehemently against any kind of birth-control or abortion. The population has thus boomed wherever their influence was felt, and demographers expect an even sharper rise before any major decline can be expected. This is because the death-rate in parts of Africa is still fairly high—in Dahomey it is around 26, and in Upper Volta it is over 28. These countries have birth-rate figures of around 50, so it is easy to see what will happen when medical services improve. As it is, population growth is preventing economic progress in many parts of the continent. Over much of Africa there are no organized family planning associations, either private or government-sponsored; in many cases this is due to a law left over from French colonial days. In fact, almost the only countries with family planning facilities are those that have been influenced by Britain, except for the United Arab Republic and Tunisia. In Tunisia family allowances are *stopped* after the fourth child; this is supposed to act as a deterrent to parents who would otherwise have large families.

One of the main barriers to progress is that the remoteness of many communities makes it very difficult for medical services to reach them and there is

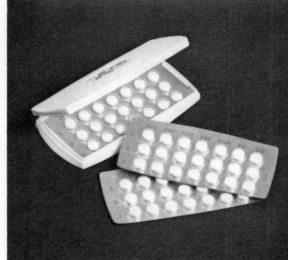

In the past, a pregnancy a year from marriage to middle age was the lot of most wives, such as the mother of 20 children, shown above. Right: "the pill"—the most effective means of birth control available. A wide variety of brands come in packs like these. When taken regularly and according to instructions, it is virtually completely effective.

therefore widespread ignorance of the advantages of family planning and maternal welfare. Where there are centers, they are usually in towns, attached to hospitals, such as the clinics set up in 1967 in Algiers. South of the Sahara, the first independent African government to adopt a national family planning program was Kenya. They began with a training scheme in 1967 and the clinic program began in 1968 at the Kenyatta Memorial Hospital, Nairobi. If Kenya's population, now about 10.5 million, continues to grow at the present rate (birth-rate 50, death-rate 20), it will double in 23 years; nearly half the entire population is under 15, and this points to an explosion if action is not taken immediately.

The United Arab Republic has a particularly difficult problem with its over-all population at about 35 million. The average birth-rate is over 40, and the death-rate is 14.8, and although there is a massive family planning program under way, the balance is heavily weighted against major improvements in the short term. The general standards of living and education are very low indeed and the impetus to reduce family size is still largely absent among the peasant farming communities that make up the bulk of the population.

Egypt in particular has created an immense problem in the effort to increase food production. The Aswan High Dam has been built to extend irrigation, by storing the flood water of the Nile for gradual release during the dry season. But this same dam intercepts the silt plus organic matter that for thousands of years has kept the Nile Valley fertile. The result is that, though there is a gain in water, there is a loss of soil fertility, which will have to be made good by manufacturing artificial fertilizers, using much of the electricity generated at the dam. Such fertilizers are no substitute for the organic matter now trapped with the silt behind the dam; and to make matters worse, in a few decades the dam will silt up, as dams always do (there are 2000 silted dams in the United States alone), so that it will no longer even do the job for which it was intended.

Over much of the rest of independent Africa, the International Family Planning Federation maintains that national governments are hesitant about taking a positive stand, but if private ventures were to get under way and persuade the population (and some of the members of the governments) that it is economically and socially desirable to limit families, then they might in time be amenable to some concerted action. Unfortunately there is no time; the economic expansion desired by these governments will fall further and further behind.

One of the problems with China (which has a population of over 787 millions), as with other socialist countries, has been that Marxism-Leninism does not recognize the existence of a population problem. Marx interpreted Malthus's 19th-century idea of "overpopulation" as a relative surplus of labor—an essential characteristic of capitalism that would be redressed by correct deployment of manpower and planned use of resources. (In simple terms Malthus held that despite every improvement in agriculture, people would always breed and multiply up to starvation level—when famine and disease would operate as a check.)

Marxists, in theory at least, see the reduction of the birth-rate not as the prime necessity, but as one of the many alternative solutions to the problem of mass poverty. Other solutions are to increase economic productivity and to improve health and education. These theories are in direct opposition to Malthus's ideas, on which much of Western population theory is based. In many other socialist countries, however, the state accepts the reality of the population problem, and some of them have extremely enlightened family planning services and policies, as we have seen.

To sum up, Malthus's theories are partly out of date now because he could not take into account the tremendous changes that were to occur because of advances in medicine and technology. But no more can we take for granted that these advances will continue to provide for us indefinitely. In the next chapter we shall take a look at some of the things that *are* being done to create an environment fit for us to live in, both now and in the future. But nothing the food scientists, sociologists, or anyone else can do or say will prevent a major disaster unless we limit the growth of the Earth's population.

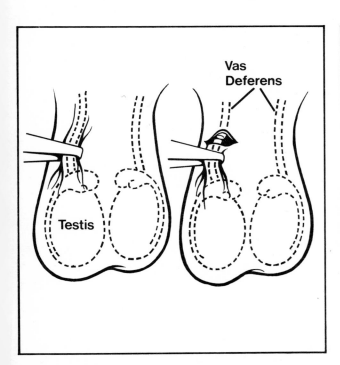

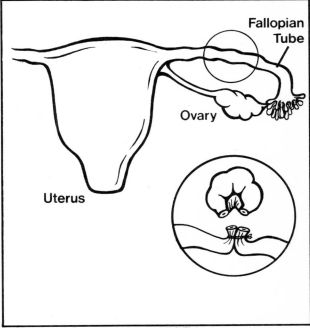

When a husband and wife are certain that they do not want any more children and wish to rid themselves of the concern over contraception, sterilization seems to be the best solution. The sterilization operation can be carried out on either partner, though it is much simpler for the male. In the man, above left, the procedure, called vasectomy, consists in cutting and tying off each of the vas deferens—or sperm-carrying tubes. In the woman (right) the operation is called a salpingectomy, or tubal ligation, and consists in cutting an inch from each of the fallopian tubes—the ducts that carry the egg from the ovary to the uterus. The ends are then tied back. Sterilization does not affect health or sexual desire, but is almost always irreversible.

20

Man Pollutes His World

It is customary in Western industrialized societies for apologists to define pollution as "something in the wrong place" and it is still widely believed that, as 19th-century industrial hygienists believed, "the answer to pollution is dilution." This is the philosophy that underlies the continuing poisoning of the Rhine and other rivers in Europe; the continuing decay of major brackish water areas such as Chesapeake Bay, the Baltic, and the Swiss lakes; and the naïve "high chimney" policy of Britain and many other countries, which simply transports pollution problems across national boundaries.

The truth, as the American biologist Professor Paul Ehrlich and many others have pointed out, is that nowhere is man's ecological ignorance more evident than in this "dilution" approach to pollution. The argument, based on the simplistic view of the physical chemistry laboratory, is that if you put a liter of poison into a million liters of water (or any other carrying medium) then the highest concentration you will eventually get anywhere is one part in a million. Thus, it is argued, if you know the rate of water flow in a river or the rate of dispersion in air, you can arrive at some "acceptable" rate of contamination from an industrial source because downstream or downwind the toxicity will be progressively reduced. Such an argument is persuasive because it is simple, appears to be correct and therefore appeals to common sense, and has the committed support of over a century of tradition in the oldest industrialized societies. Yet it is true only of *physical systems* and then only if complete mixing is possible, which is seldom the case and which, in any event, may involve time-spans of many thousands of years.

Pollutants do not enter "physical" systems in the sense assumed by those who hold the dilution philosophy, because air, water, and earth, however they may be defined chemically, are the intimate and interrelated substrate of every form of life. In biological systems not only does dilution not occur, but the longer a pollutant persists the more likely it is to be reconcentrated, in particular organisms or animals, within the biological process. Failure to understand this fundamental fact can lead to grave misinterpretation of the impact of a particular pollutant, and to serious underestimation of the

The surface fresh water, on which all life on land depends, collects in lakes and rivers. Yet this most valuable resource is too often treated by man as a convenient network of open sewers. It is only in recent years that we have begun to realize the serious need to keep pollution under control.

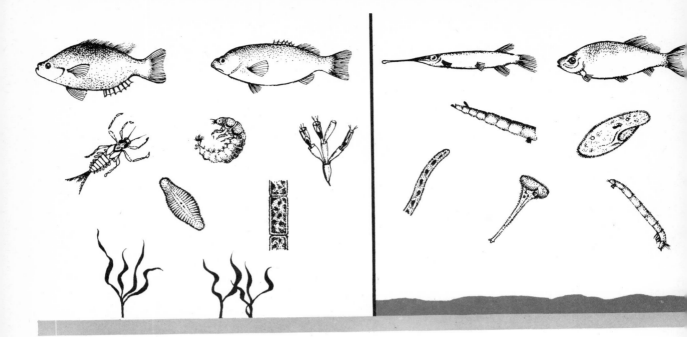

Stream with fresh, clear water. Community includes various game fish, invertebrates such as may fly and caddis fly, and many planktonic organisms.

Pollution begins, with water becoming dark and turbid, and community changing to those capable of withstanding low levels of dissolved oxygen.

importance of control. We saw an example of this in the interpretative blunder involving the decay time of DDT-type chlorinated hydrocarbons in lake water (Ch. 16). Man, suddenly aware of the enormous power of his handiwork, has since then both tried to understand and tried to conceal what was happening. It is important to understand the process of concentration in biological systems. In Chapter 4 we described how ecologists divide life structure into ecological pyramids, with the simplest organisms at the base and the higher creatures, including man, at the top. We saw that at each trophic level upward in the life chain the actual abundance of living material, the *biomass*, becomes successively smaller because of the inefficiency of food conversion. Therefore, any poison of a persistent kind distributed through the biomass at one level becomes compressed into a smaller biomass at the next and consequently its concentration rises. Whatever heads that particular pyramid may suffer a toxic insult a million times greater than the concentration at the bottom.

It is hardly surprising, therefore, that the first observations of damage caused by persistent pollutants are most often made by naturalists; nor that, over the past two decades, a number of particularly sensitive creatures have been selected by ecologists and ultimately by protection agencies as "indicator species" able to give a kind of early warning of trouble down the line. It has become painfully obvious that depletion of certain wildlife species caused by the change or destruction of their natural habitats by human activity has been more than matched by the indirect impacts of the biological systems themselves.

Indicator species can be of two kinds: those that are highly sensitive them-

Water now toxic to almost all life forms. Fish disappear; other organisms confined mainly to those that can obtain oxygen from the surface.

Pollution ends. As bacteria reduce the discharged effluent, releasing oxygen, the water gradually clears and a new community establishes itself.

selves because of something special about their ecological niche—reindeer, for example, or birds of prey; and those that happen to be unable to excrete a poison that, though not highly toxic to themselves, is a serious hazard to other creatures that prey on them, including man. There are, in face, enormous variations among different species in sensitivity to poisons—a fact most often observed when man suffers acutely, but seldom seen if the sensitive species is somewhere low down in the pyramid in a complex multi-species layer.

Poison barometers

The power to concentrate, and thus increase, both pollutants and their effects can be enormous even at a single stage in a chain. Shellfish, for example, are extremely efficient concentrators of metals and other chemicals, with the result that their tissues may contain as much as 100,000 times the concentration of poisonous substances occurring in the water in which they live. What is more, because they are exposed continually to their watery environment, they provide a long-term record of specific insults that may well be intermittent and therefore difficult to "catch" by ordinary sampling techniques. It is their *insensitivity* to a whole range of poisons that makes them potentially useful as barometers of the amount of contamination contained in their environment. They survive and tell us what is there.

Fortunately for man, if he learns to use them properly, there are also plants that concentrate toxins in much the same way, from air and from soil. Work carried out in Europe and the USA during the past few years has shown that some common kinds of moss can be used, through their long-term accumulations, to indicate air contaminants at levels well below the limits of sensitivity of mechanical sampling devices.

Perhaps the most sensitive indicators of all are the lichens, symbiotic colonies of fungi and algae that not only accumulate metals and other poisons directly, like the mosses, but also are highly sensitive to sulfur dioxide (SO_2). Unfortunately this is a sensitivity that, as with the only slightly less sensitive mosses, results in disruption and death, for the SO_2 increases the acidity of membrane surfaces and blocks the direct ion-exchange processes on which the life of these organisms depends. As recent surveys have shown, vast areas of the USA, Europe, Australia, and Japan are literally moss and lichen deserts. Sulfur dioxide, produced principally by the burning of fossil fuels, has made an unexpected but massive mark.

The importance of indicator species is this. Man's ecological knowledge is fragmentary, yet with almost every slight advance in knowledge it becomes increasingly clear that some poison, formerly regarded as tolerable, is in fact far more damaging to life processes than had been thought. Distortion or disruption of complex living webs depends on the behavior of the most sensitive links. Unless these are preserved, the whole structure is at risk.

The extreme sensitivity to contaminants of some kinds of phytoplankton, the basis of freshwater and marine life, was not seriously investigated until a few years ago, and then only in academic laboratories outside the umbrella of existing environmental control agencies. In 1971, it was found that concentrations of methyl mercury as low as one tenth of one part in 1000 million could reduce the planktonic growth-rate and therefore, potentially at least, could have a direct bearing on overall marine and freshwater productivity. Similar findings had been made earlier about various persistent pesticides. Although, in the actual circumstances prevailing in continental shelf and freshwater areas, there may be other limiting factors (such as the availability of nutrients or oxygen), the possibility of accidental interference with major natural processes remains very real. It is important to bear in mind the enormous scale of interference of this kind, and the impossibility of rectifying the situation once decline has set in. Man would have to wait helplessly until the biological system had cleared itself of contamination, and that would probably take several decades. It is relevant, and to some extent menacing, that the only consistent survey of plankton productivity in the North Atlantic— carried out over the past decade by British scientists—shows an unexplained, but real and continuing, drop in productivity. Yet man's ignorance of his own habitat is so great that it is quite impossible to assess whether this is some natural fluctuation, or the result of oceanic contaminations, or perhaps a little of both. So in considering pollution we have to bear in mind several factors that stem from the complexity of natural systems and from human ignorance, for we are still environmentally blind.

Cleaning up the mess

To many people "pollution" conjures up a picture of the excrement of technology, the effluent from chimneys and from outfalls in rivers and estuaries, the invisible toxic aerosols that pour out of vehicle exhausts and pervade urban areas, the soot on the windowsill, and the fogs and smogs of Northern Hemisphere urban winters. Yet these, like the sudden poisonous algal blooms off the coast of California or Britain, are only the obvious part of the picture. Because they are obvious and, in the industrial West, scandalously bad, they are now beginning to receive real attention by municipal authorities and governments. In many areas, it is argued, the situation is improving. Lake Erie, for instance, is not dead but returning slowly to life with an increasing productivity of coarse fish; Britain's Clean Air Acts have brought the return of songbirds to London's once near-barren parks; and the chemical giants of the USA and West Germany are now thinking twice before dumping waste.

All this is obviously to the good, and the somewhat rosy picture it provides of earnest and genuine official attempts to plug the toxic leaks of technology is encouraged persuasively by both industry and governments. Calculations carried out in the USA and Europe suggest that, for less than 2 percent of the Gross National Product, industrialized societies could clean up the worst of the mess, and—reluctantly—they are beginning to do so. Their reluctance stems primarily from the belief that expenditure on environmental protection puts industry at an "economic disadvantage" and that a dirtier industry

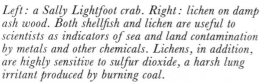

Left: a Sally Lightfoot crab. Right: lichen on damp ash wood. Both shellfish and lichen are useful to scientists as indicators of sea and land contamination by metals and other chemicals. Lichens, in addition, are highly sensitive to sulfur dioxide, a harsh lung irritant produced by burning coal.

333

somewhere else will gobble up the markets. Such a point of view, bolstered by threats of unemployment, still powers the hand of industry and distorts the efforts of agencies attempting even "facelifting" improvements.

Organic mercurials

It has taken a long time and many disasters for industrialized societies to reach this uncomfortable and inadequate compromise, but even at this elementary level of environmental protection the lessons have not yet gone fully home.

Mankind's handling of the very highly poisonous compounds called *organic mercurials* has been punctuated by one disaster after another. Within two years of their discovery in the late 19th century, technicians who had handled them died in a state of crippled idiocy, and in Germany, syphilitic patients on whom the compounds had been tried as treatment began to suffer symptoms of acute poisoning. Public health and medical authorities, whose criteria on the toxicity of mercury had been built up empirically through the industrial and medical use of the much less poisonous *inorganic* mercury compounds, appeared not to notice that something new was on the scene.

By the 1920s the powerful fungicidal properties of organic mercurials were being investigated and exploited by industry, and in spite of confirmation of their extremely high toxicity in animal tests, large-scale manufacture began. So did the long and growing toll in disablement and death of those handling

Above: underwater oil tent being lowered into the water at Santa Barbara, California. A "tent city" is being built under the ocean off the California coast to catch oil seeping from a leaking oil well before it contaminates the waters of the Santa Barbara Channel.

the compounds. Before 1940 it was perfectly clear to anyone who cared to examine the evidence that ethyl and methyl mercury compounds were so insidiously and highly poisonous that they should never be manufactured at all. Yet, even though it was known that other, less poisonous, compounds could serve equally well, use of these most poisonous compounds escalated in both agriculture and paper manufacture. Only 30 years later, after a whole series of mass poisonings (caused principally by the accidental use of mercury-treated seed as food), after large-scale depletion of seed-eating wildlife, and after the discovery that by other routes inland waters were highly and dangerously contaminated in many parts of the world, did the authorities act. Control is still only partial, and in 1972 Iraq suffered another large-scale poisoning incident that affected not only humans but cattle, poultry, and fish. The situation was so serious that the government ordered that anyone dumping treated grain in rivers should be summarily shot.

In humans and other mammals, certain of the organic mercurials selectively and permanently destroy parts of the brain. The damage is not reversible. Because we are endowed at birth with a limited number of brain cells, which do not replicate, but decay with time to produce senility, additional damage should be avoided at any cost. But the brain is highly complex and contains many alternative pathways, so that functions may continue apparently unimpaired even when a large number of cells are disrupted. By the time symptoms of organic mercury poisoning appear, massive damage has been done. As was

A sparrow hawk photographed from a balcony in London. Their reappearance in our big cities could be a hopeful sign that their numbers are increasing (and agricultural poisons decreasing) or it could be that only those hawks that prey on the nontoxic city scavengers survive.

Left: junked cars in the British countryside. The disposal of solid waste is a serious problem in the affluent countries—burning merely substitutes air pollution for land pollution, and open dumps and fills are both expensive and unsightly.

Devastation of a beach in the heavily industrialized county of Durham, England. The need to preserve social amenities, such as beaches, has been balanced against other needs—in this case, the disposal of industrial waste.

Top: smog over midtown Manhattan. Wasted fuel; damage by corrosion and blackening to stone, metalwork, and other fabric; amenity losses; and damage to health through respiratory disease, are some of the evils of atmospheric pollution. There is also the effect on plant life: this tree (above left) has been killed by chemical pollution; a few plants (above right) struggle to survive in the red dust from the local steelworks that blankets this English town.

shown by the tragic parade of the Minamata victims in Stockholm during the UN Conference on the Human Environment in 1972, the damage is forever.

But the Minamata and other Japanese mercury-poisoning "incidents" were of a different kind. These involved the industrial discharge of large amounts of *inorganic* mercury that, in the marine ecological system, were converted to di-methyl mercury, which concentrated up the food chain. Only when birds, cats, and peasant fishermen and their families began to go mad and die was it realized that something was seriously wrong. Even then it took several years for the source to be identified and the biological route to be understood. Some estimates suggest that more than half the fishing population making its livelihood from Minamata Bay and nearby stretches of coast was affected, and from the number and nature of poisonings it became clear that infants and children were especially vulnerable. What was worse, the Minamata incident was followed by others elsewhere in Japan, and was wholly ignored by government and other protection agencies elsewhere in the world.

There is, of course, some mercury continually reaching biological systems from natural sources, because mercury is a component of the Earth's crust, but the levels are usually very low. What has now been learned is that there is no threshold below which mercury does no damage, and that the best we can do is to keep man-mobilized contributions to a minimum. Yet, because in a great burst of fear some aspects of the mercury problem were overstated during the early 1970s, especially in the USA, many apologists now say that the mercury threat is a myth. What they really mean is that there is no evidence of any massive buildup of man-mobilized mercury compounds in the oceans, and that, if we do some simplistic sums assuming complete mixing, the man-mobilized contribution is negligible. Such sums are irrelevant. The dangers are on land, in food chains, in inland waters, and in estuary and coastal regions—the most productive parts of the biosphere. The danger is real, and will not be driven away by wishful thinking.

The mercury story is important because it unmasks past incompetence and irresponsibility of industry and government in handling hazardous developments. Also, because it shows that the processes of complex biological systems are unpredictable and powerful yet tend to be ignored until a major damaging situation occurs; and because it demonstrates that in dealing with pollution we need to think not simply of industrial effluents, but also of the *products* of industry. This is the real lesson of the past 20 years, and, as we shall see later, it has many important aspects of its own. It is fairly easy, for example, to work at a list of potential pollutants cropping up in industrial effluents of one sort and another, and to say which must be rigidly controlled, which need careful regulation, and which can be safely discharged because, provided natural systems are not overloaded, they will be rapidly broken down into harmless components. Persistent poisonous substances, such as the heavy metals and many pesticides and their intermediaries, should be extremely rigidly controlled, because we already know of their injurious effects in ecological systems

One of the mercury-poisoning victims of Minamata, Japan. Altogether 67 people died and 330 residents were permanently disabled at Minamata, and elsewhere in Japan some 10,000 people were affected by what came to be known as Minamata disease. So far nearly 3½ million dollars' compensation has been granted to the victims.

and suspect—from highly suggestive evidence—that they are deeply involved as causative agents in human and animal disease. True, the heavy metals pose special problems because some are essential micronutrients for either plants or animals. But many of the most toxic of the heavy metals—lead, cadmium, and selenium, for instance—have a harmful effect wherever they occur, and yet, in large-scale industrial use, are becoming widely dispersed at abnormally high concentrations throughout the biosphere.

Lists of this sort, defining "grades" of pollutants, are characteristic of the guidelines of air and water pollution agencies, and form the basis of existing international agreements on such things as ocean dumping—the Oslo Convention, for example, and the more recent International Convention. Such lists embody traditional policies reaching back to the early days of river pollution control, and make the remarkable assumption that their effects in natural systems are neither additive nor amplified.

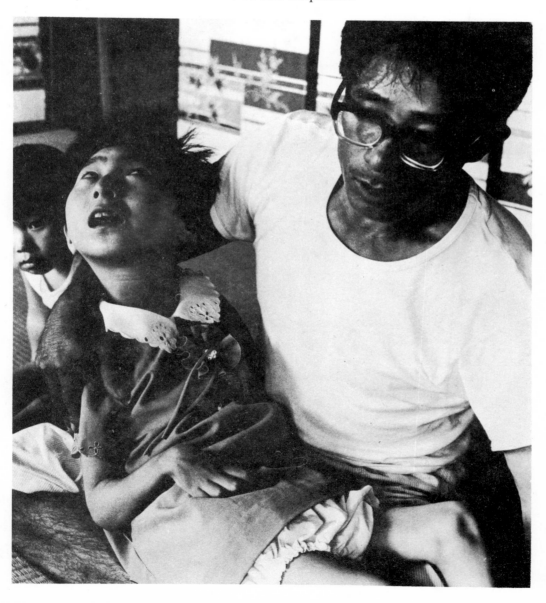

In 1967 the U.S. oil tanker Torrey Canyon *grounded on rocks off the Scilly Isles, southern England. Vast quantities of oil leaked from her damaged tanks, fouled the British and French coasts, and killed many thousands of sea birds. As the oil washed Britain's southern coastlines (bottom: oil at Porthleven, Cornwall, less than 30 miles from the Torrey Canyon wreck), the disaster was treated as a national emergency and frantic efforts were made to find ways of cleaning up the mess—one of which was to treat the oil slick with detergent (left and opposite). It is often not until oil is spilled on a nation's doorstep, as with the Torrey Canyon, or the Santa Barbara leak (p. 334), that people realize the seriousness of oil pollution. The effects of oil spills on marine ecosystems are known to be serious, and it poses questions, too, about the concentration up the food chain of carcinogenic agents within the oil. Detergents used to clean up the mess have, in most cases, made matters worse—not only do they help to disperse the slick over a wider area, but they may themselves be toxic to many forms of marine life. As the volume of oil transported at sea increases each year, efforts are being made to combat oil-spill more effectively. Methods of containing and collecting oil before it reaches shore are being studied, and efforts are being made to develop strains of bacteria that will break down the oil into components that can be absorbed harmlessly into the marine ecosystem.*

Crossing the toxic threshold

Seabirds often die suddenly in large numbers, especially from tides of black oil, and when in 1969 somewhere between 50,000 and 100,000 razorbills and guillemots died within a few weeks in the Irish Sea, it was at first assumed either that there had been some rare natural disaster—a crisis in the food chain, or an epidemic of avian disease—or that some highly poisonous pollutant had somehow reached the area in large quantities. (Oil, in this instance, was clearly not to blame.) Because preliminary examination of dead birds revealed no obvious cause of death, a very thorough investigation was mounted, perhaps the most detailed ever to follow a wildlife kill of any kind.

The findings posed a grave warning to mankind. There was no single dominant cause, but the birds' tissues were rich in a wide range of persistent poisons. Weakened after the molt and subjected to the stress of a brief spell of stormy weather, food deprivation had resulted in the mobilization of stored poisons from fatty body tissues. Lead, mercury, selenium, arsenic, cadmium, chlorinated hydrocarbons, and polychlorinated biphenyls were all present and in some cases close to known toxic thresholds. The first symptoms of poisoning are behavioral, and often involve loss of appetite. It seems that this whole population of normally vigorous seabirds crossed some unsuspected lethal threshold in which stress resulted in self-poisoning and a downward spiral to death. They starved in calm September weather, amid an abundance of food.

Several very important points emerge from this tragedy. Contamination of European coastal waters, like that of the coastal waters of the USA, is already at dangerously high levels, and it is little use trying to measure the potential hazards of the situation merely by looking at individual contaminants. It is the "toxic cocktail" that matters. The thresholds of environmental calamities are often obscure, but it is clear that present levels of environmental poisons are sufficiently high for them to work in combination to multiply the effects of natural stress many times—perhaps even hundreds of times. Vast amounts of persistent poisons have already permeated and undermined the entire life system, and are likely to set off stress disasters far into the future. We can do nothing now to prevent this happening. What is more, this population crash involved the adult, and therefore biologically most robust, members of the species. We know virtually nothing about the effects farther down the food chain, or on other marine animals that might well be more sensitive. Our biological awareness, in events of this kind, starts and ends with the ragged litter of dead animals for which it is necessary to postulate the most probable history.

These sources of contamination are clearly industrial, because some of the poisons do not exist at all in nature and others, in uncontaminated circumstances, are at levels that pose little threat of natural catastrophe. It is easiest to assume that what is being seen in and around industrialized countries is simply the result of poor effluent control. If we clean up the smokestacks and the liquid outflow, it might seem that things will rapidly get better. Unfortunately, whatever is done along those lines—and it must not be discouraged for these or any other reasons—there can be no rapid improvement because the clearance of nondegradable poisons from biological systems, and from the great water cycle of the Earth, is an extremely slow process. The insults being leveled by man at the whole web of life (including himself) is a burden that we have already imposed on generations not yet born. If we ceased using DDT tomorrow it would still be in rain a century hence.

But pollution control, in the orthodox accepted sense, deals with only a fragment of the problem. As the mercury story showed, the major impact was caused not by industrial effluent, but by an applied technology. We have to grasp the savage and salutary fact that any activity alien to evolved biological systems is likely to be damaging, and that almost every aspect of existing industrial technology is alien in that sense and will therefore provoke a biological backlash. The processes of nature are, if one excludes the energy flow from the sun, entirely cyclical. The processes of mankind, especially of industrialized mankind, are almost entirely open-ended.

Lead: versatile, deadly

The attitude of industrial societies to lead, a highly potent nerve poison, is a good illustration of the present situation. Wherever lead is smelted or in some other way involved in industrial processes, such as the casting of pipes or of lead alloy components, considerable attention is given to the control of fumes. The workers, essentially a selected group of humans who happen to have a high level of lead tolerance, are kept under continual medical scrutiny to ensure that their lead intake remains below the level of obvious clinical poisoning. From this very special context, guidelines have been laid down for what is or what is not lead poisoning in some strict, though as it happens arbitrary, sense. These criteria have no application to whole populations, in which there exists an enormous variation of individual sensitivity to particular poisons. Also, except in East European countries and Russia, they take no account whatever of the behavioral effects of lead, its ability to cross the placental barrier and thus affect the growing fetus in its most highly sensitive stage, or the general and gradual buildup of lead in the biosphere. Explicitly, the guidelines cannot and do not apply to children of any age, and they do not apply and cannot be applied to the organic compounds of lead that now make up a significant component of lead contamination in urban socieites.

Lead exists in low concentrations almost everywhere in the Earth's crust, and the natural processes amid which life has evolved mobilizes about 150,000 tons a year, which goes through the various life-cycles to end up eventually in sediments, and hence back in the Earth. At the present time, and at a growing rate, man is mobilizing about 5 million tons each year, so that even if complete physical mixing could occur, the lead burden on natural cycles—and lead has been found to be damaging to biological processes wherever it has been studied

—would be 30 times greater than normal. But thousands—perhaps tens of thousands—of years would need to pass before physical distribution could approach uniformity, and, as we know, biological systems prevent this happening. In any case, man is distributing almost one million tons of lead a year—400,000 tons in Europe and the USA alone—through car exhaust fumes as a fine aerosol that could hardly be better designed for rapid absorption through human and other mammalian lungs.

While on one hand society treats lead with great respect as a poison, and in the case of organo-lead compounds (which are so poisonous that they have been rated as chemical weapons) provides industrial control of the most stringent kind, including space suits and air lines for those handling them, on the other hand it sprays these same compounds around in vast quantities on a wholly unprotected public! Of the 5 million tons or so of lead mined, refined, and processed into products each year, only about 5 percent is industrially recycled. The rest, by one route or another, eventually enters natural cycles, animals, and plants, where it may briefly be stored but from where it is liable at any time to be released again onto some hidden but damaging pathway.

True, during the 1970s—and not always for the right reasons—new controls have been brought to bear that, in a long time, might reduce the worst excesses of the lead buildup, but they do not solve the pollution problem. The nature of our technology is such that its products, having served their brief purpose, are simply dumped or slowly leached into the biosphere. In any case, the new controls have to fight their way against the bitter opposition of long-term industrial commitment, and are strictly limited in the rate at which they can progress by the long turnover time of technological change. In the meantime, urban children throughout the Western world will continue to absorb lead in quantities that not only renders them highly vulnerable to acute poisoning in the event of release of lead stored in the tissues—such as at a time of fever—but also in a high proportion of the child population exceeds the level at which brain and behavioral changes are known to occur.

In one sense lead is a special case, for it happens to be particularly poisonous to animals at the top end of food chains—those with highly developed central nervous systems. But in another sense lead is typical, for the major characteristic of modern technology is its large-scale transition from natural products—which can safely enter natural systems—to products for which no natural degrading cycles occur.

Man against nature

The argument that, if we are to survive, we must close our technologies and integrate them with existing natural cycles, has been most eloquently and persuasively voiced by Professor Barry Commoner. Leaving aside for the moment consideration of the impact of economic growth itself (which can certainly increase pollution) and population growth (which in an important sense necessitates a rise in human consumption and pollution), the question

Genetic mutations, cancers, and stillbirths may have always been brought about by the natural background radiation (cosmic rays, radioactive substances in the Earth's crust, and natural radioisotopes) against which man has evolved. Since 1945, when the first atomic bomb was exploded, radiation from nuclear-weapon testing has increased the likelihood of genetic defects. Strontium 90, part of the fallout from nuclear explosions, is present in the bones of every living person.

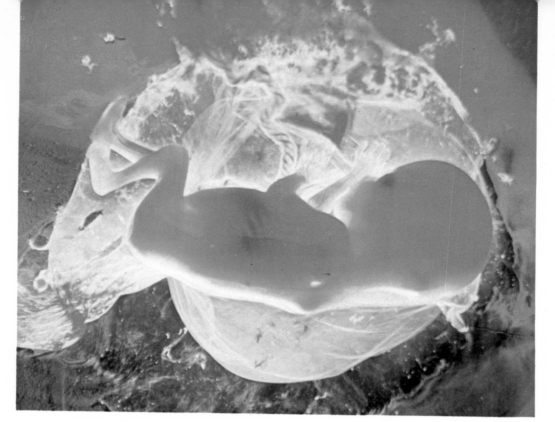

to ask is whether the massive increases in human effluent during the period since World War II can also be attributed to some other factors.

Commoner's analysis, which is American-based but has relevance to every affluent and developing society (nations comprising less than 30 percent of the world population produce roughly 85 percent of global pollution), makes some discoveries of profound importance. Economic growth in the USA during the period 1946–71, taking the Gross National Product and measuring it in the conventional way as total production, has risen 126 percent. Yet the average American now consumes just about the same amount of food (poorer in vitamins), uses about the same amount of cleaners, buys about the same quantity of clothes, occupies about the same volume of new housing, needs about the same amount of goods transport, and consumes about the same amount of coffee or beer as he did in 1946. But during the same period the amount of pollution, measured as persistent and damaging materials entering the environment, has risen about 1000 percent. In Europe, on present trends, a broadly parallel situation will have been reached by the end of the present decade.

Why has affluent man, superficially so much cleaner than at any time in his history, become so rapidly and in reality so much dirtier an animal? Barry Commoner's explanation is both simple and of worldwide significance:

> His food is now grown on less land with much more fertilizer and pesticides than before: his clothes are more likely to be made of synthetic fibers rather than cotton or wool: he launders with synthetic detergents rather than soap; he lives and works in buildings that depend more heavily on aluminium, concrete and plastic than on steel and lumber: the goods he uses are increasingly shipped by road rather than rail: he drinks beer from non-returnable bottles or cans, rather than from returnable bottles or at the tavern bar. He is more likely to live and work in air-conditioned surroundings than before. He also drives twice as fast as he did in 1946, in a heavier car, on synthetic rather than natural rubber tyres, using more gasoline per mile, containing more tetraethyl lead, fed into an engine of increased horsepower and compression ratio. . . .

True, the transition over this period has also endowed the average citizen with more refrigerators, TV sets, transistor radios, telephones, and other consumer goods than he had in 1946, but these, although linked intimately with resources use, are peripheral to the major transition. To create and meet the changes, driven by narrowly based cost-benefit analysis and arguments of consumer "convenience," industry has proliferated in other no less damaging but secondary directions. To feed the new synthetic material industries with essential feedstocks—such as the chlorine for the organic chemicals needed in the manufacture of pesticides, herbicides, plastics, and so on—other new energy-demanding and highly polluting industries have been established, so that even at the *producing* end the impact of industry on the environment has increased very sharply. But, as Professor Commoner and others explain, this is by no means the whole story. In many instances the *products themselves* have a massive impact on the environment, either because they are unknown in nature, and no natural systems exist for their breakdown and biological recycling, or because they are directly poisonous and ecologically disruptive.

"Indestructible" wastes

When Thor Heyerdahl sailed his raft across the Atlantic in 1970, mankind learned not only that human migrations had been possible long in the past by means of rafts and ocean currents, but that the Atlantic was now covered in tar-like clots that at times "stretched from horizon to horizon." More has been learned since then, about the 3–4 million tons of oil that reach the oceans directly and indirectly each year, and about how long it remains in the surface layers, and its effect on marine life. But if Thor Heyerdahl had been able also to see the ocean bottom, particularly that of the productive shelving areas around the land masses of the industrialized nations, he would have seen industrial detritus stretching from horizon to horizon. For there, intimately mixed with the cyclic processes of sea-bottom marine life, are the almost indestructible fragments of nylon, Dacron, PVC, and the hoard of other products that pile up in the environment as their usefulness to humans expires. The natural degrading "half-life" (that is, the time it takes for half of any given amount of material to be broken down in the biological system) of some kinds of plastics is said to be 10,000 million years. This means that when the Earth-life finally succumbs to the expanding sun, most of the plastics that have reached natural systems will still be there; whereas the oil, like the men who squandered it and thereby robbed future generations of its most valuable benefits, will have long gone.

So the major sources of pollution are not the effluent pipes and smoke-stacks of ill-controlled industry, but the technologies on which industrial man now leans in a posture of growing affluence. This is a fundamental issue that will be looked at in more detail later. But even those aspects of advanced society that seem to have no connection with pollution can have effects that, quite strictly, lead to ecological disturbances that are exactly parallel to the slow, accumulating effects of poisoning. There are also forms of pollution that, although brought about by the nature of industrial exploitation in a "free market," are not widely regarded as pollution at all, but are nevertheless corrupting and need careful scrutiny.

The pesticide spiral

The case against persistent pesticides does not rest on the dangers posed by poisons that reach into and threaten every corner of the biosphere. Nor do they rest on the conservationists' very proper concern for wild creatures, which have suffered accidental and crippling impact. It is based on a much more powerful argument. The broad-spectrum-insect-killer answer to pest control has come about through a failure to understand the basic problem.

The problem is not, nor ever will be, to kill "pests" so efficiently that they are exterminated, for this is not possible. "Pests" are only insects that, because of some ecological distortion, have been allowed to multiply and to occupy a much larger niche than is normal.

In using pesticides man does two things. He creates conditions in which pressures lead to the emergence of resistant strains, thus necessitating an ever-upward spiral in the amount of pesticide that has to be used to destroy a particular insect, while at the same time he destroys even more effectively than the target insect those species of insect that in natural circumstances keep the "pest" population down. Because the controlling insects are normally higher up in the food chain, their life cycle is longer, so their rate of selection toward resistance is lower, and the pesticide impact is individually greater. The technology not only ensures that resistant species develop rapidly, but also ensures that the ecological system is so disrupted that the "pest" can sweep back unopposed by all its natural enemies. This is not simply inefficient or dangerous, but downright stupid.

True, during the past few years some enlightened agronomists and ecologists have been promoting new and sensible notions about "integrated" pest control, such as introducing natural predators, but the urgent need for the widespread adoption of such systems is by no means widely recognized. On the contrary, as broad-spectrum pesticides are being controlled *for other reasons* in advanced countries using intensive agricultures, industry is unloading its ecological poisons on developing and ecologically naïve nations.

Antibiotics—use and abuse

All this, perhaps, is well known but inadequately digested. What is less well known and not digested at all, is the ecological error underlying the present use of, and impact of, the antibiotics. Antibiotic use in relation to food production will be discussed in the next chapter, but these substances are also pollutants in the sense that, as biologically highly active compounds, they have become widely dispersed in the environment and can have effects that are both dangerous and unexpected. Although it would be wholly wrong to suggest that antibiotics are not of great value in the treatment of disease, the circumstances of greatest value are probably very limited, and the present patterns of use pose very real threats to humanity that stem wholly from the *disruption of bacterial ecology*. When discussing antibiotics we have to remember several important facts. The broad-spectrum drugs are exactly analogous to broad-spectrum pesticides, in that their effects are not confined to target organisms, but hit all sensitive organisms. The proliferation of bacteria indicated by symptoms of disease is exactly analogous to the infestation of a crop by pests. It indicates not that, in some separate way, a pest has emerged, but that an ecological equilibrium has been disturbed. To achieve any major successes in disease control it is not to medicine that we should look, but to public health, which, like good husbandry, creates conditions in which

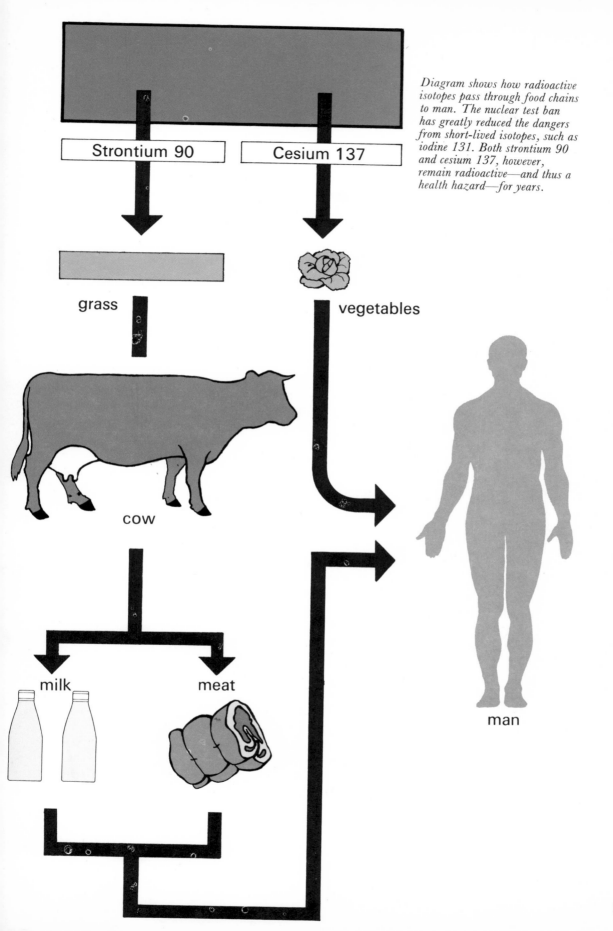

Strontium 90

Cesium 137

Diagram shows how radioactive isotopes pass through food chains to man. The nuclear test ban has greatly reduced the dangers from short-lived isotopes, such as iodine 131. Both strontium 90 and cesium 137, however, remain radioactive—and thus a health hazard—for years.

grass

vegetables

cow

milk

meat

man

damaging organisms are at a disadvantage; we must realize that all but a tiny percentage of all bacteria are not simply harmless, but absolutely essential to the continuing well-being of life processes.

Emergence of the "R" factor

It was discovered very early in the use of antibiotics that naturally resistant strains existed among bacteria, and that prolonged use of a single antibiotic in the treatment of a particular condition—syphilis, for example—resulted in the emergence of infections almost wholly immune to treatment with that antibiotic, whatever the dose used. This was and is a serious thing, but not insurmountable, because new antibiotics with different molecular structures keep coming along to enter the battle and score new local successes. True, the infamous "hospital staphylococcus" arose as a highly resistant strain that led to extensive human suffering, but, by closing hospital wards and taking all medical precautions, the threat is kept under some kind of control.

In the late 1950s, however, first in Japan, then in Britain and the USA, bacteriological research revealed that patterns of resistance to single antibiotics had extended to resistance factors covering several of the most widely used antibiotics. What was more, the multiresistance genetic factor (or factors) could be transmitted directly from one bacterium to another, whether or not these were of the same genus. From the early 1960s onward the threat to public health by these transmissible "R" factors has loomed like a shadow behind antibiotic use and is now seen as one of the most important and dangerous side-effects.

Most of the many kinds of bacteria that inhabit the mammalian gut and enable it to operate properly are a component part of the everyday environment, and there is a continual movement of bacteria into and out of the larger living creatures. Everything that lives, whether an amoeba or a man, is nothing more than a leaky bag of cytoplasm and nuclei, with a process continual exchange going on between it and the environment in which it lives. It is quite impossible to say where environment ends and creature begins, for the two are inseparable. Bacteria are part of both, and are essential to both, just as both in turn are essential to bacteria.

With the emergence of transmissible "R" factors, we thus have a situation in which bacteria essential to life are capable of passing on to pathogenic, or disease-causing bacteria high levels of resistance to a whole range of antibiotics at a single contact. The transmission can be very rapid, as laboratory experiments have shown, and in the case of diseases such as typhoid—for whose treatment only one useful antibiotic exists—there is every possibility of the disease emerging in a highly resistant form at any time. The drug involved—chloramphenicol—is used in animal husbandry in most parts of the world, and thus necessarily produces resistant strains of bacteria carrying a transmissible "R" factor. In the ordinary course of events this must reach typhoid bacteria, which sooner or later will emerge in man in epidemic proportions.

During the 1960s and early 1970s public health studies of enteric bacteria in the USA, Britain, Israel, Japan, and many European countries confirmed that, even in people who had never been under antibiotic treatment, there existed gut bacteria possessing a whole range of transmissible resistances to antibiotics. In Europe, public and government attention was focused on the problem by two incidents in Britain in which 39 infants died of an acute enteritis that failed to respond to all the drugs normally valuable in the emergency treatment of disorders of this kind. The source of the resistance has not in these cases been conclusively identified, but is irrelevant to the general lesson. Antibiotics, as they are at present being used, may constitute one of the most serious hazards to the maintenance of public health.

This is not evident from the torrent of persuasive and often expensively glossy literature that arrives in your doctor's mail from drug companies, but it is very evident from the series of local emergencies—often in hospitals— where the pressure of antibiotic use has led to an undermining of the natural ecology of bacteria. In burns units and a number of surgical wards it has been found that the only way to suppress a rampant pathogenic organism is to withdraw all antibiotic treatment. The decline in the population of pathogenic bacteria, measured by the reduction of infections as other nonpathogenic organisms repopulate their normal niches, can be dramatic. Even more important, there is no possibility whatever that such improvements could be achieved by antibiotic use, because, like pesticides, the antibiotics themselves create the distortion that permits the explosive increase of pathogenic bacteria.

Antibiotics as anti-immune factors

It is now becoming evident that antibiotics have other effects that may be seriously disadvantageous. It was noticed in the 1960s that people treated for a specific condition with antibiotics did not develop as complete a natural immunity to reinfection as might be expected. This was thought to be because the elimination of the disease-causing bacteria reduced the opportunity for normal immune reactions—the development of antibodies, and so on—to take place. By the beginning of the present decade another view had been voiced. The pattern of the emergence of what are known as "immune deficiency diseases" appeared to match exactly the increasing use of antibiotics. Indeed, all the evidence suggests that diseases of this kind, which involve the failure of the intricate and powerful natural defense systems that have evolved over many millions of years to protect man and animals from disease, did not appear on the scene until 1952, after the widespread introduction of antibiotics. The implication, which is of very great importance, is that in some way (as yet unknown) the use of antibiotics is interfering with and destroying the natural immune response. One hypothesis is that permanently damaging interference may occur if antibiotics are used during infancy, or at the fetal stage if the antibiotic crosses the placenta, as most of them will.

Although it may not be conventional to regard these apparently powerful

A rapid rise in mental illness is seemingly one of the penalties of technological "progress." Many of the actual tools of progress, such as aircraft, are themselves major sources of stress. *Above:* one of the frequent and frustrating traffic jams on the runway at La Guardia Airport, New York. *Left:* the occupants of this house by a major European airport suffer a 100 dB roar every 2 minutes during the day in summer. *Right:* a crowded London street. Congestion, added to noise and dirt, are destroying the quality of life in most of the large cities of the developed world.

disease-curing weapons as "pollutants," the time may well have come for us to think very carefully along those lines and to reject, very firmly, the notion that in the discovery of antibiotics mankind has liberated itself from the ravages of bacterial disease. The reverse may be much closer to the truth.

Information and manipulation

Getting at the truth about antibiotics, like getting at the truth about persistent pesticides, is extremely difficult, partly because of the time needed for negative aspects of the technologies to reveal themselves, and partly because of another kind of pollution. This is the corruption of information, deliberate or unconscious, that results from the differing commitments of powerful segments of society. The public is often overwhelmed by a barrage of conflicting or, more generally, confusing sets of evidence that surround particular controversies. It is extremely important to remember that where lobbying is involved, or where commitment has been made through, for example, permissive legislation, the truth is most likely to come from the least committed expert source. In the case of antibiotics this is neither industry nor government, but the public health laboratories, particularly those involved internationally in the investigation of transmissible "R" factors and the emergence of immune-deficiency diseases. And for an understanding of pesticides it is no use listening to the expert who points out that human levels are low and therefore harmless, because he has obviously failed to see, or is disguising, the real problem.

Nowhere is the evidence more confusing and apparently conflicting than in analyses of the global effects of pollution, such as the injection of nitric oxides and water vapor into the upper atmosphere by supersonic transport aircraft (SSTs) or the increase in carbon dioxide in the atmosphere resulting from the burning of fossil fuels. Arguments about SSTs, which have raged predominantly in the USA, rest on imperfect knowledge of the chemical dynamics of the upper atmosphere, on empirical observations of the effects of pollutants introduced into the upper atmosphere by weapons testing, volcanic eruption, and industrial activity at the Earth's surface, and on whether it would be possible to reverse an adverse effect if one was observed after the large-scale introduction of SSTs. The unfortunate and salutary truth is that at the present time knowledge of the natural variations of upper-atmosphere chemistry and their effects on the Earth's energy budget and climate are so fragmentary and of such recent origin that it would be extremely difficult, if not impossible, to disentangle a man-induced change from a natural variation, even after long and careful study. The question we really need to answer is whether man should embark on a new and potentially hazardous technology while in a state of ignorance that will prevent him measuring its impact.

The "greenhouse effect"

The carbon dioxide question is fraught with similar difficulties. Present estimates suggest that about half the carbon dioxide given off into the atmosphere remains in it. Carbon dioxide, like water vapor, increases the "greenhouse effect" of the atmosphere by selectively blocking the outward infrared radiation from the Earth, thus increasing the Earth's surface temperature. It is estimated that a doubling of the CO_2 concentration in the atmosphere would result in a surface temperature increase of about 2°c. Since 1940 man's activities, mainly through the burning of fossil fuels, have increased the atmospheric concentration of CO_2 by 30 percent. The surface temperature should therefore have increased globally by about 0.6°c.

Pollution and thermal change

Detailed records of the surface temperature, which, at least in the Northern Hemisphere, have been fairly comprehensive for more than 50 years, tell us that in fact the surface temperature has slowly fallen since 1940, a trend that can easily lead us into a position of great vulnerability. It is easy to argue, for instance, that the increased cloudiness resulting from additional water vapor released by man's activities is more or less compensating for any changes that would result from the higher atmospheric concentration of CO_2. That is certainly a comforting thought, because it implies a kind of massive protective stability in the Earth's processes, which mankind may distort a little but which is too powerful to be disrupted.

The fragmentary evidence of long-term cyclic changes of the Earth's climate—gleaned from written records, shifts in agricultural practice, studies of glaciation, and the fossil record—points firmly to other possibilities. During about 90 percent of its history the Earth has not been glaciated at the poles, and between periods of high glaciation there appear to be a number of short-term cyclic changes, each with major climatic impact but involving only small changes in the Earth's overall energy budget. The switch from glaciation to non-glaciation involves a change of only about 3 percent of incident radiation, and in either state the presence or absence of highly reflective ice cover leads to a condition that is quasi-stable.

In this quasi-stable state the lesser cyclic variations are too small to trigger a major glaciation change, although they may seriously affect the biological productivity of large regions by reducing, for example, the growing season or the temperature of surface waters. Recent studies have suggested that the Earth is at the moment on the downswing of one of these short-term cyclic changes, and that early next century Europe can expect surface temperatures about 1°c lower on average than those prevailing today. This means that any effect being produced by increased CO_2 in the atmosphere is being masked by a natural variation. Because the CO_2 implies an upward increment in surface temperatures *its effect will not be apparent until the next cyclic peak is approached*, which could be about 2050. The fear is that the additional warming effect of CO_2 will be sufficient to amplify the effects of a small cyclic climatic change so that a major glaciation change is triggered. The melting of the ice-

Freshwater resources are failing to keep pace with the increasing demands made upon them by growing populations and ever-expanding industrialization. It would seem insane, therefore, to continue to overload our freshwater lakes and rivers with a burden of pollution so great that their self-purifying properties break down. Left: Lake Geneva, in one of the reputedly cleanest countries in the world, carries a large burden of trash. Right and below: the crisis in freshwater supplies in America is largely due to the pollution of her lakes and rivers by the dumping of vast quantities of domestic and industrial effluent. A particular hazard is the presence in fresh water of large quantities of nitrates and phosphates, which are washed off agricultural land where they were used as fertilizers. If they find their way into drinking water they can lead to serious illness, even death, because they cause oxygen starvation of the tissues of the body.

caps from the polar land masses would result in a rise of about 24 meters in the general level of the oceans. It requires only a glance at the contours of the major continents to see that this would be catastrophic.

Is there a real cause for worry? Many climatologists believe that there is. Pollution, as we have seen, is not just an unavoidable product of effluent: it is the result of technological policy, and the new and massively growing technologies of the advanced nations are driven almost entirely by fossil energy. All the artificial energy that we use ends up as additional thermal energy at the Earth's surface, where it distorts the natural thermal budget. In 1970 man was releasing each year, in the burning of oil, natural gas, and coal, an amount of energy that had taken one million years to lay down through the natural processes that capture and store solar energy. By 1980 he will be burning the equivalent of two million years each year. Because of the gross inefficiencies inherent in thermal processes, only a small proportion of this energy ends up in transient useful work, such as propulsion, light, or new chemical bonds in plastics, fertilizers, or other products. By far the greatest proportion is dumped directly into the atmosphere or into surface water, together with the pollutants it has gathered or released during the burning process.

So energy production is not only a measure of industrial activity and of pollution; it is also a major source of pollution through the production of direct thermal changes. Before looking at the graphs of energy use and projections for the future, with their implications of a massive transition to nuclear energy and pollution burdens of a new and sinister kind, it is important to get a picture of the level of man's direct energy impact at the moment. It is often argued that this is negligible compared with the solar input, but this is not the case.

The energy budget

In many of the most intensely industrialized areas, or the high population density areas of advanced nations, man-released energy already equals, or far outstrips, the net energy input of the sun. Even when averaged over large areas, such as Central Europe or the eastern states of the USA, the man-released contribution is between 1 percent and 2 percent of the solar input. On the present projections of energy demand for Western and Central Europe, assuming a growth rate of about 5 percent a year, the human input to the energy budget will pass 4 percent of solar input by the end of the century and will add 50 percent to solar input by half way through next century. On similar projections, the global man-made energy input will pass the 4 percent mark about 2020. We need an increase of only 3 percent to trigger a glacial change. The local "hot-spots" already created by areas of intense energy and dissipation have been seen to create climatic changes such as increased cloudiness, higher fog incidence, and lowered solar input, so it can be argued that the natural balancing system is, on a larger scale, likely to protect mankind from disaster. There is, however, no evidence that this will be the case.

Existing knowledge of the dynamics of climatic change is inadequate for prediction: the direct input of thermal energy at the Earth's surface is additive to any "greenhouse effect" that may arise from CO_2, and because the major thermal pollution impact will first be felt in the Northern Hemisphere—thus creating an overall global imbalance—fundamental but unpredictable changes are almost certain to occur there first.

Energy demand (or consumption) is generally given as "millions of tons coal equivalent" (mtce), and according to UN and OECD statistical projections the present rate of growth in demand (5 percent a year globally) means that the 1953 overall demand of 2500 mtce will grow to 27,500 mtce by the year 2000. In a projection assuming a linear demand growth related to human per-head income, the figure for the year 2000 is 19,500 mtce. Either way, the demand is enormous and cannot be met by fossil fuel reserves, which, even on present rates of consumption, will be seriously depleted before the end of the century.

Estimates of reserves are notoriously unreliable and it is always pointed out that reserves of coal far exceed those of oil and gas (which are now given a 30-year lifetime at present rates of use), that the shale and tar sand reserves of oil are large and have yet to be exploited, and that any deficiencies can be

COMPARISON OF POWER INPUT AT THE SURFACE OF THE EARTH
(1970 data)

Solar energy (average)	Power (watts)	Area (1000km²)	Power density (watts/m²)
Total flux	1.75×10^{17}	5×10^5	350
Net radiation at surface	5.00×10^{16}	5×10^5	100
Man released energy			
Nordheim-Westphalen industrial complex	1.0×10^8	10.103	10.2
Benelux countries	1.2×10^{11}	73.0	1.66
Fed. Republic of Germany	3.3×10^{11}	246	1.36
Eastern US (14 states)	1.0×10^{12}	932	1.11
Central USSR	2.2×10^{11}	256	0.85
Central Western Europe	1.1×10^{12}	1665	0.74
USA total energy production	1.6×10^{12}	7760	0.24
World energy production	8.0×10^{12}	5×10^5	0.016
Weather phenomena			
Thunderstorm	1.0×10^{10}	0.1	100.0
Cyclone (typical)	2.0×10^{14}	1000	200.0

Energy in ever greater amounts is needed
to sustain the expanding economies of the
industrialized countries. The developing
countries, too, will need vast amounts of
energy if they are ever to become self-
supporting. The future, then, would seem to
be one of an insatiable, spiraling demand
chasing a rapidly diminishing supply.
At present we depend largely on reserves of
fossil fuels: oil (top left), coal (left)
and gas—both manufactured coal gas and
natural gas, usually found in association
with oil. Falling water is another natural
resource that has been harnessed (above)
to drive electricity turbines, though this
method of obtaining power is restricted to
areas where quantities and location of
water are favorable. It cannot possibly be
expanded to fill the gap that will come as
the fossil fuels run out. Nor will atomic
fission from a radioactive isotope of
uranium (right), which at one time was
thought to hold the key to unlimited sources
of cheap power. As these power stations have
multiplied, so have the worries about
accidents and doubts about their ability to
contain radioactive contamination of the
environment. The storage of their long-
lived radioactive waste is already a
serious problem.

made up by increased investment in nuclear power. If we run out of oil, optimists point out, we can always use coal as a starting-point for distillates to run vehicles and industry, and, as in the recent past, the world's energy industry will somehow cope with the ever-rising demand. Leaving aside for a moment the practicality of proposals for new (and highly polluting) technologies to nurture the growth of ever larger (and highly polluting) industries, what of the future power demands of the emerging nations?

In 1970 the annual consumption of electricity per head of population in the developed world, which comprises roughly 30 percent of global population, was 4000 kWh. The remaining 70 percent of people on the Earth utilized less than 150 kWh each. It can be seen from this that even if the intention of humanity was to provide the whole world with electrical energy at the present level of the developed countries, production would have to be increased by 6300 percent. To raise global per-head consumption to present USA levels, the increase would have to be about 10,000 percent. Even in terms of climatic impact, these are impossible figures. In terms of fossil fuel reserves they are nonsense. From the mid-1980s onward a severe gap begins to develop between what is possible, and current projections of demand, even if these are of the self-centered and globally blinkered kind normal in advanced industrial societies. True, in the charts that show projecting energy demands, that gap is generally labeled "nuclear" or "nuclear + fusion" and, for the past two decades, Western societies have been subjected to an avalanche of propaganda extolling the clean and energy-rich nuclear future.

Nuclear power

Yet, for economic, technical, and safety reasons, the first two decades of development of nuclear fission as a source of electrical energy have resulted in only partial success. By 1972 public and scientific opposition to further exploitation had driven the US nuclear power industry to new and highly expensive safety standards, and had resulted in a demand from the high-powered Pugwash Conference for a total moratorium on fast-breeder reactors —the technology essential to meeting the energy gap.

Two main criticisms have been voiced: the fission route to nuclear power is "clean" only in the sense that it involves no discharge of chimney effluents such as CO_2. Instead, it necessitates the storage or disposal of radioactive wastes whose potential biological hazards are far greater. Because some of the active waste has a long half-life it imposes the need for custodianship stretching many centuries into the future. Further, there is no safe answer to storage. Such "high-level" waste can either be stored where it can be seen and properly monitored in tanks on the surface of the Earth, but where it is both vulnerable to accidents and an ever-present threat through mistakes or sabotage, or it can be pumped into some kind of underground storage area—such as salt strata—where it cannot be seen, monitored, or controlled. If it began to leach into aquifers and reach the biosphere there would be nothing that man could do to defend himself. By the middle of the next decade global high-level storage would, on present projections, amount to an activity equivalent to many thousands of Hiroshima-type bombs, so this is not a negligible problem.

It may be highly significant that the most trenchant opponents of nuclear power include many scientists involved with its early development, including Dr. George Weil, who operated the world's first nuclear reactor with Enrico Fermi, David Lilienthal, first chairman of the US Atomic Energy Commission, Dr. John Gofman, co-discoverer of Uranium 234 and U233 and one-time head of the US AEC's radiobiology and safety division, and a growing army of biologists and ecologists who understand the genetic and other hazards of exposure to ionizing radiation. Their view, bitterly opposed by the nuclear industry, is that the technology has gone too far too soon, and offers very little in exchange for enormous risks. That was also the concensus view of the 1972 Pugwash Conference, although this international meeting of scientists was looking rather further ahead. Their major concern was the proposed rapid development and proliferation of fast-breeder reactors, which utilize plutonium as fuel. The dangers stemming from waste storage would still exist, but in this technology there is reason to believe that the plants themselves would be unstable and dangerous, and that in any case the increased traffic and processing of plutonium—a basic nuclear bomb component—would expose mankind to real and very serious threats from diversion, illicit bomb-making, and sabotage. Plutonium of the kind involved is probably the most hazardous pollutant now entering the environment, as was demonstrated by the frantic cleaning-up efforts made by the US authorities after the accidental release of an H-bomb led to plutonium contamination at Palomares, Spain, in 1970. The Pugwash view is that mankind must abandon the fission route to electrical power, concentrate research on controlled fusion (which offers fewer insidious hazards), and in the meantime learn to use energy much more efficiently than at the present time.

Historically both governments and the energy industry have regarded it as their responsibility to see that the energy demands of society are met. Efficiency of energy use has had zero priority in an energy-rich world where competition between fossil fuels has pumped demand up in the interests of ever-greater turnover. Yet those analyses that have been made, by OECD and other organizations, of the efficiency of present energy used make it quite clear that there is an enormous capacity for saving in the existing system, which could provide the breathing space that advanced societies need to get their priorities right. Pressure to meet demand remains seemingly inexorable, however, and as nuclear waste proliferates the commitment to an expansion of open-ended and wholly wasteful technologies becomes further entrenched. This self-perpetuating upward spiral, involving ever greater quantities of waste and dumped energy, is seen by many to be the greatest threat that mankind has ever faced.

21

Can Man Feed Himself?

If food production depended upon artificial sources of energy, then the world would starve tomorrow. The basic driving power on our planet comes from the sun, and, because of the immutable physical laws of energy and entropy, this dependency is as permanent (or as transient) as the sun itself. In the laboratory, under ideal conditions, the photosynthetic mechanisms of green plants may convert 3 percent of the light energy from the sun into the carbohydrates and proteins the animal kingdom requires for food. In the field, conversion efficiency is probably less than 1 percent, even under the best conditions, but is generally much lower because of other limiting factors, such as the availability of nutrients in the soil, soil structure, availability of water, and leaf condition. If one adds the inevitable inefficiencies of artificial energy production to those natural systems, it follows that to support arable food production artificially would require the generation on Earth of roughly four times the light energy received from the sun, a thermal burden that would be more than 100 times greater than could be borne by the planet without considerable climatic changes.

As an optimistic generality it is assumed that, under good natural conditions, it takes one acre of land to feed a single human being. It has been estimated that the amount of farmable land on Earth is between 7000 and 8000 million acres and, on this basis, it is often assumed that the Earth will be able to support a stable population of 7000 or 8000 million people. Indeed, demographers and agronomists now tend to base their views of the future on what they call an "asymptotic world of 7000 million people," implying that this number will be approached but never quite reached. As we saw when discussing population, there is at the moment no worldwide indication that population growth will in fact be brought under human control in time to achieve a stable and supportable population. It is now certain that there will be a massive overrun in some parts of the world and that population control in India, Pakistan, the Far East, South America, and even parts of Africa is liable to be natural and locally disastrous.

Because it takes relatively marginal lands as "arable," the figure of 8000

Harvesting wheat in Washington State. At the moment, many countries depend on United States exports for their grain. With the growth of her own population, and subsequent absorption of her grain surplus, the rest of the world will have to become self-supporting or starve.

million acres is misleading. At the present time only one quarter of that area is actually farmed. The remainder depends for its usefulness on major technological investments such as irrigation, the provision of desalinated water, massive increases in the production of fertilizers and agents for the control of pests and diseases, and the overall application of farming practices that require careful tailoring to individual regions, and even to particular localities. Almost every major attempt to open up new farming lands during the past two decades has failed because too little was known of the natural limitations prevailing in the region under development. The soil-stabilizing effect of forests has been underrated, erosion and desert areas have been promoted by accident, water requirements have not been met, the economic structure necessary for marketing often does not exist, and the overall costs of schemes have been greatly underestimated.

The shrinking Earth

Each week, at the present time, there are about 1·5 million additional human mouths to feed. The cost of opening up new lands at the rate needed to match this growth would be staggeringly high, and is certainly well beyond the capability of either the developing countries that are in most urgent need, or existing aid programs financed through the UN agencies. Even if the costs could be met, there would be a time lag of at least a decade between the initiation of projects and their achievement of useful food production. In any case this method of attempting to cope with rising demands would be doomed to ultimate failure because of the finite nature of the planet. Those who live in the richest and most favored societies need to remember that Europe's population and industrial expansion was possible mainly because large underpopulated areas existed in the world in the 19th and early 20th centuries. There are no more Australias, North American corn belts, or Africas to open up and exploit. The trap is closing, and closing fast.

We can say that solutions to the problem are being sought along two complementary routes. The developed countries are applying new and highly intensive methods of farming, which should mean much higher production of traditional food sources—especially livestock—from existing farmed areas. In developing countries, partly through the ministrations of the UN Food and Agriculture Organization, productivity is being raised by the introduction of high-yield crops.

"High-yield" is, as it turns out, something of a misnomer, but before examining what it really means it is worth looking at realistic figures for the amount of sustainable productive farmland available to mankind. In 1969, after some six years of deliberations, the FAO published a very complex study called "Provisional Indicative World Plan for Agricultural Development," which soon became known simply as the "IWP." This made it clear that in many parts of the world—South and East Asia, North Africa, parts of Latin America, and the Near East—there was no real hope of expanding the

GRAMS PER PERSON PER DAY

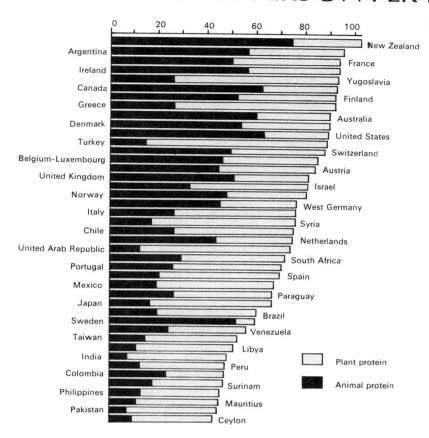

Tables showing protein supplies in 43 countries. The vast inequalities of distribution of vital protein between countries is evident in both tables. Top table shows distribution of plant and animal protein. Where animal foods occur rarely in a diet, even more of the lower quality vegetable protein is needed to make up the deficiency. Bottom table shows how the animal protein part of the diet is constituted.

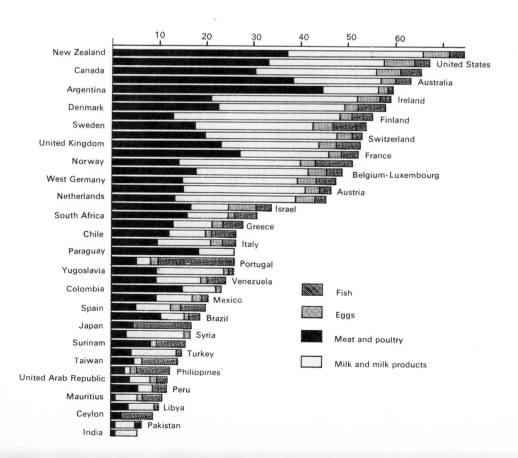

area of useful farmland. Indeed, in some regions it would soon be necessary, said this report, to return recently cultivated areas to grassland as permanent pasture if they were not to become desert. True, in Africa and in some parts of South America there was scope for expansion, but the costs of development would be prohibitive and "it would often be more economical to intensify utilization of the areas already settled."

According to the British agronomist Robert Allaby, mankind had its back against the wall by the mid-1960s. By then, neither the USSR nor China had any agricultural reserve apart from forest or pasture, and Europe had in reserve a mere 1·4 percent of its total land area that was tillable. Room for expansion elsewhere was almost equally limited: South America 2·9 percent, Asia (other than China) 5·2 percent, North and Central America 4·7 percent, and Africa 3·7 percent. We can forget about expanding into new lands, says Allaby, for they do not exist. As if to clinch the point, Professor George Bergstrom of Michigan, who published a careful study of the Earth's limitations in 1969, said flatly that it was a safe prediction that "for a population twice as large as at present, which experts agree we will have prior to the year 2000, the output will not fill minimum vital needs."

Proteins—haves and have-nots

Land constraints and economic constraints together determined the shape of the IWP, which aims, in the first decade, to increase cereal production on existing cultivated land. The major global food deficiency is protein, so this makes sense for in most of the hungry world the protein source is directly vegetable. In the wealthy regions it is indirectly vegetable, in the sense that the prime protein source is animal fed on vegetable materials sometimes enriched by other proteins. As a broad generality it is true to say that people in the rich parts of the world eat far more protein than they need for normal growth and health, while those elsewhere never quite get enough. But precise figures do not exist. The FAO, by a kind of balancing act, decided that the minimum protein requirement was not, as many nutritionists believed, around 55-60 grams per person a day, but 48·8 grams.

Nutritionally speaking, animal protein is to be preferred to vegetable protein, for man has evolved as a carnivore, and animal protein has components that do not occur in the vegetable form. A balance is therefore necessary, but this varies very widely in different parts of the world. Although it is true that if all the food produced on the Earth could be distributed properly there would be enough for everyone at the present level of population, the discrepancies are enormous. In Australia, New Zealand, and the USA, for example, consumption of protein per head is around 100 grams a day, and well over half of this is animal. In Europe the figure is almost 80 grams a day, of which 40-60 grams is animal. India, Pakistan, Sri Lanka (Ceylon), and large parts of East Asia have average protein consumptions that are very close to the FAO minimum figure, with an animal protein content of 15 grams or

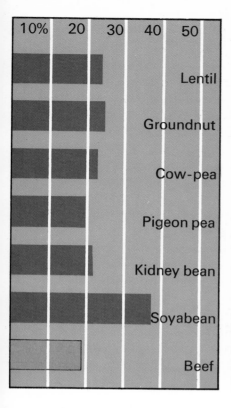

10%	20	30	40	50	

Lentil

Groundnut

Cow-pea

Pigeon pea

Kidney bean

Soyabean

Beef

Above: comparison of the protein content in various legumes and in beef. Carbohydrate content is almost as high in cereals (about 60 percent), and fat is low except in soya and peanuts.

Above: after the saiga was hunted to near extinction, the Russians bred it in reserves, restocked marginal land, and now crop it for meat with great success. Below left: wildebeest, farmed on savanna. Right: the capybara, not yet farmed but possibly a very practical new protein source.

less. In the early 1960s, for example, the average protein content of the daily diet was 50 grams, with only 6·4 grams coming from animal sources, which of course include fish. In Sri Lanka and Mauritius consumption per head was at the same borderline level, and it was only a little better in large areas of South America.

But, quite clearly, if vegetable protein production could be substantially increased and at a rate greater than that of population expansion, then improvements would follow. In the 1960s this did not happen. By 1968 the peoples of India, Sri Lanka, Mauritius, Central Africa, and South America were, in general, worse off than they had been five years earlier. In India protein consumption per head was down to 47 grams a day and, with the migration of hopeless masses of humanity to shantytowns, nutrition-deficiency diseases were more widespread and more serious than at any time in the past.

Dwarf cereals

The hope for the future, as was indicated earlier, rests on the so-called "high-yield" varieties of grain and rice that have been produced under the research programs supported by the FAO, the Rockefeller Foundation, the Mexican government, and the Philippines, and that won the 1970 Nobel Peace Prize for Dr. Norman Borlaug. Rather optimistically, the Nobel committee attributed to Dr. Borlaug a "breakthrough in wheat production that will make possible the removal of hunger in developing countries within a few years." The key was the selection of varieties that, in the right conditions, would produce much greater harvests than in the past.

Breeding of new varieties of staple cereals for higher yields has been the standard practice in Western agriculture for almost a century, and the trend is steadily toward shorter-stemmed varieties in which more of the total plant mass goes into the seed-head that is used for food; thus the plan for the developing countries followed that of intensively farmed temperate regions and

Right: a genetics and plant-breeding laboratory where scientists work on producing improved rice strains by irradiation. Research bringing new and sophisticated techniques is replacing the more traditional methods of producing hybrids by crossing. Far right: dwarf hybrids—wheat, in this case—stand up to wind and rain much better than the long-stemmed variety (on the left of the picture), which has lodged because its stalks are too slender. Lodging causes tremendous losses in cereal yields.

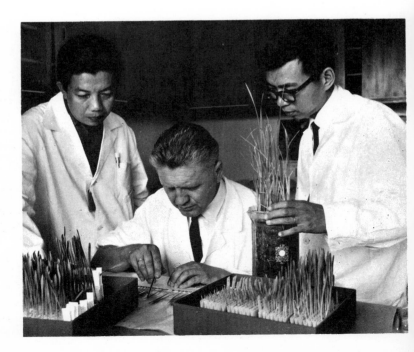

was not, basically, "new." The speed and direct purpose with which the "high-yield" varieties were produced was quite new, however, and represented a response that was unique in history to the desperate plight of humanity. There is no doubt that Dr. Borlaug deserved his prize. But the major question is whether the new varieties will work, for they possess major disadvantages that arise directly from their ability to produce heavyweight seed-heads. They are not only "high-yield" varieties of cereals; they are also high-demand varieties, and the indications by the early 1970s were that these high demands were not being met and never could be met by the hungry countries themselves.

The "Green Revolution"

The situation in the 1960s and the transition to the so-called "Green Revolution"—a term that is misleading because it suggests that the battle has been won—was eloquently summed up in *Only One Earth*, a book written for the 1972 UN Conference on the Human Environment at Stockholm by Rene Dubos and Barbara Ward:

> Between 1960 and 1965, the alarming fact emerged, in the wake of worldwide population census, that while population in the developing countries had grown by 11·5 percent, their food supplies had increased by only 6·9 percent. The gap was actually widening and with unchanged agricultural practices would continue to do so. In 1962 an average farming family of between five and six people had about 6½ acres on which to feed themselves and about two and a half other people. In 1985 they would be down to 5 acres and the others dependent on them would be just over four. This is the inexorable mathematical progression of population growth. It is virtually impossible for traditional agriculture to go on supporting such an increasing load of people on limited amounts of land.

Top left: although some fish are farmed by rearing the young in tanks and releasing them later into lagoons, oceanic fish farming is not yet practicable on a large scale. Fish provide only two percent of world food. Center: mollusks are staple food in parts of S.E. Asia, but because they rot quickly they are difficult to market economically. Top right: algae, once thought to have good potential as food, need expensive processing. There is, however, much research being done into farming the sea.

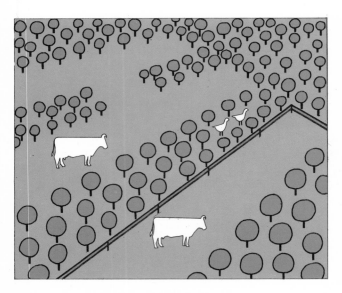

Above left: 3D-farming brings tropical marginal land into use. Tree legumes in rows provide shade; their roots contain bacteria that return nitrates to the soil. Their fruit is cropped; chickens eat windfalls. Livestock graze on controlled pasture between trees; their manure acts as a fertilizer, increasing pasture yields. The system is stable and self-perpetuating. Above right: eland antelopes with cattle on a farm in Rhodesia. Game animals do better than domestic breeds on poor land.

Above: a selection of different legumes that are used as foods. Their importance is partly in their high protein content. Below: "Incaparina," a meal synthesized from Indian corn, sorghum, and cotton-seed; it is marketed in Central America. "Pronutro" is another, made from soya, peanuts, and fish.

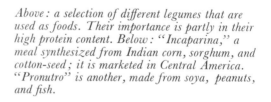

Right: tremendous strides are being made in conventional agriculture. This is a typical factory farm of the type needed if the world's people are to be fed. New cereals, too, are revolutionizing agriculture in the developing countries (p. 370).

Pointing to the rapid decline that occurs when fallow areas are reduced to increase local yields—and, incidentally, confirming that Africa's nomadic tribes are not the sons but the *fathers* of the desert—this gentle and persuasive book seizes upon high-yield varieties of cereals as the solution. These short-stemmed plants can tolerate over 120 pounds of nitrogen fertilizer per acre, can mature in 120 days instead of 150, and, theoretically, can provide a yield growth of 2·7 percent a year in those parts of the world that are in most urgent need. This looks very promising, and in trial plots one new variety of rice increased yield almost tenfold over traditional varieties. This was achieved by cropping three times in one year, however, and with an input of fertilizer of about 270 pounds an acre. That figure, well above the average for Europe, is beyond the reach of developing countries, and without external aid will remain so for the rest of the century.

Nevertheless, the areas put down to high-yield crops increased rapidly in the late 1960s and early 1970s. In 1965-6 the total acreage of new varieties was about 40,000. By 1969-70 it had reached almost 44 million, of which 24·5 million acres were under the new wheats. Even these figures represent only a small proportion of the total acreage under cultivation in the developing world. Locally, however, they have produced very significant increases in annual production. Total yields in India and Pakistan, for example, rose by over 30 percent above the previous record yields in 1968, which happened to be a good growing year. But taking the long-term trend rather than this local peak as an indication of prospects, the situation there is grave.

The fertilizer quandary

There are three basic problems related to the implementation of Green Revolution plans, plus a number of peripheral problems that may be very significant locally. First, as has been mentioned, is the production of fertilizer. Forgetting for the moment that land everywhere, if it is to remain productive, requires careful husbandry, and that the practice of multiple cropping and high fertilizer use implicit in the Green Revolution eventually destroys the soil, can even the most industrially capable of developing countries produce the fertilizer it needs to increase grain yields?

In India, where five-year agricultural plans have been operated for two decades and where both the administrative structure and the transport systems are relatively good, the goal for fertilizer production by the start of 1972 was 2·4 million metric tons. The actual achievement was less than half this amount, and the barriers to production—such as shortages of essential feedstocks, power and engineering failures at plants, and a pervasive social apathy—cannot and will not be quickly remedied. On the basis of the achievements in India, Pakistan, and the emerging African nations, it is reasonable to assume that for at least one decade, and probably two, targets will be undershot by half and hence increases in grain yields will be much smaller than expected.

It has been estimated, for example, that if India and Pakistan were to convert all the acreage at present cultivated to the new high-yield varieties this would require a fertilizer input of roughly two thirds of world production at 1972 levels. At the moment, of course, most of this production goes into the intensively farmed land of the developed countries. Even if this enormous quantity of artificially produced nitrogen fertilizer was available, there is no certainty that it could be distributed, applied, and utilized efficiently. As we saw in the last chapter, only about 30 percent of fertilizer actually reaches the plant even in favorable circumstances. What would happen, through the enrichment of surface waters with nitrates in tropical and semitropical countries caused by fertilizer leaching, is simply not known. But the consequences, particularly in regions where inland waters provide protein through fishing, could be extremely serious. Fertilizer use requires skill.

In both India and Pakistan, where subsidy systems operate to encourage the use of the new grains and to help farmers apply the necessary additional fertilizer (without the fertilizer, of course, yields from the new cereal varieties do not increase at all), it has been found that socio-economic conditions tend to favor the "progressive" farmers who enlarge their estates and set their production to provide maximum economic returns, not maximum yields. Sudden local gluts simply destroy the existing market, and, as everyone knows, starving, poverty-stricken millions are not a significant purchasing force. Hence production becomes geared, not to human requirements nor even to desired national targets, but to levels that approach the *best economic return.*

What is more important, perhaps, is that high-yield grains and fertilizers tend first to be concentrated in the best farming areas. This means that any deleterious effects stemming from changes in soil structure will first affect the best farming land. Without intensive studies of climatic, soil, and ecological conditions in individual areas—and because of shortage of skilled manpower these cannot be carried out adequately at the present time—the penalty of long-term losses of productivity seems inevitable.

Water—the vital factor

The second major problem related to the implementation of Green Revolution plans is the provision of enough water for the increased yields. Water requirements for green plants are massive, and productivity of areas already under cultivation, and especially of marginal lands that will have to be brought into use, is strictly limited in many cases by the availability of water. Many estimates exist for the amount of water required for a single healthy cereal plant to reach maturity, but the variation is large and inevitable because transpiration rates vary widely with climate and soil structure. It can be generally assumed, however, that in practice one pound of dry rice as food will require over 3000 pounds of water for its production. In the case of wheat the requirement is considerably lower, and sorghums are lower still, roughly 1000 pounds and 600 pounds respectively. We tend to forget that only a

It is estimated that more of our freshwater supplies lie within the top half of the Earth's crust than in all the lakes and rivers combined. Only a fraction of this water is at present economically usable. In many arid and semiarid parts of the world it is the only water available the whole year around.

One such country is Malta, where (above) a traditional method of drawing water from a nonflowing artesian well for irrigation purposes is the horse-powered bucket wheel. Below: livestock farmers in western India water their animals at a well.

In vast areas of the world, rainfall is scanty or confined to one short period in the year. In these regions farmers for centuries have had to resort to controlling the distribution of water on which their crops depend—the method they evolved was irrigation. Great river valleys present the easiest method of irrigation, as in the Vale of Kashmir (above), where the river is allowed to flood its banks onto the land, which is divided into basins by low earth banks. The seed is then sown by scattering broadcast. In the Sudan (above right), furrow irrigation is possible by releasing water stored in the Sennar Reservoir of the Blue Nile. Crops are planted in the earth ridges. Technology comes to the aid of Pennsylvania farmers by pumping piped water to sprinklers (right).

portion of the plant is useful as food, that the enormously complex root structure of a plant has a surface area considerably greater than that of a tennis court, and that only a very small fraction of the water absorbed and used by the plant is actually incorporated in plant tissue. Its major purpose is to get nutrients to the right place, and to provide the right conditions for photosynthesis to take place.

So, although there is plenty of water on our planet, the water demands of high-yield agriculture are met naturally over only a very limited proportion of the land surface. Elsewhere—and this includes large parts of Africa, the Near East, India, the Far East, and South America, all of which are in desperate need of increased agricultural yields—the naturally available supply of water has to be increased dramatically if land is to be cultivated in a systematic way. But, as the history of the Punjab and of North Africa has shown, the traditional techniques of providing irrigation by channels not only requires great initial effort, and expenditure on maintenance, but can rapidly lead to waterlogging and surface salt enrichment that renders land infertile.

Drought and flood

As we saw in Chapter 11, the enormous productivity of tropical forest areas is assumed to be an indication that they can easily support high-yield agriculture. But they have evolved in such a way that they make the most of climatic and soil conditions. Once the overhead cover of trees is removed to make way for crops, the typically sudden, very heavy tropical rainfall damages

Godavi River in flood, western India. Too much or too little water is a recurrent problem in many of the developing countries. Storing of water during the monsoon season would make possible the irrigation of larger areas of land for crop growing.

surface plants, and where it falls on soil, is highly erosive. It has been estimated that the loss of surface soil from an acre of tropical rain forest is about 2·5 tons a year, which is almost entirely returned from minerals in the rain and is, in any case, enriched by leaf fall. Once the cover is removed, however, the loss through erosion can reach 250 tons per acre a year. This means that the surface soil, even under the best management, will have a productive life of less than 20 years. In many areas it may be less than 5 years.

It is therefore quite obvious that equatorial agricultural techniques cannot be applied in such regions, although fruit- and nut-tree agricultures might be developed in such a way that the enormous growth potential of high rainfall and high temperatures can be exploited. Very little research has, as yet, gone into such developments. The major attention has been directed toward increasing water supplies through storage and irrigation in these parts of the world where monsoons provide an apparently adequate rainfall but where, at the present time, most of this rainfall rapidly runs off and is lost. The problems are not those of rainfall, but of ensuring that the water is where it is wanted for the length of time needed to produce mature crops. And, in the monsoon countries, one year of low rainfall is the harbinger of certain famine.

In highly developed but largely arid countries, such as Israel, economic and technological conditions permit the development of water distribution systems that largely avoid the wastage through evaporation and leakage that are inherent in traditional ditch and dike irrigation engineering. Storage can be in large surface or underground tanks, and, distributed by closed conduit, water can be delivered by pipes to the root areas underground. Plastics and silicones can be used to reduce water-loss both at the surface of the soil and through the plant itself, and can also be used to direct and improve runoff and its collection. With skill, though at high cost, the water demand for a given type of plant can be reduced dramatically, perhaps by as much as 90 percent. But such techniques, although invaluable for the relatively small arid areas being transformed to highly productive agricultural land in Israel, are too complex, demanding, and expensive for more widespread application at the present time. Yet they possess the great advantage of by-passing the natural consequence of traditional irrigation—a buildup of surface salts. If such techniques can be afforded, they are certainly an acceptable solution.

Israel is also in a position to exploit the very expensive technologies of desalination, thus leaving her free to supplement natural rainfall to a significant extent, because the benefits of agricultural payoff outweigh, in the long term, initial capital expenditure. Although no large-scale schemes of this kind have yet been introduced, it seems probable that they will form an integral part of her development program during the late 1970s and 1980s. Again, however, this kind of technological solution to water and agricultural deficiencies is not among the options open to the developing countries in the world.

But none of this should lead to the conclusion that a high rainfall is essential for agricultural productivity. Even with rainfall below 2·5 inches a year,

Piping water and establishing communes has turned the potentially fertile soil of the semidesert Negev into an important economic factor in the life of modern Israel. Left: picking tomatoes on a kibbutz struggling to survive in the Negev. Above and below: farming on the Kibbutz Yotvata, with the Edom Mountains in the background. The blue pipes in the upper picture drip water to keep the roots of the young corn plants moist. Green peppers (below) are growing under plastic sheeting that catches and recycles moisture otherwise lost through evapotranspiration—water that is transpired into the air through pores in the leaves.

acceptable yields can be obtained if the water penetrates the soil and is held by a suitable humus-rich soil structure. This suggests that good husbandry, the maintenance of soil structure, and genuine farming expertise are at least as important in the hungry developing countries as new varieties of plants and massive irrigation schemes.

Plant breeding—gains and losses

But let us return to the "high-yield" plants. The genetic characteristics that result in short stems and high nitrate tolerance have not been achieved without other genetic changes. One, is a lower resistance to disease and to crop pests; the other, no less significant in terms of world food production, is that the structure and protein content of the grain both suffer.

New varieties are constantly being bred, of course, but one of the rice strains produced by the International Rice Research Institute in the Philippines during the 1960s, a major prop of the Green Revolution during its early phase, turned out to have a poor texture when cooked, to mill badly, and to have a protein content 3 to 4 percent lower than traditional varieties. Similar deficiencies have characterized some high-yield wheats. Although, taken as yield-per-unit-area, the protein value of the new crops is higher than that of traditional varieties, the amount of protein in a given weight of grain is less. Because diet is generally assessed in calories—that is, carbohydrate content—with an assumed proportion of protein, the new varieties produce on a weight-for-weight basis a diet that is actually of poorer quality.

Both the Philippines Institute and the International Maize and Wheat Improvement Center in Mexico, with the aid of the Ford and Rockefeller

Left: plan of the National Water Carrier in Israel. Completed in 1967, it seeks to improve local distribution of irrigation water and to supply the Negev with water. From the Eshed Kinrot pumping station (A) water flows to Rosh Ha'ayin (B) in central Israel. With additional supply from the Yarkon River, the water continues to Beersheba and the central Negev. Right: laying part of the pipeline to Rosh Ha'ayin.

Foundations and the blessing of the FAO, are seeking alternative varieties that will produce high yields with an improved protein content. The road may be very long indeed, and because every genetic gain implies some new genetic deficiency, it is by no means certain that there will be an early success. It is normal for a new variety to take about 10 years to reach the proven stability and yield characteristics that make it acceptable for commercial use. A decade looks very long when measured against the rates of population growth in the hungry parts of the world. For it means that it will be almost two decades before a genuinely protein-rich cereal agriculture could become reality after the appropriate varieties have been developed.

Plant genetics and disease

Resistance to disease is another important criterion. The varieties traditionally grown in developing countries have become established largely because, in the absence of pesticides or other chemical means of pest and disease control, they remain tolerably productive. Such inbuilt resistance is genetic, and, almost on an empirical basis, the genetic strength of varieties has been maintained by crossing and by learning from direct experience in the areas of cultivation. It is a very different matter to set up breeding and testing stations in Mexico and the Philippines in the hope that what does well there will also do well in quite different circumstances in other parts of the world.

The sad truth is that very little is known about the insect and plant disease ecologies of many of the areas in which the new varieties will eventually be grown, and that there is a high likelihood of failure because of infestation. True, infestations can be controlled for a while by the intensive use of broad-spectrum fungicides and pesticides. Experience has already shown, however, that such techniques are environmentally and ecologically damaging, and present no adequate answer to the problem.

In Europe and most of North America the seasons themselves are the most valuable of existing crop-protection systems, because the cold winters ensure that the rate of survival of pests and fungal spores is low. This is not the case in many parts of the world, and is seldom true in those areas most in need of higher crop yields.

It may sound attractive to have a climate that permits, say, three croppings in 14 months, but this means that the climatic and biological conditions promote the survival and proliferation of pests and diseases. Experience of fungal infestations of cocoa plantations in Africa has shown, however, that with adequate field knowledge a minimal application of chemical control agents can allow a large increase in yields without opening up the way for epidemics of highly resistant pests or diseases. But the development of acceptable crop protection regimes takes many seasons, and requires high expertise and very careful management. It also necessitates a technical vigilance not yet available to the peasant farmers on whose shoulders the major burden of producing more food will rest.

In advanced countries it is easy to forget that high productivity of agriculture and livestock often depends on importing valuable materials from less-favored countries. The Netherlands is often quoted as an example of highly efficient agriculture, and certainly the yield per acre is very high. But the Netherlanders are not self-supporting, nor are the Danes. The Danish pork and bacon industry is founded on imported fish protein from hungry South America, and much of the oil cake cattle feed of advanced and rich nations comes from areas where the human population is close to starvation. The increasing pressure of population in developing countries must inevitably be felt on the food industries of the rest of the world. The population and agriculture problems of the hungry world are equally the problems of the rich.

The search for solutions

The question facing the whole world is whether solutions may exist that do not involve the impossible costs and techniques that seem inseparable from the Green Revolution as it has so far been conceived.

In 1972 the Ford Foundation began to look very hard at an entirely different route to high agricultural productivity. It rests not on the pumping-up of available nutrients by means of fertilizers and pesticides, but on a better understanding of natural processes. One of the grave but seldom mentioned disadvantages of artificial nitrate fertilizers is that their presence depresses the activity of natural nitrogen-fixation mechanisms in plant roots.

Unfortunately, there have been few measurements of the nitrogen-fixing ability of natural areas, so that reasoned comparisons cannot be made with nitrate-fertilizer application. But studies in the 1970s at Rothamsted in England revealed, surprisingly, that in "wilderness" areas nitrogen fixation could amount to the equivalent of 1 kilogram of nitrate per hectare a day (kg/H is roughly equal to pound/acre). In the least productive areas fixation was in the range 20-40 kg/H a year, which (because of their inefficiency) equals an application of artificial fertilizers in the range 50-100 kg/H a year.

The question is whether this natural fixation rate can be increased without involving penalties. The possibility that such an advance might be feasible emerged in the late 1960s in Brazil during studies of a kind of grass, *Paspalum notatum*. In natural soil fixation of nitrogen many kinds of algae, bacteria, and fungi are at work, often working particularly actively in and around root fibers and in many cases depending on mutualistic relationships with each other.

Such complex biological systems often require special conditions to achieve their highest productivity, and some plants appear to possess as yet unknown characteristics that encourage the right conditions for particular organisms that happen to be efficient nitrogen-fixers. The important point about *Paspalum notatum* is that it has an association with mutualistic bacteria capable of fixing between 70 and 90 kg/H of nitrogen a year. This is the equivalent of between 200 and 250 kg/H a year of artificial fertilizer, well up to the levels needed for high-yield crops.

Genetic manipulation—possibilities

Paspalum is a grass of great importance for livestock-grazing in South America, but one of the intriguing points about it is that it possesses some cereal-like characteristics. Because the root association with nitrogen-fixing organisms is specific, it is presumably also genetic, and the possibility immediately opens up of transferring the genetic factors associated with high nitrogen fixation of *Paspalum* to new varieties of high-yield grains. The prospect of grains that behave like legumes is, to put it mildly, startling and immensely promising. Yet, at the present time, no success has been achieved in transferring the bacterial characteristics of one plant to another, and, in spite of the obvious importance such an ability would have on future agricultural yields, the amount of research going into this kind of work remains derisory.

Whether or not genetic manipulation of this kind, achieving greater yields without an inherent fertilizer demand, is possible without some as yet unknown loss is a moot point. But the fact remains that it is only through investigations of local soil and insect ecologies that a basis will ever be found for the expansion of agriculture in developing regions at a rate that can hope to match increasing population and at a cost that, because it may largely exclude the need for both fertilizers and pesticides, can be met by the countries involved.

The second but no less important point about agricultural improvements that embrace a greater understanding of natural processes and their manipulation to greater yields, is that such manipulations are much less liable to damage soil structure, to lead to pest population explosions, or to result in long-term lowering of productivity.

Eat now, pay later

There is, however, little time for the necessary investigation into local soil and insect ecologies. And there is a desperate shortage of expertise in the areas most in need of such research. What is more, when—in 1972—the American biologist Professor Barry Commoner said that "agricultural science has led us into a very dark narrow alley," he was not thinking specifically about the way Green Revolution techniques were mistakenly mimicking the agricultural practices of the advanced countries—he was looking hard at agriculture in the advanced countries themselves. Trends everywhere during the past decade, driven by the "economics" of large-scale production and the demands of the deep-freeze, "convenience"-food market, were toward monoculture production, increased fertilizer use, and, inevitably, greater pesticide use.

Monocultures, as has already been mentioned, provide an ideal substrate for the emergence of pest and disease epidemics. The USA in 1969 lost about 17 percent of its wheat yield to rust. In some areas of Britain, where large monoculture areas are more common than in the rest of Europe, the losses were up to 40 percent in 1970. New regimes of fungicide spraying on an enormous scale have been developed, and these peaks have not been repeated. But they are symptoms of a deep sickness.

Right: maps of the distribution of a few of the world's important non-renewable resources. They show that the Northern Hemisphere is richer than the Southern in energy resources—which could be balanced if the vast quantities of deuterium in the sea could be economically harnessed.

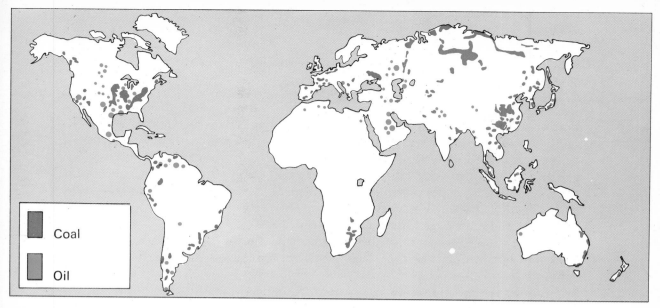

Coal

Oil

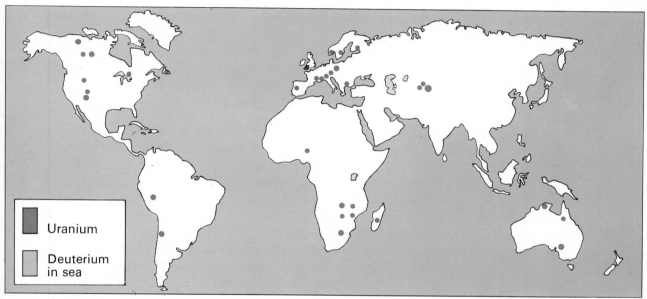

Uranium

Deuterium in sea

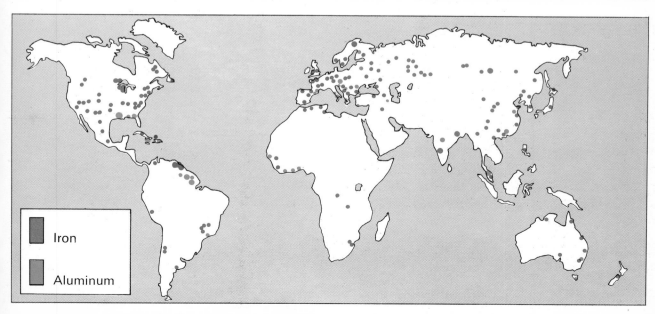

Iron

Aluminum

It is as though the harsh lessons of the US dust bowl (a product of greed, loss of grass and forest-cover, and poor husbandry) and the Irish potato famine (a result of monoculture practices as much as of lack of agents to control fungal disease) have not yet been learned. True, with new varieties of plants and with enormously increased fertilizer application—which often results in a plant product of poorer structure, quality, and flavor than those it replaces—yields in the advanced agricultural countries have been increased dramatically. In the USA the higher yields are actually produced from a smaller acreage of land than 20 years ago.

But the yield growth curve is beginning to flatten off, there is increasing evidence of soil deterioration, and pest problems are becoming increasingly difficult to contain. Taken broadly, the deleterious side effects of established techniques are now looming larger than the initial benefits, and, in countries as far apart as Canada, Australia, and Switzerland, surface-water burdens of nitrates and biocidal chemicals are being seen as penalties that cannot be sustained for long.

Paradoxically, the parallel development of intensive husbandry techniques for livestock has highlighted some aspects of the overall deterioration of soil structure and added to the demand for fertilizer. Studies in European countries suggest that the fertilizer value of organic wastes that do not go back to the land is just about equal to artificial fertilizer demand. But fertilizers are easier to spread than manure. What is more, many of the manure slurries produced in intensive units are contaminated with antibiotics, metals, or steroids, used to increase growth rates and to control disease; and mildly permissive legislation has made it cheaper for farmers to pour organic waste into rivers, and hence into the sea, than to devise ways of conserving it and returning it to the land.

Agriculture, the energy consumer

It is not that the idea of boosting available soil nitrate to increase yields is fundamentally wrong; it is simply that the technologies that have so far been established are inapplicable to developing countries and are also intrinsically unsound, partly because they tend to encourage poor agricultural practice and partly because they necessitate massive industrial development and consume large amounts of energy. The energy demand of agriculture in developed countries is already enormous and in 1972 was outlined succinctly by Michael Perelman in a publication by the US Committee for Environmental Information. This is the American picture:

> Electricity use by farmers accounts for about 2·5 percent of all energy used. Thus agriculture consumes the equivalent of 350 trillion BTU (British Thermal Units) of fuel, or an equivalent of almost 2 million BTU for each inhabitant of the United States. (2 million BTU = 14 gallons of gasoline approx.) Total energy use is much higher. Delwiche, in 'Research for the World Food Crisis,' estimates that more than 10 million

BTU of energy are used each year for each acre of land we cultivate. In 1964, at the time of the last US Census of Agriculture, the US devoted about 319 million acres to the production of food crops. That much land, which excludes the acreage under cotton, tobacco and grazing pasture, would therefore require the equivalent of about 150 gallons of gasoline a year for each American we feed or *about five times as much energy as we consume in food*.

This estimate does not take into account the energy required to manufacture the equipment used in agriculture, nor that required for storage and transportation. Nor does it include the energy needed to produce fertilizers. In the USA, says this report, the fertilizer industry requires

> about 10 million calories for each kilogram of nitrogen fertilizer produced. In 1969 US farms consumed about 7·5 million tons of fertilizer which required the equivalent energy of more than 1·5 billion gallons of gasoline, or about 8 gallons for each American fed.

These figures are not high when measured against some consumer demands, but they make agriculture the largest single industrial consumer of fossil fuels. Agriculture should be an *energy-producing* industry, in that basically it is capturing energy from the sun and transforming it into food energy, and the present technological trend is wholly unsound.

It is possible that the looming energy crisis will force agriculture in the developed countries to begin to weigh its practices on the basis of output per unit of energy instead of output per man-hour, and to lead industry into developing new—perhaps microbial—techniques for production of nitrogen-rich fertilizers. Certainly, microbial techniques can be used for the production of protein from organic wastes, a possible feedstock for animals to replace supplies that at the moment come from protein-hungry parts of the world. Because microbial processes are best driven by the sun, they are energy capturing, not energy wasting, and are therefore environmentally sound.

Indeed, the possibility of meeting the global protein gap through processes that either extract hitherto unused protein from tree leaves, or generate proteins through the use of organic waste for microbial cultures, is going to loom ever larger during the next two decades, for both offer major new sources without creating large demands for industrial capital or artificial sources of energy. But whether such proteins will ever be acceptable as more than a prop for animal feeds, or—at best—as a means of topping-up otherwise deficient diets, is a question that only time can answer.

Experience up to now has suggested that many of the microbially produced proteins, unfiltered (as it were) by the metabolism of a higher mammal, are not suitable for human food. In 1972 the ever cautious British Medical Journal went so far as to publish a strong warning that foods of this kind, unknown during the evolution of man, needed to be given much more searching long-term scrutiny for ill effects than any drug. This does not mean that such processes will never be used for the large-scale production of human

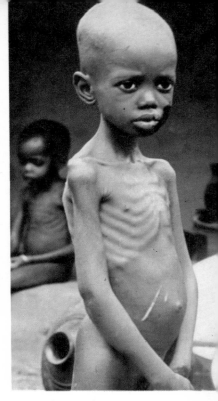

Above: a British farm technologist in Gambia supervising the plowing of land for rice cultivation by local smallholders. The developing countries of the world look to the richer industrialized nations for development funds and technical expertise rather than for food aid, which does not help them to be self-supporting. In times of crisis, however, food aid is the only immediate remedy. Below: flour being distributed during the Congo Civil War.

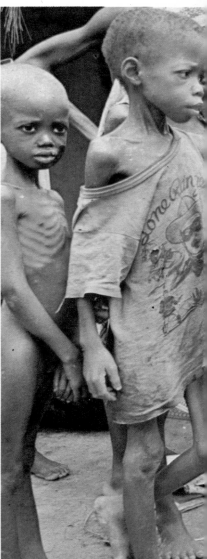

For millions of people throughout the world, hunger is a daily fact of life, needing only drought, flood, pestilence, or warfare to push whole populations over the borderline into famine and starvation. *Right*: a familiar sight in and around the cities of India. For many families the whole cycle of life from birth to death is lived in desperate poverty in the streets.

The Nigerian civil war resulted in a drastic reduction of the food supply for the breakaway eastern region of Biafra, with widespread deprivation (*left* and *below*). Children are especially vulnerable to the effects of long-term hunger, and may suffer permanent physical retardation, the younger ones in particular may also suffer mental retardation from a lack of protein during the vital stages of brain development.

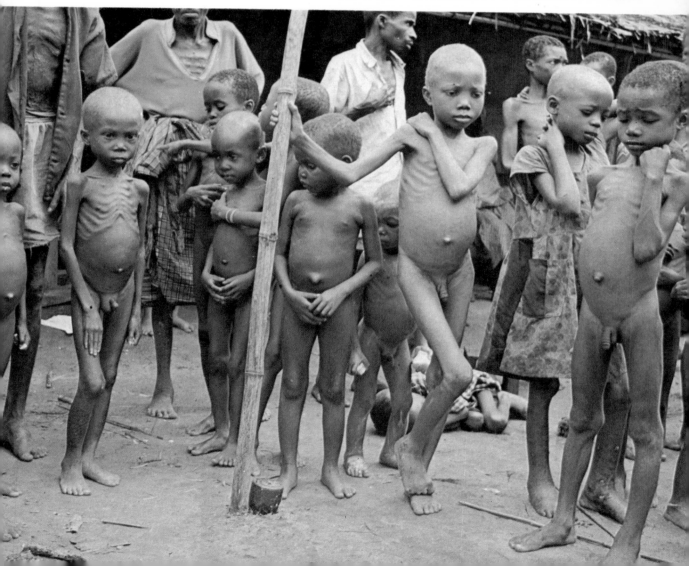

food. But it does mean that they cannot safely be introduced on a global scale in time to meet the food gap. At best they will protect the intensive husbandry in the developed countries when the developing countries reach that point of economic self-sufficiency that will permit them to retain the proteins that at present they must export to offset economic starvation. That, of course, is the basic dilemma faced by the developing countries, and (as it turns out) it is highly advantageous to the rich parts of the world.

The harvest in the sea

Remembering that some of the protein used for intensive husbandry foods in the developed parts of the world comes from the sea (the South American anchovy harvest feeds Danish and US pigs, for example), it seems reasonable to suggest that the protein gap might be most easily filled by making greater use of the fish-producing capability of the oceans. They are, after all, vast and seemingly inexhaustible. Unfortunately, the commonly held belief that the marine harvest can be substantially increased is false. The situation has been neatly summed up by the marine biologist J. H. Ryther, who wrote in 1969 that

> the open sea—90 percent of the ocean and three quarters of the earth's surface—is essentially a biological desert. It produces a negligible fraction of the world's fish catch at present and has little or no potential for yielding more in the future.

The basic problem lies in the physical nature of the marine food cycle. On land, trees produce foliage, and in the fall, the foliage is returned more or less directly to the soil, which in turn feeds the tree. Gravity provides the link for an important part of the nutrient cycle. In the sea, things are different. Photosynthesis can take place only in the uppermost layer—the top few meters of water—because at greater depth there is not enough light. There is a steady rain of marine detritus, like leaves during the fall, which ends up as sediment on the sea bottom, below the productive layer. This means that only where there are upwelling currents or sufficient natural turbulence to lift nutrients from the bottom to the surface layer do conditions exist for the support of abundant life.

In the open sea, nutrients are therefore scarce, which means in turn that photosynthetic organisms are small and also relatively scarce. Because these organisms are small, several steps—trophic levels—are necessary between photosynthesis and edible fish. This means, as we have already seen, that the efficiency of productivity is much reduced. Natural turbulence ensures, however, that most coastal waters have a higher productivity, and in a few parts of the world there are offshore and coastal regions where natural currents lift nutrients to the surface.

It is important to remember that the maximum sustainable harvest is well below the total productivity of the oceans, because a large proportion of the fish stock has to remain in the ocean to reproduce. To obtain the maximum sustainable yield, fishing would have to be controlled in such a way that the age structure of the population in the sea was at its most advantageous both for reproduction and for average efficiency of nutrient conversion at that trophic level. In fact, the fishing industry overexploits every fishing ground it enters. The result is that productivity falls, and total yield, in spite of more "efficient" techniques, is well below the sustainable maximum.

Not all species of fish are useful as food, and not all useful fish can reach the human food chain, because a proportion—perhaps about one quarter—is taken by other predators, such as seabirds. At best, men might eventually achieve a world harvest of 75 million metric tons. Compared with global protein demand, this is a small amount.

The present world fish harvest is not known with certainty because of the absence of reliable information from many parts of the world, and the total absence of figures from the Chinese mainland, but is estimated to be about 50 million metric tons. Trends at the present time are not, as might be expected, toward higher harvests. By the end of the 1970s harvests of some major food species were falling. Although one major cause is undoubtedly overfishing, we may be beginning to see an actual decline in marine productivity caused by other human activities. It happens that some of the most productive marine areas are estuaries, and, for relatively trivial economic reasons, the estuaries in many parts of the world are either at the receiving end of effluent dumped into rivers or the sites of industrial developments because they provide an apparently cheap source of water and a free dumping ground for waste.

The dangers of these shortsighted practices are well known to both govern-

It is unlikely that the natural marine harvest will increase by as much as one fifth—the extra protein that the world will need by the year 2000. Much more encouraging is the prospect of automated rearing tanks built on the sea bottom, as in the artist's impression above

ment and industry but the trend continues. In Europe, the Far East, and North America, estuarial use is in fact accelerating. At the same time, with public awareness of the problems that might be entrained, and growing resistance to the degradation of the environment, it seems possible that good sense might prevail before the end of the decade. By then, however, marine nurseries may well be disrupted and once-productive areas so contaminated that it may be decades before they can recover. Worse, because the fishing industry is still driven by a hunting policy that pays little attention to the global need to allow stocks to recover to the point at which maximum yields can be sustained, the overall situation could deteriorate rapidly.

There is already serious evidence of falling productivity in large marine areas such as the Mediterranean and the Baltic, and overfishing has affected the yields of the North Sea and the North Atlantic. But at the present time it is impossible to distinguish between the effects of fishing policies, long-term climatic variation, and changes wrought by other human activities. The absence of conclusive and therefore persuasive arguments provides latitude for the continuation of activities that, in the long run, must be damaging.

There have, however, been a number of positive suggestions for increasing the marine yield, and for hauling the fishing industry out of the paleolithic hunting age into that of controlled culture. It has been suggested, for example, that man now possesses in the form of nuclear reactors a means of creating artificial areas of marine upwelling. Perhaps this is possible, but the costs would be enormous and the overall economics and ecological effects uncertain. Much more important are the studies being made in many parts of the world of fish farming that utilizes, perhaps, both waste fish offal and the excess heat from power stations.

Fish farming

Fish farming is not in any sense a new idea, but even in countries such as Japan, where the techniques of freshwater farming make a valuable contribution toward the protein requirements of the national diet, the scale of operation is relatively small and confined to a few freshwater species. Of marine species only shellfish such as oysters and mussels have in the past been farmed in any significant way.

Recent research, carried out largely in Britain but also in the USA and Sweden, has suggested that several species of marine fish, all high-class table foods—such as turbot, lemon sole, and plaice—may be farmed with advantage. It should be added that all attempts to increase the yields of sea fisheries by introducing large numbers of young fish have failed. If farming is to succeed at all it can do so only on a very carefully controlled and confined basis

Experiments have shown that the major practical problems attached to fish farming stem principally from the differing and hitherto undefined food requirements of a fish as it develops from the egg through various stages of larval growth. Achieving the right physical conditions for survival and rapid

Top: fishing boats at Keflavik, Iceland. Already many formerly rich fishing grounds around Europe are overfished and there is natural concern for the future in a country such as Iceland, almost the whole of whose economy is at present based on sea-fishing. Below: Japanese fish farm used for rearing tuna. The young fish are caught in the open sea and transferred to the farm to be fattened for marketing, thus by-passing depletion by predators.

growth is a much easier problem to solve, although there may well be new and unsuspected problems of epidemic disease when large marine fish farms first expand beyond the pilot stage. Yet it has already been found that one expected problem with flatfish farming—the need to provide a very large area of tank to achieve a worthwhile scale of production—is not a problem at all. It has been found that under intensive cultivation flatfish grow fastest when they are in conditions that require that they "stack up" when at rest, and that a density calling for them to layer 6-12 deep is quite practicable. In simplistic terms, because cultivated fish reach table size in about two years instead of the four years required in the wild this suggests that a tank area covering about 15 square kilometers would be able to equal the table species productivity of the entire North Sea.

The obvious problem is food. An overall conversion efficiency of 10 percent might be possible, but that means that even if *all* fish offal could be processed for fish-farm food the total yield could be only small—one tenth of present-day waste. But it is not inconceivable that planktonic foods could be artificially cultivated on land in solar-driven systems utilizing other organic wastes as nutrients. Such possibilities have not yet been explored, but they appear to be feasible. It has been suggested, for example, that the first pilot-scale fish farms would be developed with three major production components: breeding stations and food production units on land, and marine cultivation cages in

protected, and perhaps artificially warmed, coastal regions. The marine cages, in which fish would develop from a length of a few centimeters to table size, would have three separate internal levels enclosing surface-feeding fish, flatfish, and—at the base, utilizing the waste nutrients from above—shellfish.

Fish farming: prospects

Alternative suggestions include the artificial enclosure of shallow bay areas and the construction of permanent but sea-fed fish-farm tank areas on coastal land. None of these possibilities has been fully explored, although small-scale marine-cage experiments are being carried out off the west coast of Scotland and in Norwegian waters. Larger freshwater farming experiments are already under way in the USA, Canada, and Africa.

Curiously, one of the most important possibilities from the point of view of providing protein in the most needy areas, the use of freshwater fish in paddy fields both for harvest and for control of mosquito larvae (on which some species of fish selectively feed), has been almost totally neglected in favor of control by pesticides.

Yet it is clear that there may well be biologically sound routes toward increasing the worldwide fish harvest, provided that research and pilot-scale experiments are sufficiently supported, and also provided that the governing philosophies of food production escape the straitjacket imposed by existing traditions and by the practices of chemical fertilizers and chemical pest control. These factors, both old and relatively new, are slowing down the development of more advantageous and productive methods of manipulating natural systems undamagingly and to man's advantage.

Too many mouths

But it has to be conceded that, even in the best of all possible worlds in which the developing countries would be given enormously enlarged help by the rich nations, the physical limitations of the planet mean that food production cannot increase for long to match the present rate of population increase. It may be possible to feed the 7000 million humans who will inhabit the Earth in the early 21st century, but the natural productivity of the Earth, unaided by the input of artificial energy, is only about enough to support a human population of roughly 2000 million. Unless men devise ways of making much greater use of the natural input of energy from the sun—and as yet there is no real trend in that direction—then mankind will go increasingly hungry in the decades immediately ahead. Indeed, although Western politicians continue to grace banqueting halls with after dinner speeches implying that all will be well and that Malthus was wrong, all the evidence suggests that they are wrong and that Malthus was right. The period immediately ahead poses for man threats of a kind that has not, and could not have, existed in the past. Their resolution will require changes of philosophy and practice of the most profound kind.

Opposite: hungry peasants in Hunan province, China. The sheer number of people threatens to defeat our efforts to feed the hungry world. More than half the world's population lives in Asia, where it is predicted there will be from 3000 to 4800 million mouths to feed by the year 2000.

395

22

Suicide or Solution?

From what we have been saying during the course of the last two chapters of this book, we have all the ingredients of an impending ecological crisis. The ingredients of that crisis seem, at first examination, to be so complex and so confusingly interrelated, that it is hardly possible to sort them into an easily understood or logical form. But if we keep in mind the fact that the living Earth, however complex, is a single, interrelated and living whole, then it becomes possible to see that the innumerable patterns of action and reaction group themselves into a few major dominant factors. These factors are all supranational and therefore arguably outside the control of individual governments. But, as could be seen at the 1972 United Nations' Stockholm conference on the Human Environment, this *need* for global cooperation is at least understood, if not practiced. What is more, the machinery for its eventual practice already exists, if only in an embryonic and highly fragile way.

Perhaps the most important single event toward the hastening of the recognition of worldwide ecological unity was not the Stockholm conference itself, but the view of the blue and misty planet Earth brought back by America's Apollo astronauts from lunar flights. If men had not realized it before, there could be no doubt afterwards that survival depended solely on maintaining the biological integrity of this small, unusual, and most beautiful planet. That is the beginning of wisdom. The next step is to identify the main factors on which maintenance and biological survival depend.

Ingredients for crisis

We have seen in earlier chapters that, taking into account the physical limitations of the Earth and the basic laws of both thermodynamics and ecology, the web of life is itself limited by indefinable but very real constraints. Two of the pressures now stretching life's supporting structures to the limits of planetary constraint are population growth and the increasing burden of pollution—measuring pollution not simply as effluent from industrial smokestacks and waste pipes, but also as the outpourings of the supposedly beneficial final products of industry.

A further and equally obvious constraint is the availability of natural

The small blue planet Earth, as seen from Apollo 17. Man must realize that he is only part of a complex web of life spread over the surface of his planet. Survival depends on all the parts living together in equilibrium.

resources. These can be either essential fossil fuels needed to drive industry as it is at present organized, or the minerals and other materials that industry and society need for the continuation of their existing life style. But other important factors and general interactions need to be added to these three basic cornerstones. Population growth and the growing expectancy of "the good life" in the largely materialist sense understood by developed consumer societies (capitalist and communist) is geared to capital investment in both industry and agriculture—with the necessary tradeoff between these investments, for what goes into industry cannot go into agriculture.

We now have five major factors: population; natural resources; capital investment in industry; capital investment in agriculture; and pollution. Considered on a global basis, these components make up the major features of the current growing pressure on the supporting biological system. Although appearing simple and generalized, they interact in a quite complicated way. To take one example: capital investment in industry produces more goods, reduces the availability of both land and capital for agriculture, increases the pollution load, and thereby reduces natural biological productivity. This in turn may reduce the availability of food, and result in population curbs through increasing the death-rate in infancy, which in its turn reduces the need for capital investment to meet rising expectations.

This series of interactions, taken at its simplest level, should be seen as a chain of events with ranging time lags between action and interaction. If it seems difficult to follow, how much more difficult it is to try to grasp what the outcome is likely to be if there are changes in two or more of the components (or *variables*, as mathematicians would call them). The human mind, although remarkable in its ability to analyze systems and to make deductions and intuitive choices, is extremely poorly equipped to understand the internal

Above, left to right: population; food (and natural resources); capital investment in industry; capital investment in agriculture; pollution. All are linked and all interact with each other and with other components in the life-support systems. If we can stabilize population growth, and bring the other four components into equilibrium, then we shall gain time to pursue the urgent problems of a fairer distribution of the Earth's riches and raising the quality of life.

workings of *multivariable* systems such as those that control our world.

The Forrester model

The difficulty of predicting the behavior of multivariable systems has been understood by mathematicians for generations, but it was first connected directly to the problems of our planet by Professor Jay Forrester at the Massachusetts Institute of Technology. During the 1960s, working very much on his own, Forrester attempted the first logical analysis of the world's main variables. As a model of the real world, he designed a mathematical structure within which the variables could interact, and embodied the model in a computer that would predict the outcome of differing movements of the variables involved.

It is often argued that models of this kind bear no relation to the real world, and that their predictions are necessarily false because they cannot take into account the finer detail of the real situation. Yet normal human thought processes are almost entirely dependent upon the use of models—constructed, to be sure, from experience, not from mathematics, but models just the same. We accept these without question, yet regard as suspect the much more complex interactive models that computers can handle with ease, because our own minds cannot themselves take in such complex dynamic structures.

Because we are unable to follow the interactions, and therefore reach their logical final outcome, we firstly admit their complexity, but then go on to apply the well-known human stamp of optimism, arguing tacitly that, having survived for so long, humanity will not passively watch as the real world pressures it has created turn into monsters of self-destruction.

It is also argued, defensively, that any computer model of the real world must contain assumptions about the nature and the extent of interactions. It

must also oversimplify the complicated chain-reaction mechanisms that are characteristic of life structures. This is true enough, but because humanity is already approaching some worldwide constraints, it is in fact the *general* trends that matter most. Although it is absurd to suggest what direction humanity will choose to go in at some time in the future, it is crucial to put as many facts and trends as possible into a model of man's present actions to see where these are leading. Given a model that is a broad but accurate mirror of the major components regulating global behavior, it then becomes possible to test the outcome of various policies to see whether mankind could achieve a better future for himself than the one indicated by his present behavior. The very fact that at this time of enormous global stress man possesses the tools with which to fashion a policy for the future is one of the most hopeful facts in the human predicament.

Professor Forrester was perhaps the first to realize that the tools for survival already existed as mathematical and electronic raw materials, and was certainly the first to attempt to hammer them usefully into shape. Needless to say, his work has been attacked and even savaged, especially by economists, whose notions of control through shortages and rising prices are seen in the context of Forrester's model to be inadequate and hazardous.

The runaway exponential

Perhaps the most important characteristic of major human trends at present

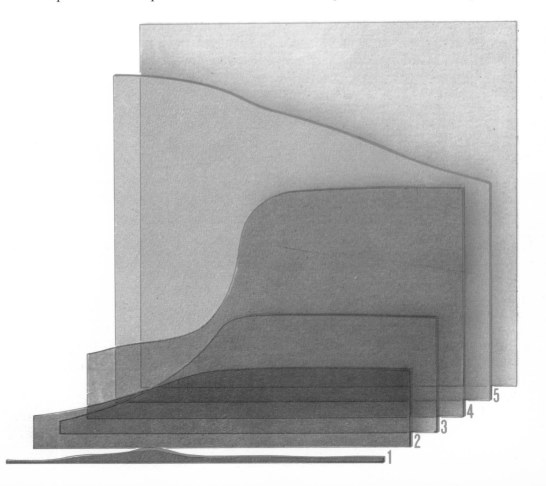

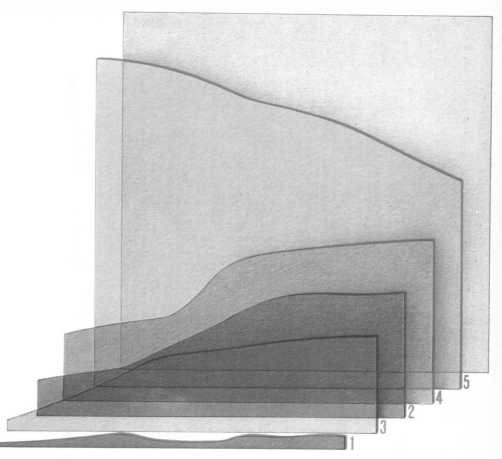

Carefully selected information about present world ecological problems was fed into the Meadows interactive model (p. 404) and processed into the three graphs shown here. Far left: a state of equilibrium sustainable far into the future is produced when technological policies are added to growth-regulating policies and the population is stabilized at 1975 levels. Technological policies include the recycling of resources; devices for controlling pollution; an extended lifetime on all forms of capital; and methods to restore eroded and infertile soil. Changes in social values include emphasis on food and services rather than on industrial production. Births are set equal to deaths and industrial capital investment set equal to capital depreciation.

Top right: if the strict controls on growth are removed, and population and capital are regulated by the natural delays in the system, the equilibrium level of population is higher and the level of industrial output per head is lower. Here it is assumed that perfectly effective birth control, and an average desired family size of two children, are achieved by 1975. The birth-rate only slowly approaches the death-rate because of delays inherent in the age structure of the population.

Bottom right: if policies for stabilization are delayed until the year 2000, the state of equilibrium is no longer sustainable. Population and industrial capital reach levels high enough to create food and resources shortages before the year 2100.

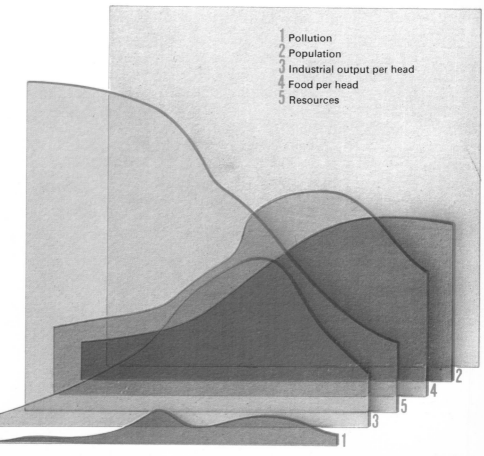

1 Pollution
2 Population
3 Industrial output per head
4 Food per head
5 Resources

is that they are growing in a manner resembling the *mathematical exponential*—that is, they are doubling their value in a constant time interval. To offer an interest rate of, say, 5 percent a year is merely another way of saying that the capital sum involved will grow exponentially with a doubling time of 14 years. Any steady growth rate measured as a fixed proportion over a fixed time interval—and both politicians and economists at the moment are obsessed with growth at some notional fixed rate, such as 4 percent a year, for the Gross National Product "to be healthy"—is in fact exponential growth.

Now the surprising thing about exponential growth is the way in which it accelerates—a dangerous characteristic when the growth is contained by some absolute physical limitation. As an illustration of this Professor Forrester, in his book *World Dynamics*, took population as an example, choosing an arbitrary world crisis level of 8000 million people (which may well be too high) and an arbitrary doubling time of 50 years (which was rather lower than the actual global population doubling time during the 1950s and 1960s). As the diagram on the opposite page shows, for many hundreds of years the 50-year doubling time is no problem at all; but then, quite suddenly, the population soars into crisis in less than a single doubling interval!

There is one factor connected with doubling time of even greater importance to mankind. Growth is largely determined, as we saw in Chapter 19, by the age structure of the population at any given moment. The expansion of a growing population tends to be given continually increasing impetus because the proportion of young, and therefore potentially fertile, members becomes larger with increased growth rate. So it follows that it is quite impossible to change suddenly from growth to stability. Even assuming an educated desire of a society to stabilize its population, it can only do so (without tyrannical legislation controlling births and a grotesque distortion of its age structure) if stabilization is sought over several generations. This means that there is an enormous overrun of numbers from the time at which a decision is made to aim at stability. It is no use waiting until the crisis appears, for by then it is much too late to avoid disaster.

True, although the *global* picture of population growth at the present time appears to be exponential, the *local* picture in Europe and North America is slightly different. In the advanced countries, taken as a whole, population growth is around 1 percent a year and apparently slowing down. Yet, even though these societies are undergoing a demographic transition involving a shift from large to small families, there is a slight growth of around $\frac{1}{2}$ percent a year in even the slowest growing group of advanced nations. In the rest of the world, where the same kind of demographic transition has not yet started on any significant scale, the growth rates are in the range $2\frac{1}{2}$–$3\frac{1}{2}$ percent.

It will take at least five generations for the transition to small families to be accomplished, which means that, if even present living standards are to be maintained, there must be a matching economic growth rate over the same period, and a similar agricultural growth to maintain existing dietary standards.

Right: how population doubling time accelerates according to the model set up by Professor Forrester. With a doubling time of 50 years, it takes 600 years to reach half *the crisis level; then, in just one more 50-year period, the curve surges upward to crash through the crisis level.*

These requirements feed back into two other factors that Forrester saw as major components of the life-supporting system: capital investment and agricultural investment. We know from published statistics that these components are in fact expanding on an exponential basis at the present time. The question therefore arises of what will happen to the existing world system if these parallel and interrelated growths are allowed to continue. We have already seen that industrial investment and agricultural investment lead inevitably to greater pollution by making possible the massive use of energy and resources. The urgent questions are where will current trends take us, and, if the results of our investigations look intolerable, what action can be taken to change the global system to one that has some chance of long-term existence? Remembering the population pattern of growth that appears to leave man-

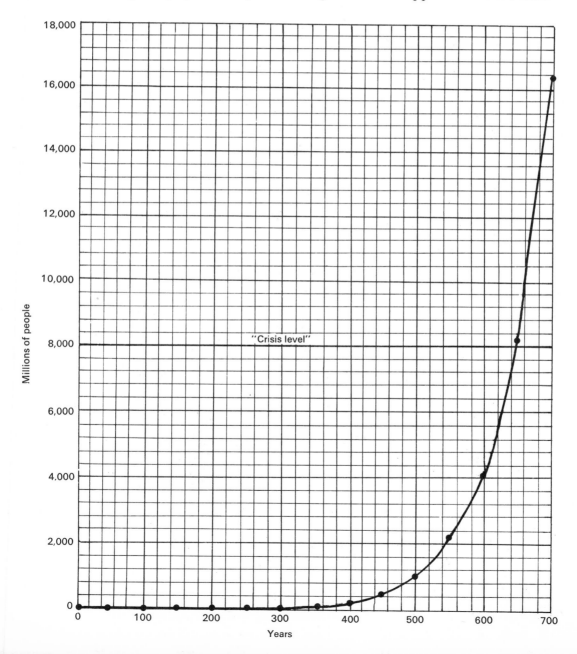

kind less than 50 years to achieve stability, what other factors will influence his chances of survival?

The Forrester model threw up some surprising answers. It was found, for example, that an assumption of five or ten times the known resources of the planet merely delayed the onset of stability by a few decades. Intuitively, you might guess that larger investments in pollution control, of the kind now being integrated into the social and industrial habits of the advanced world, would resolve the pollution problem. The model says not.

The Club of Rome

Forrester's work is directly opposed to the laissez-faire philosophy of industrial and economic growth that motivates all the advanced nations of the world. The conclusions and the implications he draws from them, although not yet even starting to be integrated into political and industrial thinking, were grasped rapidly by intellectual groups in Europe and the United States. In particular, and against a background of growing disillusionment about the real effects on society of the continuing acceleration of throwaway consumer production, Forrester's work made an immediate impact on an organization called the "Club of Rome."

This informal organization, a kind of invisible but very high-level college, grew out of a meeting of some 30 individuals—scientists, economists, humanists, educationalists, and international civil servants—that was held in the Accademia lei Lincei in Rome in April 1968 at the instigation of Dr. Aurelio Peccei, an Italian industrialist and economist. They met to consider a topic of enormous complexity and importance—the predicament of mankind. At this meeting, from which the Club of Rome emerged, it became clear that in spite of widely differing backgrounds everyone present shared an overriding belief that the major problems facing mankind were so complex, so little understood, and so closely interrelated that no solution could be reached through existing institutions and policies.

The Meadows model

Their early meeting included one in 1970 at Bern, during which Professor Forrester presented his work, and this led to an ambitious undertaking called the "Project on the Predicament of Mankind." Financed principally by the Volkswagen Foundation, the project's first phase amounted to an enlarged investigation of global environmental problems along the lines established by Forrester. This was carried out at the Massachusetts Institute of Technology by a group under Professor Dennis Meadows. The resulting interactive model was much larger than the one used by Forrester, and the information finally plugged into it was selected with the greatest possible care. To no one's real surprise this new and enlarged model confirmed, quite unambiguously, all the warnings that Forrester had first voiced.

Presented technically at meetings in Moscow and Rio de Janeiro, and non-

The population of the world could all stand on the Isle of Wight, England. Or, to put it another way, the inhabitants of the world could fit into the United States at a density per square kilometer of 300 people—less than the density of people per square kilometer in the Netherlands. Even so, as the text explains (p 402), the arguments against continual growth of population remain unaltered.

technically in 1972 in a book called *The Limits of Growth*, the outcome of this initial project drew some sharp criticism, but nevertheless shook the least blinkered of governmental advisory agencies out of their complacency. The study was, however, widely misinterpreted and criticized as if it were an attempt to predict the future—in spite of the fact that its clearly stated purpose was to pick out those elements that have most influence on the long-term behavior of the worldwide industrial system, and to analyze the interaction of current trends.

Some aspects of criticism were constructive. Although allowing some latitude for technological advances, such as the harnessing of virtually limitless energy, the model may have seriously underestimated the extent to which technological societies could delay the onset of crisis. In substance this criticism is valid, but it affects merely the timescale on which change can take place. There may be not three but six generations during which transition to a state of equilibrium with nature can be painlessly achieved. Measured against the timescale of human evolution the difference between three and six generations is marginal: yet considered in relation to the rate of change of society during the past century in the industrialized nations, an additional half-century in which to make a social adjustment to conditions of equilibrium could be of very great importance. As the Russian chairman of the Club of Rome meeting in Moscow put it in 1972 "man is no mere biocybernetic device." He possesses the power of thought and of calculated action that can override the vectored drive of the pressures upon him, and his goals are not exclusively materialistic. But man's economic and political systems *are* fundamentally materialistic, and this study shows beyond doubt that such fundamental values must change and change quickly. In the conclusion to this first report, which is arguably one of the most important documents in the history of man, the executive committee of the Club of Rome have this to say:

> We are unanimously convinced that rapid, radical redressing of the present unbalanced and dangerously deteriorating world situation is the primary task facing humanity. Entirely new approaches are required to redirect society toward goals of equilibrium rather than growth. Such a reorganisation will involve a supreme effort of understanding, imagination and political and moral resolve. *This supreme effort is a challenge for our generation. It cannot be passed on to the next.*

It is necessary to give this conclusion special emphasis because it is not the outcome of an unsupported value judgment by detached idealists, but a carefully weighed summing-up by men of industry, economics, science, and international relations.

Groping toward action

It is not surprising that, while economists and mathematicians were working toward a computer model of man's predicament, other scientists—

Picture on two following pages
The sea is the one last place left on Earth that man can expand onto, and into, as populations grow. It can also be exploited for its abundant minerals and metals—oil, gas, coal, diamonds, sulfur, cobalt, uranium, and many others. These are just two of the reasons why we should explore the possibilities of undersea cities. Overleaf: artist's impression of a deep-sea city that the U.S. giant General Electric

principally biologists and ecologists—had reached almost identical conclusions to those of the Club of Rome. Underlying the wave of rising concern about environmental problems in the USA and Europe, and recognizing campaigns to "clean up" industry as little more than sticking plaster and bandage on gaping wounds, was a far more radical wave of thought that culminated in documents and ideas as radical as the Communist Manifesto, and perhaps more important to the world. Loosely, it could be argued that student unrest throughout the Western world, the creation of communes of young people who opt out of materialist societies, and the upsurge of interest in *soft* technologies (that is, industrial and agricultural techniques designed to utilize and harmonize with natural systems, rather than operate against them), are all symptoms of this redirection of human aspirations. But one of the most important statements emerged in Britain in 1972 through the magazine *The Ecologist* in the form of a manifesto, supported in principle by a powerful group of scientists, and called "A Blueprint for Survival."

The Blueprint based its argument on the hard realities of global limitation, and on the impossibility of continual and growing exploitation of our resources, coupled with an ever-expanding population. It attempted to formulate a new philosophy of life that would lead to a transition toward a stable society, working in true equilibrium with natural systems, and involving the least traumatic routes to social change. As such it was a more openly political document than earlier statements about the world situation—which the Club of Rome had neatly described as man's "predicament." That it involved a flat and complete rejection of the current philosophies of industrial societies meant, inevitably, that to those committed to those philosophies it appeared to be a dangerously subversive document. Yet what it really said was that change is inevitable and that it is far better to initiate industrial and social changes in the right direction as early as possible than to wait for crises or local disasters to force changes on us. Its introduction was quite specific on this point:

> The principal defect of the industrial way of life with its ethos of expansion is that it is not sustainable. Its termination within the lifetime of someone born today is inevitable—unless it continues to be sustained for a while longer by an entrenched minority at the cost of imposing great suffering on the rest of mankind. We can be certain, however, that sooner or later it will end (only the precise time and circumstances are in doubt), and that it will do so in one of two ways: either against our will, in a succession of famines, epidemics, social crises, and wars; or because we want it to . . . in a succession of thoughtful, humane and measured changes.

The Blueprint drew criticism not, perhaps, because of its general philosophy, but because its solutions and proposals for future social structures appeared to many people to be regressive. One of the most dangerous and subversive delusions imposed on the thinking public of industrialized societies is that a move toward an increasingly complex technology is "progress," and that it

hope to start constructing in 1980. The city will be built of 12-foot-diameter glass spheres fitted together to make a complete self-contained city 12,000 feet down in the Atlantic Ocean. In preparation for man's new environment, scientists have been experimenting with an artificial gill—a synthetic membrane that extracts air from water while at the same time keeping water out. With such devices built into undersea dwellings, or eventually even implanted into our lungs we shall be able to breathe under water.

therefore implies an advance in social achievement. Conversely, any move toward deep-rooted natural processes is "backward," and represents a retreat to some former primitive state. This delusion has enormous power and is inherently corrupt.

Its champions are, however, persuasive and highly respected, believing as they do that man's present technological route constitutes progress and must continue unchanged, whatever its consequences. But the *real* situation is one in which real progress—in the sense of change, expansion, and contraction—will inevitably continue and is truly unstoppable. "Progress," in the sense of unlimited development of technologies, of piling technology upon technology in the hope that the new will cure the ills of the old, is not inevitable, and can be stopped. It can be stopped by making decisions in favor of other human needs. That is the "progress" for which the world is hungry.

The most important point about the Blueprint was that it looked far beyond the scrabbling dollar-centered industrial societies of our time, wracked by the stress, distortion, and emptiness of life in engulfing conurbations, to decentralized societies based on technology in harmony with nature, and to largely self-supporting community groups of about 10,000 people each. It stated four basic criteria for a stable society capable of giving the greatest individual satisfaction to its members: (1) minimum disruption of ecological processes; (2) maximum conservation of materials and energy—that is, an economy based on stock rather than output; (3) population stability at a level naturally sustainable by the productivity of the Earth; and (4) a social system *in which the individual can enjoy, rather than feel restricted by, the first three criteria.*

Who would argue that the goals implicit in these criteria are not "progress"? They take in the separate problems that, as components of the environmental crisis, have torn the advanced societies for a decade, while also reaching out to the much deeper problems of spiraling demand for resources, energy, and urban land. Indeed they contain one expression of the new spirit that, very slowly, seems to be changing industrial man's view of his world. In its rejection of a policy of massive consumption of resources, the Blueprint philosophy is directly opposed to that of the existing industrial and social order. In its cry for cultural development, and the integration of the individual in society, and in its emphasis on the fundamental art of "cultivation" (of human sensibilities, of social advance, and of food), it is the reverse of the view of progress expressed by the "big" technology propagandists.

Yet what is new is not perhaps the spirit so much as the carefully reasoned expression. At first glance we might consider the heartfelt cries of the Blueprint to be contained in these sentences:

> I know not why it should be a matter of congratulation that persons who are already richer than anyone needs to be, should increase their means of consuming things which give little or no pleasure except as representative of wealth. . . .
>
> It is only in the backward countries of the world that increased (in-

dustrial) production is still an important object: in those most advanced, what is economically needed is a better distribution, of which one indispensible means is a stricter restraint on population. . . .

The density of population necessary to enable mankind to obtain, in the greatest degree, all the advantages both of cooperation and of social intercourse, has, in all the most populous countries, been attained. . . .

If the earth must lose that great portion of its pleasantness which it owes to things that the unlimited increase of wealth and population would extirpate from it, I hope, for the sake of posterity, that (societies) will be content to be stationary, long before necessity compels them to it. . . .

It is scarcely necessary to remark that a stationary condition of capital and population implies no stationary state of human improvement. There would be as much scope as ever for all kinds of mental culture, and moral and social progress: as much room for improving the art of living and much more likelihood of it being improved, when minds cease to be engrossed by the art of "getting on."

These words were written over a century ago at the height of the first Industrial Revolution in Britain, by John Stuart Mill. It can be argued that, at least in some ways, the changes foreseen in his vision of the world have in fact taken place. But if you separate the social and cultural advances of the past century from the industrial expansion and its impact on society over the same period, it can be seen that the shortcomings of industry with its ever-accelerating consumption of resources outweigh the social and cultural advances that have been made. The root cause of this imbalance is that the factors that govern technological proliferation are not the same as those that govern cultural and social advance. There is, to be sure, an area of overlap, but the main effects of directing society to goals of economic and technological expansion for its own sake—which has been and still is the driving motive of industrialized societies—are antisocial and anticultural.

It is not simply that social and cultural priorities become secondary to the built-in priorities of growing technologies, but that our technologies, by their very size, impose a special kind of uniformity on human purpose by narrowing and limiting expectation. Industrial technology as we now experience it conditions men almost wholly to greed, so that the aspirations of even the most affluent societies are directed toward ever greater material abundance. That no route has yet been found for a fair redistribution of material wealth both within individual societies and between societies at different levels of affluence is one of the major causes of social and political stress. It is in default of a humane redistribution of wealth that the materialist scramble both continues and accelerates, and this makes it the basic cause of environmental and ecological stress.

Economic growth
It is only very recently that some economists have begun to look critically

at the phenomenon of economic growth and escalating consumption of resources that seems to characterize and determine human activity. The questions being asked, principally in relation to something called the "quality" of *human* life (not in relation to the health of the whole fabric of life on Earth, which is a far bigger problem), are whether present economic standards properly measure human advancement, and whether, in any case, they produce the right answers. An expert science and technology committee of the Organization for Economic Cooperation and Development (a political association of the developed nations of most of Europe, of the United States, and Japan), in a document on technology and economic growth prepared for a meeting of ministers in October 1972, suggested firmly that the economic goals that had motivated governments for centuries were no longer valid. It stated that the attention of governments must turn toward social and environmental improvements and the encouragement of technological developments that did not have a detrimental effect on the environment, even if this meant a substantial cutback in economic growth.

The transition implied in this OECD document echoed a series of attempts to identify certain aspects of a nation's social system so that they can be analyzed and used to provide, in parallel with Gross National Product and Net National Product (GNP and NNP), some measure of the social health of the nation. In the United States a highly ambitious project began in 1966 to attempt to create a new kind of social accounting that would enable government to measure such things as the advantages and disadvantages to

The explosive gaps and inequalities between the richer and poorer nations are destined to become even larger unless the lot of the so-called developing countries is substantially improved.

413

society of economic and technological innovation. These included the state and trends of social ills, such as the breakup of families and pressures toward crime; the progress toward particular goals; such as better housing and medical services, or why progress had not been achieved; and finally to seek indicators for social improvement, the state of mobility of populations and, also, individual achievement. This culminated in a document called "Towards a Social Report," published by the US Department of Health, Education, and Welfare in 1970, which in turn led to the establishment at the White House of a research staff dedicated wholly to the identification of "social indicators" and the definition of national social goals.

All this is symptomatic of a transition in thinking, albeit with fairly limited aims and at a purely national level. Its corollary is that the international league tables of growth of GNP, which are used as a big stick to drive societies up the scrambling spiral, have no real direct meaning in terms of human well-being. It is possible for a nation of very slow growth to be socially healthier, richer, and more satisfying than one of fast growth. But this is true only when that society has advanced sufficiently to provide the basic material needs and social services for all its members. There is, as it were, a cutoff point beyond which capital investment aimed at what is known as the "creation of wealth" needs to be translated into social investment for the creation of human richness and environmental health. The United States and parts of Europe have reached that cutoff point. The snag, in Europe and the USA, is that the patch-work of society is uneven and still contains many deprived segments of the community, and that a very significant proportion of national productivity is absorbed into meeting the demands of population growth. Further, no reliable means exist to measure or ensure investment for social benefit.

It follows, however, that even measured locally in the advanced societies, one of the most important steps toward social advancement, quite apart from bringing about a change of direction away from material possessions in the most favored segments of society, is *stabilization of population*. That there are signs that governments are beginning to recognize the need for population control, somewhat in the wake of strong public pressure, is important because it coincides with an absolute global necessity. That it is being driven largely by nonglobal forces and ahead of any obvious crisis might seem to be a reason for optimism. It suggests, as the most committed advocates of the existing setup would have us believe, that the mechanisms necessary for the control of human activity within safe limits really exist. This gives rise to the further hope that, if the developed countries recognize the need for population stability, accompanied by a fairer redistribution per head of wealth and opportunity, then the less-favored nations will follow their example.

A chink of light
Unfortunately, there is nothing in the history of human behavior to suggest that the constraints applied to itself by one community will be adopted by

414

others. Nor does it follow that the stabilization of population will automatically result in a reduction of technological pressure on biological systems or on resources. Although linked, these three areas of human impact on the natural world are capable of quite independent movement. To suggest that a direct linkage exists is rather like suggesting that the abandonment of an oversized space program will make available earmarked resources, and that these will automatically find their way into the improvement of social services. This could happen only if the value judgments of society ensured that it did so.

Gordon Harrison, in his excellent book *Earthkeeping*, points out that there is as yet no valid *scientific* evidence that crisis will ensue as a result of either the depletion of the Earth's resources or population growth.

It is hard to answer assurances of this, because they mislead. After all, there was no valid scientific evidence of disaster before the Great Famine in Ireland, and the only *evidence* we possess of present world shortages is social fact, in the form of gross deprivation, instability, and widespread malnutrition. Where such conditions are an almost permanent way of life it is just not conceivable that they can be influenced merely by a shift of values in other and more favored parts of the world. It is equally misleading to suppose that the disturbing, indeed frightening, view of the future seen from the peaks of affluence will have much meaning to those far below the peak. At and below subsistence level, crisis is already a fact of life, and all the signs available to those at the peak point to crises of various kinds unless far-reaching changes are made in social values and technological systems. The sane goal is equilibrium, and while some parts of the world seek to avert crisis through restraint, the whole range of human conditions must, if justice is to have any meaning, eventually find some way of leveling out the gross inequalities.

At the present time there is evidence of a deterioration of world life support systems, and we do not know at what point deterioration will, through additional stress, become collapse. In the absence of this knowledge the only prudent course is one that avoids any increase of stress. This is not a scientific argument, but one of good sense, and it tells us that a stable population is less likely to increase stress than a growing population, and that the softening of technology will also tend to reduce rather than increase stress.

This "gentle" argument may well be applicable in advanced countries where population growth is around 1 percent a year. At the same time it needs to be remembered that growing populations imply a continuation of growth for several generations after a decision is made to achieve stability. But, taking the world as a whole, there is no room at all for gentle argument. The application of medicine and public health measures in regions where a high birthrate was formerly balanced by a high infant mortality rate has destroyed the "natural" mechanism for population control. Unless some kind of voluntary control is instituted, runaway population growth is inevitable and will continue until it hits natural controls at some higher population level. The alternatives to voluntary control involve high mortality rates, through epidemics, starvation, and other privations, more devastating than anything that has, to our knowledge, occurred in the past.

If we look again at the Club of Rome world model, we find that population control (and ultimate stability) is no answer to our predicament. Taking the absurdly overoptimistic assumption that population could be stabilized by 1975, the model shows that continuation of the existing exponential growth of capital investment and of industrial output leads to a short-term peak of human prosperity, followed by a fairly rapid crisis and collapse through using up nonrenewable resources.

The longer the delay in introducing policies to conserve our natural resources and to stabilize our populations, the lower the eventual standard of the human condition and the fewer the options open to future generations. If stabilizing policies are not achieved until around the year 2000, then according to the model stability cannot in fact be achieved. Collapse of the economic and industrial world would then be inevitable, a conclusion that, to say the least, is regarded by industry as highly controversial. As suggested earlier, the rate at which crisis builds up and the rate of the subsequent decline of the economic and industrial world might be more gradual than the model suggests.

Yet fundamentally it is difficult to see how such refinements would effect the general conclusions. As Dennis Meadows has written, there are three apparent courses: continued growth on the existing pattern; a self-imposed limitation of

Above: battle-scarred Vietnam, scene of conflicting materialist ideologies. Wars, famines, epidemics, and social crises will continue to be the lot of mankind unless supranational policies are agreed that will recognize basic human spiritual and material needs. Many young people are searching for solutions in this direction—like the New Mexico community pictured on the right.

growth; or a naturally imposed limitation of growth. Of these, *only the last two are possible*. The real argument is whether naturally occurring restraints that will come with the adoption of a new set of social values will be adequate to slow down and finally halt present trends.

At the present time there is no evidence whatever that, in an international sense, restraint of any kind is being applied except through the taking-over of mineral resources by developing countries for their own use or for increased profit. Nor is there any evidence of any real will on the part of developed and highly industrialized countries to limit their own growth in such a way that investment in the less-favored regions will eventually lead to an equitable leveling out of affluence. It is so frequently pointed out, for example, that the gap between the affluent and the developing countries is widening, that it tends to be forgotten that this situation is by no means inevitable. If economic growth per head in the affluent world was damped down to a doubling time of, say, 60 years, while that in the developing world was encouraged to a doubling time of 20 years, the balance would be redressed within two centuries. Whether such an immensely demanding transition could be sustained is another question. But it is not inevitable that the gap must widen. It is inevitable only if the present pathetically inadequate policy of devoting less than 1.0 percent of GNP to development aid is maintained. This policy makes it quite certain that differentials will not only remain, but increase.

Such self-evident statements to the effect that Calcutta today is passing through a transitional phase similar to that of a European industrial city 150 years ago, or that the developing countries can solve their current unemployment and urban problems only through labor-intensive developments, mask the real barriers to development imposed by the developed nations on those struggling to emerge.

The forces that are arranged against the developing countries are industrial and economic and pegged firmly to the dedication of the developed world to growth. This is one of the major sources of global tension, which if not changed will finally explode. Yet there are some indications that change is under way. In his final year as President of the Common Market Commission (an organization that from the point of view of the undeveloped world looks and behaves like a hostile and repressive closed shop), Dr. Sicco Mansholdt recognized the power of the arguments for redesigning technology in order to reduce its ecological impact, and seemingly accepted the logic of the Club of Rome analysis. His specific proposal for certification of consumer goods that have minimum environmental impact (the CR system), and modification of the tax system to give these goods preference, is likely to get off the ground, and his proposals for research into the achievement of "ecological and biological equilibrium" involving, for example, closed-circuit industrial production, are at least being examined. Whether they can be examined seriously enough in a political and social setup demanding an increase of consumer output, and economically crippled by inflation, is another question.

It is equally open to question whether United States industry will take seriously a resources report of profound importance made in 1972 by the Science Research Council of the US Academy of Sciences. This was an examination of the requirements of United States industry, and trends in the availability of world resources. Its conclusions were not cheering. World shortages of some industrially crucial minerals will become so great during the next two or three decades that, if some sections of industry are to avoid collapse, it will be necessary in the near future to begin a transition to the mining of large stocks of native minerals. The terrible scars left by mineral extraction on the scale needed by even the present level of industrial productivity are one measure of what is in store unless techniques are changed. Yet at present, priorities are such that exploitation remains more profitable, and seemingly more desirable, than conservation.

An awakening

In a single decade, there has emerged a whole series of related reports, statements, and new attitudes that together comprise the first deep rumble of an approaching revolution. In industrialized societies there are growing pressures for less noise, cleaner rivers, and less wasteful industries. In March 1973, for instance, under the aegis of the Council of Europe, ministers from 17 major countries produced guidelines to combat the pollution threat to Europe's natural environment. However polarized and parochial these decisions may appear in relation to world ecological problems as a whole, they have a significance larger than can be measured by their purely local impact. The voices raised against population growth in the continuous debate about national policies for population control have at least assured that there will be a candid examination of the problem. It seems probable that, before the end of the 1970s, most countries will have acquired figures for their own fertility, population size, and growth rate—the initial ingredients of policy. Further, the fossil fuel and power difficulties now being clearly seen for the first time will lead to new attitudes to the efficient use of power and the advantages of conserving resources. But it would be an overstatement to suggest that such pressures have yet changed the course of human development, though there are signs that they may do so in the near future. As was revealed by the UN Conference on the Human Environment in 1972, mankind has not yet reached the stage of thinking in terms of one Earth, of vulnerable life-supporting systems, or fo eventual equilibrium. But such ideas are being taken seriously by affluent societies, and that the worst ecological wounds are being recognized and patched is a hopeful sign. The future of mankind depends absolutely on the crystallization of a scale of values to which his existing institutions, economy, and culture are inherently opposed. Predatory man is locked up in his own systems of self-destruction, and global man is still inside him. We know that because we have heard his voice. It may be faint and distant, but it does not speak the gibberish of greed.

Picture on two following pages
With solar radiators like this one, NASA hope to avert the threatening world-energy shortage. They will be designed to orbit in space, 36,000 km away from the Earth, and will be of enormous size. The two "paddles" shown in the construction have solar cells to catch the light from the sun. From a central antenna the sun's energy will be beamed down to the Earth in the form of a radio signal, where it will be converted to electricity. The preparatory stages of the project have already started.

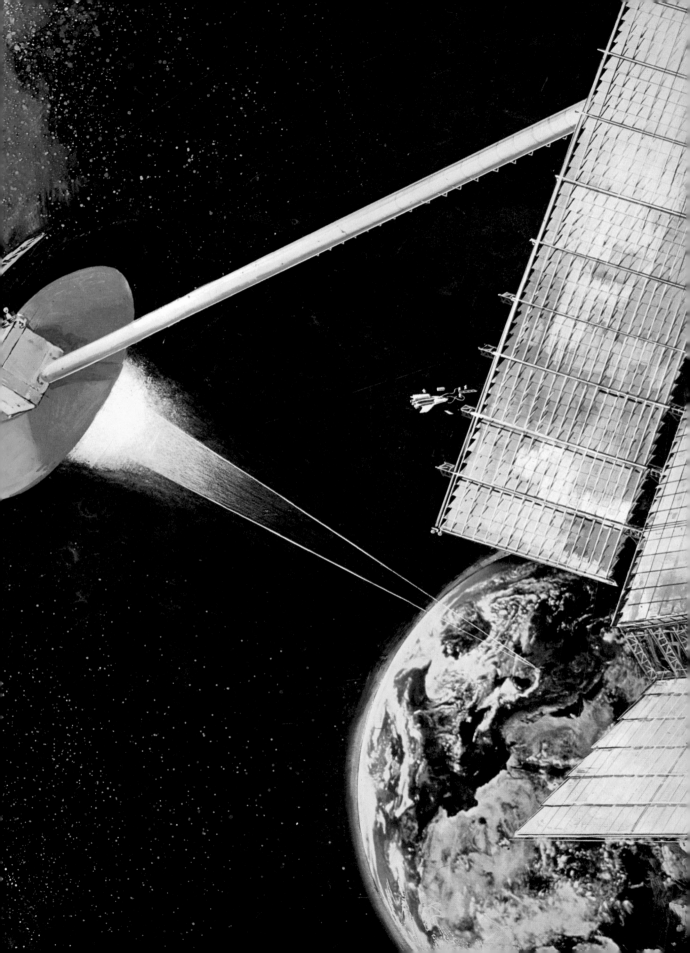

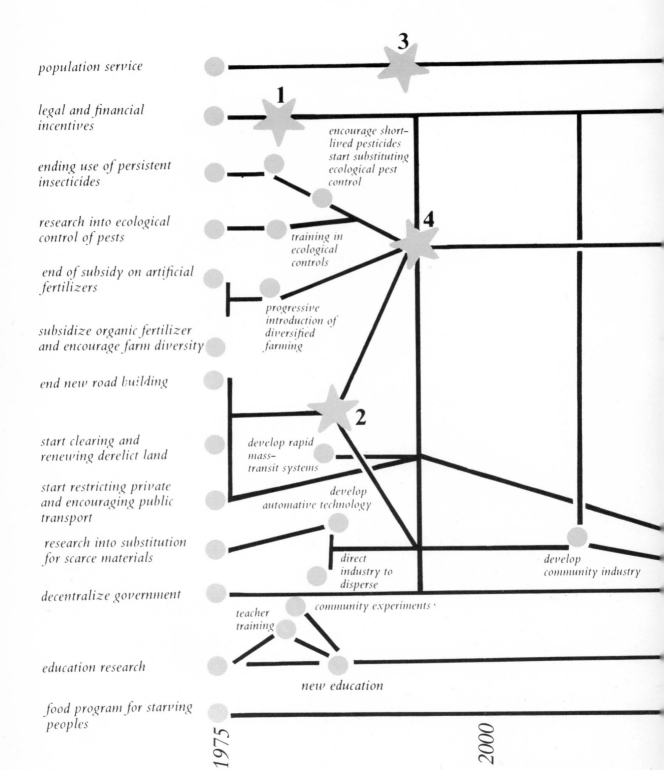

population service

legal and financial
incentives

1

3

ending use of persistent
insecticides

encourage short-
lived pesticides
start substituting
ecological pest
control

research into ecological
control of pests

training in
ecological
controls

4

end of subsidy on artificial
fertilizers

subsidize organic fertilizer
and encourage farm diversity

progressive
introduction of
diversified
farming

end new road building

start clearing and
renewing derelict land

develop rapid
mass-
transit systems

2

start restricting private
and encouraging public
transport

develop
automative technology

research into substitution
for scarce materials

develop
community industry

decentralize government

direct
industry to
disperse

community experiments·

teacher
training

education research

new education

food program for starving
peoples

1975

2000

A blueprint for a new kind of civilization by 2075

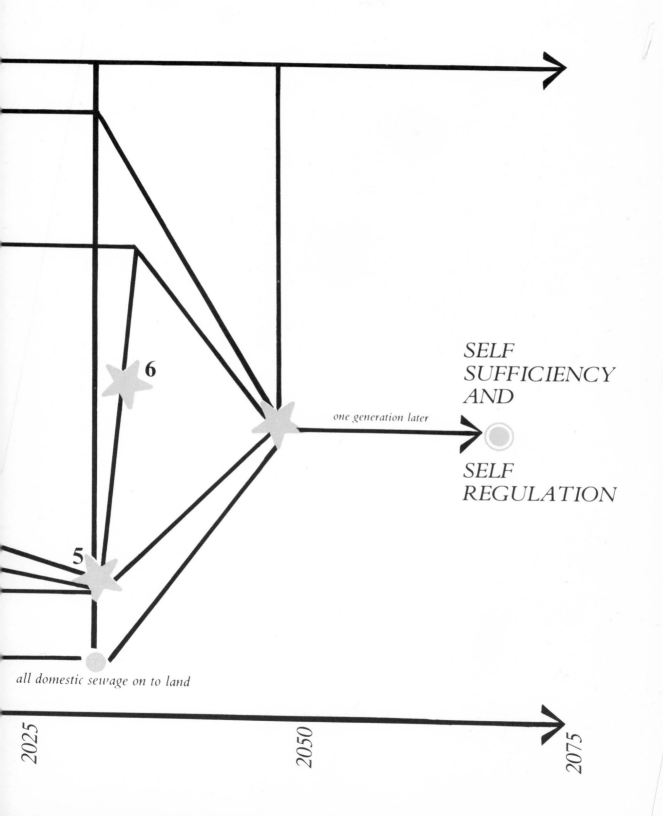

6

5

one generation later

SELF SUFFICIENCY AND

SELF REGULATION

all domestic sewage on to land

2025

2050

2075

Blueprint for the Individual

Family

Stop at two children; adopt if you want more.
Limit your reproductive capacity nonchemically (e.g. by vasectomy).

Buying

Use durables rather than disposables: no paper napkins, paper underwear, paper kitchen cloths or towels, paper plates.
Boycott overpackaged products, especially in plastics and aluminum and anything in nonreturnable or aerosol containers; if they are not available in any other form, take the "non" returnable container back to the store and explain your action to the manager.

Food

Keep pesticides out of your home.
Boycott convenience foods.
Exploit the wild foods that grow near you.
Grow and preserve as much of your own food as possible.
Eat vegetable rather than animal protein.
Buy insect- or virus-blemished foods, they are less likely to be polluted by pesticides.

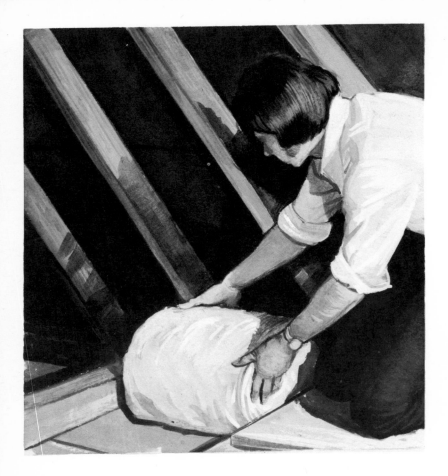

Building

Use building materials made from reclaimed industrial waste.
Insulate to the highest possible standards and so cut your energy consumption.

Water

Shower instead of bathing; wet yourself, turn the shower off while you lather, wash it off.
Wait until a dish- or clothes-washer is filled to capacity before you turn it on. Make sure poisons, medicines, or fertilizers are never flushed in the W.C.
Use minimum detergent quantities; presoak all washing; boycott enzyme detergents.

Power

Use the lowest-wattage lighting that will serve your purpose.

Use fluorescent rather than filament lamps.

Make the most of natural light: decorate with light colors and keep windows free of clutter and translucent curtains.

List all the things you now do by machine, strike out all those you could do by hand, and give the redundant machines to your local recycling club.

Transport

Buy the smallest, lowest-power car you can.

Learn the tricks economy-run drivers use—and include them in your own driving.

Use a bicycle—and remember that bicycles can accompany passengers on trains.

Give lifts; join a school-run/shopping-run car pool.

Avoid lead-additive gasolines.

Daily Life

Use both sides of the paper in letters.
Compost all organic waste.
Pass on outgrown toys and clothes.
Use cloth diapers instead of paper ones.
Buy secondhand wherever possible.
Garden without pesticides and herbicides—use derris, pyrethrum, quassia, bordeaux and burgundy mixtures, and lime sulfur.

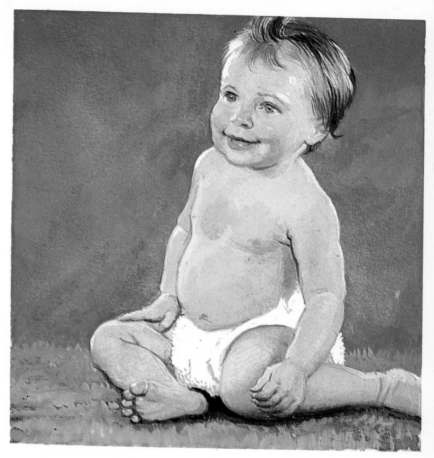

Community

Organize a local recycling club for glass, paper, cloth, iron and steel, and aluminum.
Monitor all local-government plans for their environmental effects.
Press for population policy—nationally and locally; specifically support planned parenthood, abortion reform, and sex education.
Press for maximum recycling, especially of water, in local industries.
Press your local authority to treat sewage properly.
Support Friends of the Earth, the World Wildlife Fund, and other pressure groups and conservation societies.
Write to your political representatives at all levels of government on every issue of environmental concern: keep the pressure on them.
Picket and organize boycotts of local organizations guilty of flagrant degradation of the environment.

Glossary of Ecological Terms

Abiotic environment The non-living environment consisting of inorganic nutrients, climatic factors and topography.

Absolute humidity The weight of water vapor in a given volume of air. It is usually measured in grams per cubic meter.

Antagonism (between soil ingredients). The inhibitory action of one substance on the use of another by plants when both substances are present in the soil.

Aquifer The layer of rocks, saturated with water, lying below the water table.

Assimilation The conversion of foods ingested by an animal into protoplasm.

Autotroph (primary producer) An organism that synthesizes organic compounds from inorganic ones (such as water, carbon dioxide, and salts) with the aid of an external supply of energy. The energy comes either from light (in photosynthesis) or from the breakdown of inorganic substances (in chemotrophs). Includes some bacteria, algae, and green plants.

Benthos The plants and animals that live on the bottom of the sea or of lakes.

Bioenergetics The study of the energy flow in an ecosystem.

Biogeochemical cycle The transfer of chemical materials from soil to a living organism and back again.

Biological control The use of natural predators to control pests.

Bioluminescence The production of light by living things.

Biomass The weight of organisms per unit area. It may be expressed either as gross weight, including the weight of the water content of the organisms, or as dry weight.

Biome A large area of vegetation of basic similarity, such as a tropical rain forest. It includes several different ecosystems such as rivers, marshes, clearings, and trees.

Biosphere The thin layer at the Earth's surface that can support life; it consists of rivers, lakes and oceans, the soil, and the lower atmosphere.

Biotic climax An association of plants that is maintained as climax vegetation by a biotic factor: for instance, much grassland is prevented by grazing animals from passing into woodland.

Biotic environment The living environment; the relationships between different plants, between different animals, and between plants and animals.

Calorie The calorie is defined as the amount of heat needed to raise 1 gram of water from 15°–16°C. The kilocalorie (also called the Calorie with a capital C) is 1000 calories and is the unit used in measuring the energy content of food.

Carnivore (secondary consumer) An organism that feeds on herbivores (primary consumers): for instance, a lion that feeds on a zebra.

Carotenoids Orange pigments in plants, which respond to the violet, blue, and green part of the light spectrum. They control phototropism (*See* Tropism).

Chemotroph An autotroph that obtains energy for food production by oxidizing inorganic materials. For instance, the soil bacterium *Nitrosomonas* obtains energy by oxidizing ammonia to nitrite. Certain bacteria and blue-green algae are chemotrophs.

Climatic climax The climax vegetation that appears to be most naturally suited to a particular area. It is dictated by the regional climate.

Climax The stable end-product of succession, consisting of plants and animals in equilibrium with each other and with the environment. Unless some factor in the ecological conditions changes, a climax association is capable of existing indefinitely.

Commensalism An association between two or more individuals of different species from which one derives feeding, or other, benefits, without significantly affecting the other.

Community The populations of different species, plant and animal, within a given area.

Compensation point The point at which a plant's loss of carbohydrate by respiration is balanced by its rate of photosynthesis.

Decomposers Organisms that feed on dead plant and animal material, by breaking it down physically and chemically.

Demography The study of population numbers and their variation with time.

Denitrification The breaking down of nitrogenous compounds by bacteria with the release of free nitrogen to the atmosphere.

Desert An area where the rainfall

is less than 25 cm per year, or where much of a higher rainfall is immediately lost by evaporation. A typical desert has large patches of sand between any vegetation.

Detritus feeders   Organisms that feed by ingesting small pieces of dead plant and animal material.

Disclimax *see* Plagioclimax.

Dominant   A species of plant that exerts a major influence on an ecosystem, and whose removal would radically alter the whole association. In a particular succession or climax, one—or several—of these dominant species is the most prominent plant, and the succession or climax may be called after it.

Ecological pyramid   A diagram showing the numbers, or mass, of the individuals in the different trophic levels of an ecosystem.

Ecology   The study of animals and plants and the interrelations between them, considered in relation to their nonliving environment; the study of ecosystems and biomes.

Ecosystem   All the communities in a given area considered together with their nonliving environment. In general, an ecosystem is a reasonably recognizable area, although adjacent ecosystems can overlap.

Ecotone   An area of transition and competition where one biome merges into the next. It shares some characteristics of the biomes on either side, but also contains its own unique species.

Ecotype   A group or "strain" of a species with requirements for nutrients, or tolerance of climatic factors, that differ from those of other groups of the same species.

Edaphic climax   A permanent type of vegetation within a climatic climax; it differs from the rest of the climax because a local abiotic factor overrides the influence of the regional climate.

Edge species   A species that inhabits the edges of two biomes (an ecotone) and is not found in either of the neighboring biomes.

Endemic   Naturally inhabiting a particular region.

Epilimnion   The top, warm layer

of water in freshwater lakes during summer.

Epiphyte   A plant that grows on another plant purely for support. It derives nourishment from rain and dust.

Erosion   The wearing away of the land surface by water, wind, and gravity.

Evapotranspiration   The loss of water by transpiration plus the loss oᶠ rain water by evaporation from plar and ground surfaces.

Exotic   Not naturally inhabiting a particular region, but imported by man from another one.

Feral   Reverting to a wild life from the domestic state.

Fertility (of soil)   A soil is fertile if it contains all the essential ingredients that enable a plant to complete its life cycle.

Fire climax   An association of plants that is maintained as climax vegetation by periodic fires (started by lightning as well as by man). These fires destroy plants that would otherwise become the dominant plants.

Food chain   A linear chain of organisms in which each link in the chain feeds on the one before and is eaten by the one after. At the start of the chain are the primary producers; at the end, the carnivores.

Food web   All the interrelated food chains in an ecosystem. The sum total of all the feeding habits of all the organisms in an ecosystem.

Gene pool   All the genes in a population of a species, which, through potential interbreeding of the members of the population, are available to produce a new generation.

Grasslands   Regions where the climax vegetation is grass and there are few trees. Tropical grasslands are often known as savannas; temperate grasslands include the prairies, pampas, steppes, and veld.

Gross efficiency   A measure of the efficiency of an animal in converting the food it consumes into protoplasm. Measured efficiencies vary from 4 percent to 37 percent. A high gross efficiency shows that relatively little of the food is lost as feces.

Gross primary production   The amount of protoplasm formed per unit time.

Gullying   The formation of deep gullies by the action of rainwater runoff on bare soil, and land being denuded of soil by sheet erosion.

Habitat   The home of an organism.

Herbivore (primary consumer)   An organism that feeds on primary producers (autotrophs).

Heterotroph (consumer)   An organism that obtains its organic food from other organisms. All animals, some fungi, and most bacteria are heterotrophs.

Holozoic   Feeding on other organisms. A holozoic organism is capable of ingesting undissolved organic material and breaking it down into soluble materials within its body. All animals that are not parasites are holozoic organisms.

Humidity   Water vapor in the air. *see* Absolute humidity *and* Relative humidity.

Humus   Decaying organic material in the soil.

Hydrarch   A succession that develops where water is abundant.

Hydraulic agriculture   Farming dependent on irrigation: for instance, the rice paddies of Southeast Asia.

Hydroponics   The cultivation of plants in nutrient-rich water. The plants are often grown in beds of sand or other inert material, which are flooded with nutrient-laden water at intervals.

Hydrosphere   The water on the surface of the Earth.

Hypolimnion   The cold bottom layer of water in freshwater lakes.

Indicator species   Species that are the first to suffer under polluted conditions, and to give warning of contamination.

Internal fertility control   The response in some animals to favorable or unfavorable conditions by a high or low birth-rate.

Leaching   The removal of soluble compounds by the downward percolation of water through the soil.

Ley (or **lea**) Land used as temporary pasture, or for growing hay. The farmer plows it up after a few years and sows another crop.

Limiting factor A factor of the abiotic environment that limits optimum growth because it is in short supply.

Loess A fine clay, formed from wind-blown particles; when mixed with humus, it forms a very fertile soil, light and easy to till.

Macronutrients The nine chemical elements that make up the bulk of living matter. Oxygen, carbon, and hydrogen are the major constituents; other macronutrients are nitrogen, phosphorus, potassium, calcium, magnesium, and sulfur.

Mesarch A succession that develops where there is adequate, but not excessive, rain and soil moisture.

Micronutrient A chemical nutrient needed for plant growth but in very small quantities; sometimes called a trace element.

Monoculture A farming system based on a single crop, grown year after year.

Mutualism In this book, mutualism means a close relationship between individuals of two or more species for their mutual benefit.

Mycorrhiza A mutualistic association between a fungus and a higher plant, most often consisting of an intimate relation between the roots of the higher plant and the mycelium of the fungus.

Nekton Animals that swim actively in water.

Neolithic ("New Stone Age") A stage in human history, characterized by a cultivator economy, with the start of village and town life, but with stone tools and weapons.

Net efficiency A measure of the efficiency of an animal in converting the food it assimilates into protoplasm. Measured efficiencies vary from 5 percent to 60 percent. A high net efficiency indicates that relatively little of the assimilated food is lost in respiration.

Net production The gross primary production minus the amount of protoplasm used by respiration in the primary producers. Net production is the amount of food available for the primary consumers.

Niche The habitat of an organism and the role it plays in the ecosystem.

Nitrogen fixation The production of inorganic nitrogen compounds from atmospheric nitrogen; performed by certain bacteria and blue-green algae.

Nomadism A way of life involving continual movement by herders with their herds over large areas, in search of food for the animals.

Omnivore An animal that eats many kinds of food, both plant and animal.

Paleo-ecology The study of ecosystems of the past.

Paleolithic ("Old Stone Age") A stage in human history, characterized by stone tools and weapons and a hunter–gatherer economy.

Parasite An organism that makes, at some stage in its life history, some connection with the tissues of an individual of a different species, from which it derives food. The parasite may harm the host to a greater or lesser extent but does not usually kill it.

Pastoralism A way of life involving the herding of animals.

Peat Dead marsh vegetation in various stages of alteration. It forms in cold, stagnant, acid water, lacking oxygen, where bacterial decay is reduced.

Permafrost Permanently frozen ground below the surface soil; found in polar regions and most of the Arctic tundra.

Photic zone The top layer of the deep water of the Earth, where enough light penetrates for photosynthesis.

Photoperiodism The response of an organism to the relative lengths of the day and night.

Photosynthesis The production of organic compounds, such as glucose, from carbon dioxide and water, liberating oxygen. The energy required for this reaction is obtained from light, with the aid of the green pigment chlorophyll.

Phototroph An autotroph that uses the energy from sunlight to make organic food from inorganic materials by photosynthesis.

Phytochrome The pigment that controls flowering by responding to the amount of light received by the plant. Phytochrome responds to the red and infrared part of the spectrum.

Phytoplankton The plant component of the plankton.

Phytotron A complex of greenhouses in which a sophisticated control system can simulate a great number of natural conditions by varying the light intensity, temperature, water, humidity, wind, and so on.

Pioneer community The plants, microorganisms, and small animals that first establish themselves on bare ground at the start of a primary succession.

Plagioclimax A subclimax maintained by continuous human activity.

Plankton The multitude of small living things that float in surface waters.

Population A group of individuals of the same species living in a given area.

Precipitation Rain, hail, and snow.

Predator An animal that kills other animals for its food.

Prey Animals killed and eaten by predators.

Primary consumer see Herbivore.

Primary producer see Autotroph.

Primary succession Community changes that begin on a sterile area, or on sites not previously occupied by organisms.

Pyramid of energy A diagram showing the energy available per unit time in a trophic level. Usually expressed as kilocalories per square meter per year ($kcal/m^2/yr$).

Racial senescence The condition within a species when the gene pool has become too small to provide enough variation in the species to allow for adaptation to changed conditions, and to mask weak characteristics.

Relative humidity The ratio of the amount of water vapor in the air to the amount that would saturate it at the same temperature and pressure.

Respiration (cellular or internal) The chemical reactions by which an organism obtains energy from organic compounds. Aerobic respiration requires oxygen. Anaerobic respiration does not.

Runoff The drainage of water, usually rainwater, from waterlogged or impermeable land.

Saprophyte An organism capable of feeding only on soluble organic matter from dead plants and animals; some fungi, some bacteria, some algae, but no animals, are saprophytes.

Secondary consumer *see* Carnivore.

Secondary succession Community changes that occur on sites favorable to life, or where there have previously been organisms.

Sere All the community changes (successions) from the first colonization of bare land to the final stable vegetation (climax).

Sheet erosion The removal of topsoil by heavy rainstorms.

Soil profile A section through the soil showing the different layers from the surface to the underlying rock.

Standard nutritional unit The unit expressing the energy available at any given trophic level for the next level in the food chain. Expressed in 10^6kcal per hectare (2.47 acres) per year.

Standing crop The biomass at any particular time, or during some given period.

Static husbandry The practice of rearing animals in fenced fields, rather than in roaming, tended herds.

Stratification (in ecology) The arrangement of an ecosystem into layers, such as forest canopy, understory, shrubs, herbaceous plants, mosses, and so on. It also includes the animals that live in these layers.

Subclimax A stage of succession before the climax that persists because of a continuous arresting

factor. *See* Biotic climax, Fire climax *and* Plagioclimax.

Succession The replacement of one community by another. The progressive change in the composition of a plant community, and hence the animal one, during the development of vegetation

Symbiosis An association of individuals of two or more species living together for all or part of their lives. Includes Commensalism, mutualism, and parasitism.

Taiga The vast stretches of coniferous forest south of the timberline in the Northern hemisphere; also called the boreal forest or northern coniferous forest.

Taxis Movement of a whole organism in response to a directional stimulus.
    Chemotaxis movement in response to a particular chemical substance.
    Geotaxis movement in response to gravity.
    Phototaxis movement in response to light.

Telemetry A method of keeping track of animals by battery-driven transmitters fixed to them.

Terrace cultivation A system of cultivation on mountain and hill slopes. The farmer cuts terraces into the slopes and builds low embankments around the small fields in order to retain irrigation water and to prevent soil erosion.

Territory An area of ground within which an animal is master over others of the same species.

Tertiary consumer An organism that feeds on secondary consumers. For instance: the cod (tertiary consumer) eats herring (secondary consumer), which eat copepods (primary consumers), which eat sea-water diatoms (primary producers).

Thermocline The layer of water of rapidly changing temperature in lake water in summer.

Thermoperiodism The response of living things to variations in temperature from season to season.

Timberline The point at which trees are unable to grow. Timberlines are found at extreme north and south latitudes and at high altitudes.

Transhumance The practice of moving animals from one type of pasture to another according to the season. In some regions the animals graze on scrubland during the rainy season, but are returned to river valleys or wetter upland regions for the rest of the year. In mountains the animals may be taken to high pastures in summer but return to the valleys in winter. It is still practiced in some parts of the Middle East.

Transpiration The loss of water vapor from a plant. It occurs mostly through the stomata of the leaves.

Trophic level A division of the food chain defined from other levels by the method of obtaining food: primary producer, primary consumer, secondary consumer, tertiary consumer.

Tropism A growth response by plants and sedentary animals to a directional stimulus.
    Geotropism Growth in response to gravity.
    Hydrotropism Growth in response to water.
    Phototropism Growth in response to light.
    Thermotropism Growth in response to heat.

Tundra Vast treeless tracts of land bearing a low cover of hardy land plants: lichens, mosses, and sedges. Arctic tundra occurs north of the timberline near the Arctic Circle. Alpine tundra is a similar, but not identical, type of vegetation found above the timberline on high mountains.

Water table The surface in the ground below which the cracks and fissures in the rock are saturated with water.

Weathering The action of rain, snow, frost, ice, wind, sunshine, and so on on rocks, altering their form, colour, texture, and composition.

Wetlands Areas of shallow water, often with much vegetation growing in it.

Xerarch A succession that develops where moisture is severely limited.

Zooplankton The animal component of plankton, it includes many eggs and the larval stages of aquatic organisms.

Index

Acknowledgments

Cover Picturepoint, London: Endpapers Mansell Collection: 2 Bruce Coleman Ltd.: 6 Picturepoint, London: 8 W. Suschitzky: 15 (top) Popperfoto: 17 (left) Philip H. Evans: 20 Aldus Books Ltd.: 22 (right) United States Information Service, London: 28 Jan Binblad/Bruce Coleman Ltd.: 31 (left) Josef Muench: 31 (right) Eric Hosking: 32 (top) Jurgen Schadeberg/*Daily Telegraph* Magazine: 32 (bottom) USDA Photo: 33 Richard Harringdon/Camera Press Ltd.: 34 Hiekisch/ZEFA: 36 © Photo John Dominis, *Life* 1967, Time Inc.: 37 National Institute of Oceanography/Photos Peter David: 38 Photo Andreas Feininger: 46 USDA Photo: 48 Photos Douglas P. Wilson: 49 ZEFA: 52 Photos Douglas P. Wilson: 53 (right) Soil Survey of England and Wales: 56 Marineland of Florida: 57 (left) Jane Burton/Bruce Coleman Ltd.: 57 (right) Simon Trevor (ADP)/Bruce Coleman Ltd.: 59 (top) after T. P. Ferguson and G. Bond, *Annals of Botany*, 18, 1954: 59 (centre) Gene Cox. Micro Colour (International), Trowbridge, Wiltshire: 59 (bottom) Ida Levisohn: 60 Graham Pizzey/Bruce Coleman Ltd.: 61 (left) Photo Jack Kath, Merck Sharpe and Dohme Research Laboratories, Rahway, New Jersey: 62 S. A. Smith, The London School of Hygiene and Tropical Medicine: 64 (left) Picturepoint, London: 64 (right) Jane Burton/Bruce Coleman Ltd.: 65 ZEFA: 66 Photo Ylla from Rapho Guillumette: 69 (bottom-left) Douglas P. Wilson: 69 (bottom-right) J. Allan Cash: 79 (top-left) Radio Times Hulton Picture Library: 79 (top-right) (bottom) Photos Geoffrey Drury © Aldus Books: 80 (top) Museum of English Rural Life, The University of Reading: 80 (bottom), 81 Photos Geoffrey Drury © Aldus Books: 85 (left) Photo J. H. Johns, New Zealand Forest Service. British Crown Copyright: 85 (right) Photo J. H. Johns: 86 Gordon F. Leedale, The University of Leeds: 91 (left) Aldus Archives: 91 (right) Photo Ray Dean © Aldus Books: 93 Aero Service Division, Litton Industries: 94 (top) The Wellcome Historical Medical Museum: 94 (bottom) Popperfoto: 100 (bottom) The Forestry Commission of Great Britain: (104 (right) International Minerals and Chemical Corporation, Skokie, Illinois: 105 Chuck Abbott/Rapho Guillumette Pictures: 106 Department of Agriculture and Horticulture, University of Bristol: 107 The Grassland Research Institute, Hurley, Maidenhead: 108 (top) National Oceanic and Atmospheric Administration: 108 (bottom) United States Information Service, London: 111 (left) Radiation Biology Laboratory, Smithsonian Institution: 111 (right) Russ Kinne/Bruce Coleman Ltd.: 113 Popperfoto: 114 (bottom) U.S. National Park Service: 117 Photos Edward S. Ross: 121 Photos British Columbia Forest Service: 125 (top) Rothamsted Experimental Station, Harpenden: 125 (bottom) Photographic Section, C.S.I.R.O. Division of Plant Industry: 128–129 *Daily Telegraph* Magazine: 130 Photo Keith Reid: 137 (left) Photo Geoffrey Drury © Aldus Books: 137 (right) Sigurdur Thorarinsson: 138 U.S. Forest Service: 143 Photo J. H. Johns, New Zealand Forest Service. British Crown Copyright: 145 Robert J. Ellison/*Daily Telegraph* Magazine: 147 (left) Australian News and Information Bureau: 147 (right) Photo J. H. Johns, New Zealand Forest Service. British Crown Copyright: 148 Photo Axel Poignant: 151–153 (top) Photos Laurence K. Marshall: 153 (bottom-left) Kenneth P. Oakley, *Man the Toolmaker*, British Museum (Natural History), 1965, by permission of the Trustees: 153 (bottom-right) Photo Dr. H. J. Heinz: 154 Radio Times Hulton Picture Library: 157 (bottom) A National Film Board of Canada photograph: 158 Three Lions, Inc.: 159 Nordenskiöld, *Vegas Fard Kring Asian Och Europe*, 1881: 160 Photo Henri Lhote: 163 (left) Parc Zoologique, Switzerland/Photo Darbellay, World Wildlife Fund: 163 (right) Krönberg Zoo/Photo F. Vollmer, World Wildlife Fund: 165 Photo Sally Anne Thompson: 166 USDA photos: 168 (bottom) J. Allan Cash: 169 (top-right) Reproduced by permission of the Trustees of the British Museum: 169 (bottom) Photo Harrison Forman: 171 Photo Geoffrey Drury © Aldus Books: 172 ZEFA: 174 after *Scientific American*, Paul C. Mangelsdorf, *Wheat*, July, 1953: 177 after *The Neolithic Revolution*, British Museum (Natural History) by permission of the Trustees: 180 (left) Photo G. Tomsich: 180 (right) Photo John Webb © Aldus Books: 181 after *Lands in the Desert*, Longmans Colour Geographies, Unit 16, by permission of the publishers: 183 (bottom) WHO/Photo Paul Almasy: 184 Three Lions, Inc.: 185 (top-right) Courtesy Compañia Mexicana Aerofoto, S.A.: 185 (bottom) Photo Georg Gerster